STUDENT'S SOLUTION MANUAL

TRIGONOMETRY

FIFTH EDITION

STUDENT'S SOLUTION MANUAL

LIAL • MILLER • HORNSBY

TRIGONOMETRY

FIFTH EDITION

Prepared with the assistance of

GERALD KRUSINSKI
College of DuPage

AUGUST ZARCONE
College of DuPage

HarperCollins*CollegePublishers*

Student's Solution Manual to accompany ***TRIGONOMETRY,*** *Fifth Edition*

ISBN 0-673-46815-1

95 9 8 7 6 5 4

PREFACE

This book provides complete solutions for many of the exercises in *Trigonometry*, fifth edition, by Margaret L. Lial, Charles D. Miller, and E. John Hornsby, Jr. Solutions are provided for the odd-numbered exercises. For each chapter a sample test with answers is also provided. Solutions are not provided for some of the exercises that involve open-response answers.

This book should be used as an aid as you work to master your course work. Try to solve the exercises that your instructor assigns before you refer to the solutions in this book. Then, if you have difficulty, read these solutions to guide you in solving the exercises. The solutions have been written so that they are consistent with the methods used in the textbook.

You may find that some of the solutions are presented in greater detail than others. Thus, if you cannot find an explanation for a difficulty that you encountered in one exercise, you may find the explanation in the solution for a similar exercise elsewhere in the exercise set.

Most of the solutions that require graphs refer to the answer section of the textbook. These graphs are not included in this book.

In addition to solutions, you will find a list of suggestions on how to be successful in mathematics. A careful reading will be helpful for many students.

The following people have made valuable contributions to the production of this *Student's Solution Manual*: Marjorie Seachrist, editor; Judy Martinez and Sheri Minkner, typists; Therese Brown and Charles Sullivan, artists; and Carmen Eldersveld, proofreader.

We also want to thank Tommy Thompson of Seminole Community College for his suggestions for the essay "To the Student: Success in Mathematics" that follows this preface.

TO THE STUDENT: SUCCESS IN MATHEMATICS

The main reason students have difficulty with mathematics is that they don't know how to study it. Studying mathematics *is* different from studying subjects like English or history. The key to success is regular practice.

This should not be surprising. After all, can you learn to play the piano or to ski well without a lot of regular practice? The same thing is true for learning mathematics. Working problems nearly every day is the key to becoming successful. Here is a list of things you can do to help you succeed in studying mathematics.

1. *Attend class regularly.* Pay attention in class to what your instructor says and does, and make careful notes. In particular, note the problems the instructor works on the board and copy the complete solutions. Keep these notes separate from your homework to avoid confusion when you read them over later.

2. Don't hesitate to ask questions in class. It is not a sign of weakness, but of strength. There are always other students with the same question who are too shy to ask.

3. *Read your text carefully.* Many students read only enough to get by, usually only the examples. Reading the complete section will help you to be successful with the homework problems. Most exercises are keyed to specific examples or objectives that will explain the procedures for working them.

4. Before you start on your homework assignment, rework the problems the instructor worked in class. This will reinforce what you have learned. Many students say, "I understand it perfectly when you do it, but I get stuck when I try to work the problem myself."

5. Do your homework assignment only *after* reading the text and reviewing your notes from class. Check your work with the answers in the back of the book. If you get a problem wrong and are unable to see why, mark that problem and ask your instructor about it. Then practice working additional problems of the same type to reinforce what you have learned.

6. Work as neatly as you can. Write your symbols clearly, and make sure the problems are clearly separated from each other. Working neatly will help you to think clearly and also make it easier to review the homework before a test.

7. After you have completed a homework assignment, look over the text again. Try to decide what the main ideas are in the lesson. Often they are clearly highlighted or boxed in the text.

8. Use the chapter review exercises at the end of each chapter as a practice test. Work through the problems under test conditions, without referring to the text or the answers until you are finished. You may want to time yourself to see how long it takes you. When you have finished, check your answers against those in the back of the book and study those problems that you missed. Answers are referenced to the appropriate sections of the text. As an additional review, use the sample test with answers at the end of each chapter of solutions in this Student's Solution Manual.

9. Keep any quizzes and tests that are returned to you and use them when you study for future tests and the final exam. These quizzes and tests indicate what your instructor considers most important. Be sure to correct any problems on these tests that you missed, so you will have the corrected work to study.

10. Don't worry if you do not understand a new topic right away. As you read more about it and work through the problems, you will gain understanding. Each time you look back at a topic you will understand it a little better. No one understands each topic completely right from the start.

CONTENTS

CHAPTER 1 THE TRIGONOMETRIC FUNCTIONS

Section 1.1

1. (-4, 2) lies in quadrant II.

3. (-9, -11) lies in quadrant III.

5. (9, 0) lies on the x-axis, so it is in no quadrant.

7. (0, -2) lies on the y-axis, so it is in no quadrant.

9. $(-5, \pi)$, $\pi \approx 3.14$, lies in quadrant II.

11. $(3\pi, -1)$, $3\pi \approx 9.42$, lies in quadrant IV.

13. Since xy = 1, and 1 is a positive number attained by the product of two variables x and y, then either x and y are both positive (quadrant I), or both negative (quadrant III). The graph of xy = 1 will lie in quadrants I and III.

In Exercises 15–25, the formula

$$d = \sqrt{(x_2 - x_1)^2 + (y_2 - y_1)^2}$$

is used to find the distance between (x_1, y_1) and (x_2, y_2).

15. $(x_1, y_1) = (-2, 7)$ and $(x_2, y_2) = (1, 4)$

$$\begin{aligned} d &= \sqrt{[1 - (-2)]^2 + (4 - 7)^2} \\ &= \sqrt{3^2 + (-3)^2} \\ &= \sqrt{9 + 9} \\ &= \sqrt{18} = 3\sqrt{2} \end{aligned}$$

17. $(x_1, y_1) = (2, 1)$ and $(x_2, y_2) = (-3, -4)$

$$\begin{aligned} d &= \sqrt{(-3 - 2)^2 + (-4 - 1)^2} \\ &= \sqrt{(-5)^2 + (-5)^2} \\ &= \sqrt{25 + 25} \\ &= \sqrt{50} \\ &= 5\sqrt{2} \end{aligned}$$

19. $(x_1, y_1) = (-1, 0)$ and $(x_2, y_2) = (-4, -5)$

$$\begin{aligned} d &= \sqrt{[-4 - (-1)]^2 + (-5 - 0)^2} \\ &= \sqrt{(-3)^2 + (-5)^2} \\ &= \sqrt{9 + 25} \\ &= \sqrt{34} \end{aligned}$$

21. $(x_1, y_1) = (3, -7)$ and $(x_2, y_2) = (-2, -5)$

$$\begin{aligned} d &= \sqrt{(-2 - 3)^2 + [-5 - (-7)]^2} \\ &= \sqrt{(-5)^2 + 2^2} \\ &= \sqrt{25 + 4} \\ &= \sqrt{29} \end{aligned}$$

23. $(x_1, y_1) = (-3, 6)$ and $(x_2, y_2) = (-3, 2)$

$$\begin{aligned} d &= \sqrt{[-3 - (-3)]^2 + (2 - 6)^2} \\ &= \sqrt{0^2 + (-4)^2} \\ &= \sqrt{0 + 16} \\ &= \sqrt{16} \\ &= 4 \end{aligned}$$

25. $(x_1, y_1) = (\sqrt{2}, -\sqrt{5})$ and $(x_2, y_2) = (3\sqrt{2}, 4\sqrt{5})$

$$\begin{aligned} d &= \sqrt{(3\sqrt{2} - \sqrt{2})^2 + [4\sqrt{5} - (-\sqrt{5})]^2} \\ &= \sqrt{(2\sqrt{2})^2 + (5\sqrt{5})^2} \\ &= \sqrt{8 + 125} \\ &= \sqrt{133} \end{aligned}$$

27. $(a, b, c) = (3, 4, 5)$

$$\begin{aligned} a^2 + b^2 &= 3^2 + 4^2 \\ &= 9 + 16 \\ &= 25 \\ &= 5^2 = c^2 \end{aligned}$$

This triple is a Pythagorean triple since $a^2 + b^2 = c^2$.

29. $(a, b, c) = (5, 12, 13)$

$$\begin{aligned} a^2 + b^2 &= 5^2 + 12^2 \\ &= 25 + 144 \\ &= 169 \\ &= 13^2 = c^2 \end{aligned}$$

This triple is a Pythagorean triple since $a^2 + b^2 = c^2$.

31. $(a, b, c) = (4, 6, 10)$

$$\begin{aligned} a^2 + b^2 &= 4^2 + 6^2 \\ &= 16 + 36 \\ &= 52 \neq 10^2 = c^2 \end{aligned}$$

This triple is not a Pythagorean triple since $a^2 + b^2 \neq c^2$.

33. $(a, b, c) = (7, 24, 25)$

$$\begin{aligned} a^2 + b^2 &= 7^2 + 24^2 \\ &= 49 + 576 \\ &= 625 \\ &= 25^2 = c^2 \end{aligned}$$

This triple is a Pythagorean triple since $a^2 + b^2 = c^2$.

35. $(a, b, c) = (8, 15, 17)$

$$\begin{aligned} a^2 + b^2 &= 8^2 + 15^2 \\ &= 64 + 225 \\ &= 289 \\ &= 17^2 = c^2 \end{aligned}$$

This triple is a Pythagorean triple since $a^2 + b^2 = c^2$.

37. $(a, b, c) = (3, 4, 5)$

$$\begin{aligned} a^2 + b^2 &= 3^2 + 4^2 \\ &= 9 + 16 \\ &= 25 \\ &= 5^2 = c^2 \end{aligned}$$

Now look at $(3k, 4k, 5k)$. If $k = 2$, we have $(6, 8, 10)$ which is a Pythagorean triple since $6^2 + 8^2 = 36 + 64 = 100 = 10^2$. If $k = 3$, we have $(9, 12, 15)$ which is a Pythagorean triple since $9^2 + 12^2 = 81 + 144 = 225 = 15^2$. If $k = 4$, we have $(12, 16, 20)$ which is a Pythagorean triple since $12^2 + 16^2 = 144 + 256 = 400 = 20^2$. By this, it appears that any triple of the form $(3k, 4k, 5k)$ is a Pythagorean triple.

39. $r = 3,\ s = 2$

$$\begin{aligned} a &= r^2 - s^2 = 3^2 - 2^2 = 9 - 4 = 5 \\ b &= 2rs = 2(3)(2) = 12 \\ c &= r^2 + s^2 = 3^2 + 2^2 = 9 + 4 = 13 \end{aligned}$$

The triple $(5, 12, 13)$ is generated.

$$\begin{aligned} a^2 + b^2 &= 5^2 + 12^2 \\ &= 25 + 144 \\ &= 169 = 13^2 = c^2 \end{aligned}$$

$(5, 12, 13)$ is a Pythagorean triple since $a^2 + b^2 = c^2$.

41. $r = 3,\ s = 1$

$a = r^2 - s^2 = 3^2 - 1^2 = 9 - 1 = 8$

$b = 2rs = 2(3)(1) = 6$

$c = r^2 + s^2 = 3^2 + 1^2 = 9 + 1 = 10$

The triple (8, 6, 10) is generated.

$$\begin{aligned} a^2 + b^2 &= 8^2 + 6^2 \\ &= 64 + 36 \\ &= 100 \\ &= 10^2 = c^2 \end{aligned}$$

(8, 6, 10) is a Pythagorean triple since $a^2 + b^2 = c^2$.

43. $r = 4,\ s = 1$

$a = r^2 - s^2 = 4^2 - 1^2 = 16 - 1 = 15$

$b = 2rs = 2(4)(1) = 8$

$c = r^2 + s^2 = 4^2 + 1^2 = 16 + 1 = 17$

The triple (15, 8, 17) is generated.

$$\begin{aligned} a^2 + b^2 &= 15^2 + 8^2 \\ &= 225 + 64 \\ &= 289 \\ &= 17^2 = c^2 \end{aligned}$$

(15, 8, 17) is a Pythagorean triple since $a^2 + b^2 = c^2$.

45. Are (-9, -2), (-1, -2), and (-9, 11) the vertices of a right triangle?

$$\begin{aligned} a^2 &= [-9 - (-1)]^2 + [-2 - (-2)]^2 \\ &= 8^2 + 0^2 = 64 \end{aligned}$$

$$\begin{aligned} b^2 &= [-9 - (-9)]^2 + (-2 - 11)^2 \\ &= 0^2 + 13^2 = 169 \end{aligned}$$

$$\begin{aligned} c^2 &= [-1 - (-9)]^2 + (-2, -11)^2 \\ &= 8^2 + 13^2 = 233 \end{aligned}$$

$64 + 169 = 233$, so $a^2 + b^2 = c^2$.

Yes, the given points are vertices of a right triangle.

47. (-8, 2), (5, -7), (3, -9)

$$\begin{aligned} c^2 &= (-8 - 5)^2 + [2 - (-7)]^2 \\ &= (-13)^2 + 9^2 = 250 \end{aligned}$$

$$\begin{aligned} b^2 &= (-8 - 3)^2 + [2 - (-9)]^2 \\ &= (-11)^2 + 11 = 242 \end{aligned}$$

$$\begin{aligned} a^2 &= (5 - 3)^2 + [7 - (-9)]^2 \\ &= 2^2 + 2^2 = 8 \end{aligned}$$

$8 + 242 = 250$, so $a^2 + b^2 = c^2$.

Yes, the given points are vertices of a right triangle.

49. $(4 - \sqrt{3}, -2\sqrt{3})$, $(2 - \sqrt{3}, -\sqrt{3})$, $(3 - \sqrt{3}, -2\sqrt{3})$

$$\begin{aligned} a^2 &= [(4 - \sqrt{3}) - (2 - \sqrt{3})]^2 \\ &\quad + [-2\sqrt{3} - (-\sqrt{3})]^2 \\ &= (2)^2 + (-\sqrt{3})^2 = 7 \end{aligned}$$

$$\begin{aligned} b^2 &= [(4 - \sqrt{3}) - (3 - \sqrt{3})]^2 \\ &\quad + [-2\sqrt{3} - (-2\sqrt{3})]^2 \\ &= 1^2 + 0^2 = 1 \end{aligned}$$

$$\begin{aligned} c^2 &= [(2 - \sqrt{3}) - (3 - \sqrt{3})]^2 \\ &\quad + [-\sqrt{3} - (-2\sqrt{3})]^2 \\ &= (-1)^2 + (\sqrt{3})^2 = 4 \end{aligned}$$

Since no two of a^2, b^2, c^2 add up to the third, the given points are not vertices of a right triangle.

51. Find y such that the distance between (5, y) and (8, -1) is 5.

$$\begin{aligned} d = \sqrt{(5 - 8)^2 + [y - (-1)]^2} &= 5 \\ \sqrt{9 + (y + 1)^2} &= 5 \end{aligned}$$

Square both sides.

$$\begin{aligned} 9 + (y + 1)^2 &= 25 \\ (y + 1)^2 &= 16 \\ y + 1 &= \pm 4 \end{aligned}$$

If $y + 1 = 4$, $y = 3$.
If $y + 1 = -4$, $y = -5$.
Therefore, $y = 3$ or $y = -5$.

53. Find x such that the distance between (x, 11) and (5, -4) is 17.

$$d = \sqrt{(x - 5)^2 + [11 - (-4)]^2} = 17$$
$$\sqrt{(x - 5)^2 + 15^2} = 17$$
$$(x - 5)^2 + 225 = 289$$
$$(x - 5)^2 = 64$$
$$x - 5 = \pm 8$$

$x = 13$ or $x = -3$

55. Let P = (x, y) be a point 3 units from (-5, 6). Then, by the distance formula,

$$d = \sqrt{[x - (-5)]^2 + (y - 6)^2} = 3$$
$$\sqrt{(x + 5)^2 + (y - 6)^2} = 3.$$

Square both sides.

$$(x + 5)^2 + (y - 6)^2 = 9$$

This is the equation for all points 3 units from (-5, 6).
The graph is a circle with center at (-5, 6) and radius 3.

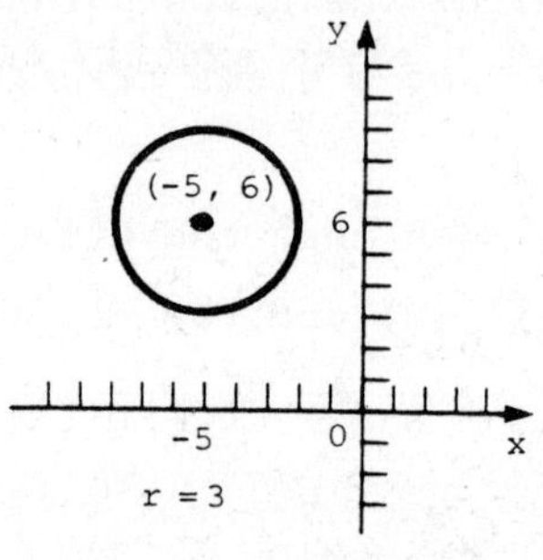

57. To find the length AB, apply the Pythagorean theorem to △ ABX.

$$(AB)^2 = (AX)^2 + (BX)^2$$
$$= 8^2 + 8^2$$
$$= 128$$
$$AB = \sqrt{128} = 11.3137$$

The distance across the top of the fabric is 20 • AB (since there are 20 strips) plus 42 in for the side of the unused triangle at the end. (Since this is a 45°-45° right triangle, it is isosceles, and its two equal sides are each 42 in.) Thus, the distance across the top is

$$20(11.3137) + 42$$
$$= 226.274 + 42$$
$$= 268.274 \text{ in.}$$

Since the price of the material is \$10 per linear yard, convert from inches to yards to find the number of linear yards.

$$\frac{268.274}{36} = 7.452 \text{ yd}$$

The cost of the material is

$$(7.452)(10) = \$74.52.$$

59. Let x = the leg of the right triangle with leg a and hypotenuse s.

Notice that $x = \frac{\ell}{2} = \frac{231.3}{2} = 115.66$.

$$a^2 + x^2 = s^2$$
$$(184.7)^2 + (115.66)^2 = s^2$$
$$34114.1 + 13377.24 = s^2$$
$$47491.34 = s^2$$
$$217.9 = s$$

The length of the edge of the pyramid labeled s is 217.9 m.

61. $\{y|y > 9\}$ This is an open interval with interval notation $(9, \infty)$.

63. $\{k|k < -8\}$ This is an open interval with interval notation $(-\infty, -8)$.

65. $\{p|p \geq 10\}$ This is a half-open interval with interval notation $[10, \infty)$.

67. $\{y|8 \leq y \leq 13\}$ This is a closed interval with interval notation $[8, 13]$.

69. This is an open interval, starting at 2, and includes all real numbers greater than 2. The interval notation is $(2, \infty)$.

71. This is a half-open interval between 1 and 4. It includes 4, but does not include 1. The interval notation is $(1, 4]$.

73. $|y| \geq 1$ is equivalent to $y \geq 1$ or $y \leq -1$. Since the interval notation for $y \geq 1$ is $[1, \infty)$ and the interval notation for $y \leq -1$ is $(-\infty, -1]$, the interval notation for $|y| \geq 1$ is $(-\infty, -1] \cup [1, \infty)$.

In Exercises 75–83, $f(x) = -2x^2 + 4x + 6$.

75.
$$\begin{aligned} f(-2) &= -2(-2)^2 + 4(-2) + 6 \\ &= -8 - 8 + 6 \\ &= -10 \end{aligned}$$

77.
$$\begin{aligned} f(3) &= -2(3)^2 + 4(3) + 6 \\ &= -18 + 12 + 6 \\ &= 0 \end{aligned}$$

79.
$$\begin{aligned} f(-5) &= -2(-5)^2 + 4(-5) + 6 \\ &= -50 - 20 + 6 \\ &= -64 \end{aligned}$$

81.
$$\begin{aligned} f(-m) &= -2(-m)^2 + 4(-m) + 6 \\ &= -2m^2 - 4m + 6 \end{aligned}$$

83. $f(2 - p)$
$$\begin{aligned} &= -2(2 - p)^2 + 4(2 - p) + 6 \\ &= -2(4 - 4p + p^2) + (8 - 4p) + 6 \\ &= -8 + 8p - 2p^2 + 8 - 4p + 6 \\ &= -2p^2 + 4p + 6 \end{aligned}$$

85. $2x + 5y = 10$

Domain: $(-\infty, \infty)$

Range: $(-\infty, \infty)$

This is a function because each value of x gives exactly one value of y.

87. $y = 2x^2 - 5$

Domain: $(-\infty, \infty)$

Range: Since $x^2 \geq 0$,

$2x^2 - 5 \geq -5$, the range is $[-5, \infty)$.

This is a function because each value of x gives exactly one value of y.

89. $y = 3(x + 1)^2 - 5$

Domain: $(-\infty, \infty)$

Range: Since $(x + 1)^2 \geq 0$,

$$2(x + 1)^2 - 5 \geq -5.$$

The range is $[-5, \infty)$.
This is a function.

91. $-x = y^2$

Domain: $-x = y^2$, so $x = -y^2$.
The domain is $(-\infty, 0]$.
Range: $(-\infty, \infty)$

If $x = -16$, $y^2 = 16$, so $y = 4$ or $y = -4$.

Since one x-value leads to 2 y-values, this is not a function.

93. $y = \sqrt{x - 2}$

Domain: $x - 2 \geq 0$

$$x \geq 2$$

The domain is $[2, \infty)$.
Range: $y \geq 0$, since a square root is always nonnegative.
The range is $[0, \infty)$.
This is a function.

95. $y = \sqrt{1 - x^2}$

Domain: $1 - x^2 \geq 0$

$$-x^2 \geq -1$$

$$x^2 \leq 1$$

$$-1 \leq x \leq 1$$

The domain is $[-1, 1]$.
Range: $[0, 1]$
This is a function.

97. Domain: $[-5, \infty)$
Range: $[0, \infty)$

This is a function because any vertical line intersects the graph in at most one point.

99. Domain: $(-\infty, -3] \cup [3, \infty)$
Range: $(-\infty, \infty)$

This is not a function because some vertical lines, for example, $x = 4$, intersect the graph in 2 points.

101. Domain: $(-\infty, \infty)$
Range: $(-\infty, \infty)$

This is a function because any vertical line intersects the graph in at most one point.

In Exercises 103–109, the domain includes all real numbers for which the denominator does not equal zero.

103. $y = \dfrac{-2}{x + 1}$

$x + 1 \neq 0$

$$x \neq -1$$

Domain: $(-\infty, -1) \cup (-1, \infty)$

105. $y = \dfrac{4 + x^2}{(5x + 1)(3x + 8)}$

$$5x + 1 \neq 0 \quad \text{and} \quad 3x + 8 \neq 0$$

$$x \neq -\frac{1}{5} \quad \text{and} \quad x \neq -\frac{8}{3}$$

Domain:

$$\left(-\infty, -\frac{8}{3}\right) \cup \left(-\frac{8}{3}, -\frac{1}{5}\right) \cup \left(-\frac{1}{5}, \infty\right)$$

107. $y = \dfrac{10 + 3x}{-9 - x^2}$

$$-9 - x^2 \neq 0$$

$$-9 \neq x^2$$

This is true for all x, so the domain is $(-\infty, \infty)$.

109. $y = \dfrac{3 + x}{\sqrt{2x - 5}}$

The domain is determined by the requirement that

$$2x - 5 \nleq 0$$

$$2x \nleq 5$$

$$x \nleq \frac{5}{2}, \text{ which means}$$

$$x > \frac{5}{2}.$$

The domain is $\left(\frac{5}{2}, \infty\right)$.

111. The area of the trapezoid is

$$\frac{1}{2}(b + a)(a + b) = \frac{1}{2}(a + b)^2.$$

The area of the two right triangles with legs a and b and hypotenuse c is $\frac{1}{2}$ ab.

The area of the right triangle with legs c is $\frac{1}{2}c^2$.

Since the sum of the areas of the 3 triangles is equal to the area of the trapezoid, we have

$$\frac{1}{2}ab + \frac{1}{2}ab + \frac{1}{2}c^2 = \frac{1}{2}(a + b)^2$$

$$ab + \frac{1}{2}c^2 = \frac{1}{2}(a^2 + 2ab + b^2)$$

$$ab + \frac{1}{2}c^2 = \frac{1}{2}a^2 + ab + \frac{1}{2}b^2$$

$$\frac{1}{2}c^2 = \frac{1}{2}a^2 + \frac{1}{2}b^2$$

$$c^2 = a^2 + b^2.$$

Section 1.2

1. $7x + 11x = 180$

$$18x = 180$$

$$x = 10$$

The measures of the two angles are

$$(7x)° = 7(10°) = 70°$$

and $(11x)° = 11(10°) = 110°$.

3. $5k + 5 + 3k + 5 = 90$

$$8k + 10 = 90$$

$$8k = 80$$

$$k = 10$$

The measures of the two angles are

$$(5k + 5)° = 5(10°) + 5° = 55°$$

and $(3k + 5)° = 3(10°) + 5° = 35°$.

5. $6x - 4 + 8x - 12 = 180$

$$14x - 16 = 180$$
$$14x = 196$$
$$x = 14$$

The measures of the two angles are

$$(6x - 4)° = 6(14°) - 4° = 80°$$

and $(8x - 12)° = 8(14°) - 12° = 100°.$

7. If an angle measures x degees, and two angles are complementary if their sum is 90°, then the com– plement of an angle of x degrees is 90° – x degrees.

9. Let x = the measure of the angle. If the angle is its own complement,

$$x + x = 90$$
$$2x = 90$$
$$x = 45.$$

A 45° angle is its own complement.

11. 75° 15′ + 83° 32′ = 158° 47′

13. 89° + 23° 42′ = 112° 42′

15. 90° – 51° 28′ = 89° 60′ – 51° 28′

= 38° 32′

17. 180° – 152° 43′

= 179° 60′ – 152° 43′

= 27° 17′

19. 90° – 36° 18′ 47″

= 89° 59′ 60″ – 36° 18′ 47″

= 53° 41′ 13″

21. 180° – 120° 42′ 37″

= 179° 59′ 60″ – 120° 42′ 37″

= 59° 17′ 23″

23. $20° \; 54' = 20° + \frac{54°}{60}$

$= 20° + .9°$

$= 20.9°$

25. 91° 18′ 54″

$= 91° + \frac{35°}{60} + \frac{54}{3600}°$

= 91.598°

27. 274° 18′ 59″

$= 274° + \frac{18°}{60} + \frac{59}{3600}°$

= 274.316°

29. 31.4296° = 31° + (.4296)60′

= 31° + 25.776

= 31° + 25′ + (.776)(60″)

= 31° 25′ 47″

31. 89.9004°

= 89° + (.9004)(60′)

= 89° + 54.024′

= 89° + 54′ + (.024)(60″)

= 89° 54′ 01″

33. 178.5994°

= 178° + (.5994)(60′)

= 178° + 35.964′

= 178° + 35′ + (.964)(60″)

= 178° 35′ 58″

35. $-40° + 360° = 320°$

37. $-125° + 360° = 235°$

39. $450° - 360° = 90°$

41. $539° - 360° = 179°$

43. $850° - 2(360°) = 130°$

45. $-985.4063° + 3(360°) = 94.5937°$

47. $30°$

A coterminal angle can be obtained by adding an integer multiple of 360°.

$$30° + n \cdot 360°$$

49. $60°$

A coterminal angle can be obtained by adding an integer multiple of 360°.

$$60° + n \cdot 360°$$

51. $135°$

A coterminal angle can be obtained by adding an integer multiple of 360°.

$$135° + n \cdot 360°$$

53. $-90°$

A coterminal angle can be obtained by adding an integer multiple of 360°.

$$-90° + n \cdot 360°$$

55. The answers to Exercises 52 and 53 give the same set of angles since 270° and −90° are coterminal. Notice that $-90° + 360° = 270°$.

For Exercises 57–71, answers other than the ones given are possible.

57.

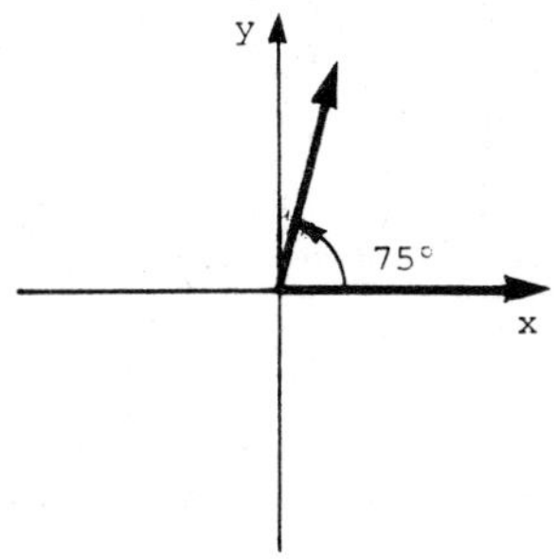

75° is in quadrant I.

A positive angle coterminal with 75° is

$$75° + 360° = 435°.$$

A negative angle coterminal with 75° is

$$75° - 360° = -285°.$$

59.

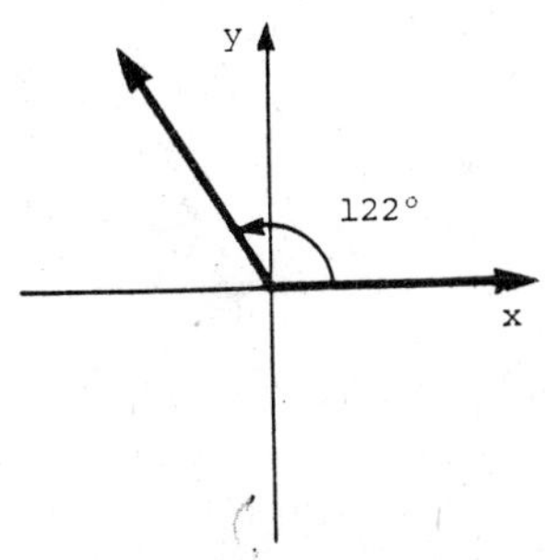

122° is in quadrant II.

A positive angle coterminal with 122° is

$$122° + 360° = 482°.$$

A negative angle coterminal with 122° is

$$122° - 360° = -238°.$$

61.

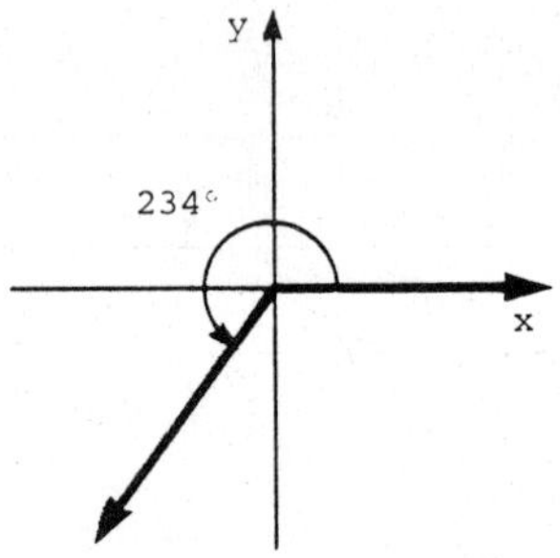

234° is in quadrant III.

A positive angle coterminal with 234° is

$$234° + 360° = 594°.$$

A negative angle coterminal with 234° is

$$234° - 360° = -126°.$$

63.

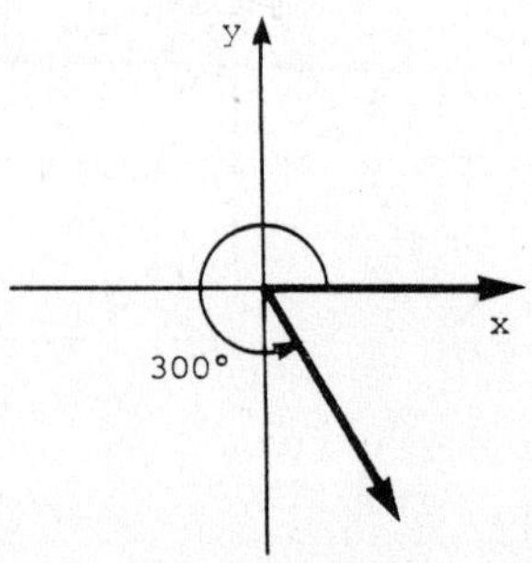

300° is in quadrant IV.

A positive angle coterminal with 300° is

$$300° + 360° = 660°.$$

A negative angle coterminal with 300° is

$$300° - 360° = -60°.$$

65.

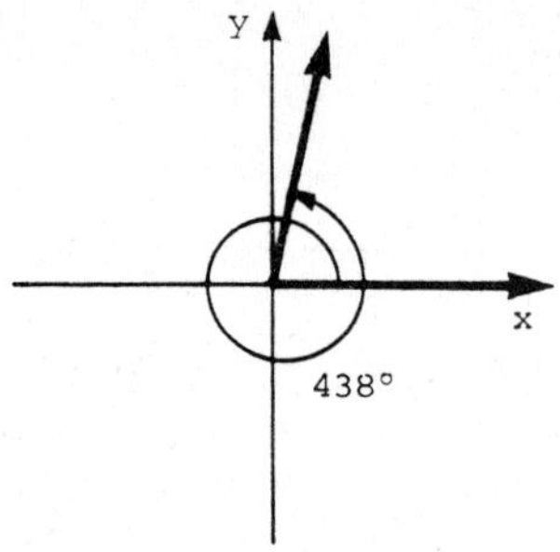

438° is in quadrant I.

A positive angle coterminal with 438° is

$$438° - 360° = 78°.$$

A negative angle coterminal with 438° is

$$438° - 2(360°) = -282°.$$

67.

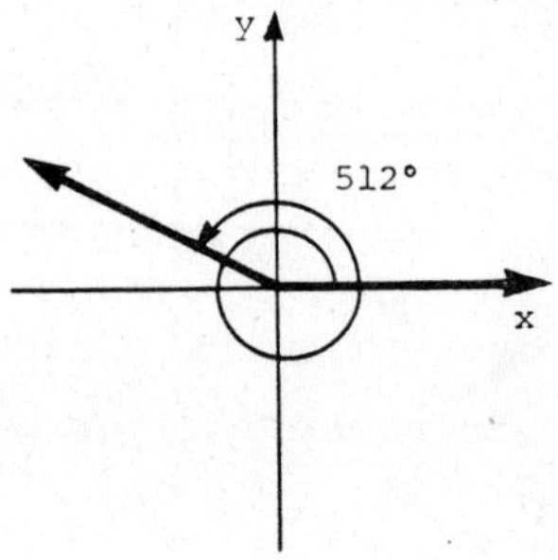

512° is in quadrant II.

A positive angle coterminal with 512° is

$$512° - 360° = 152°.$$

A negative angle coterminal with 512° is

$$512° - 2(360°) = -208°.$$

69.

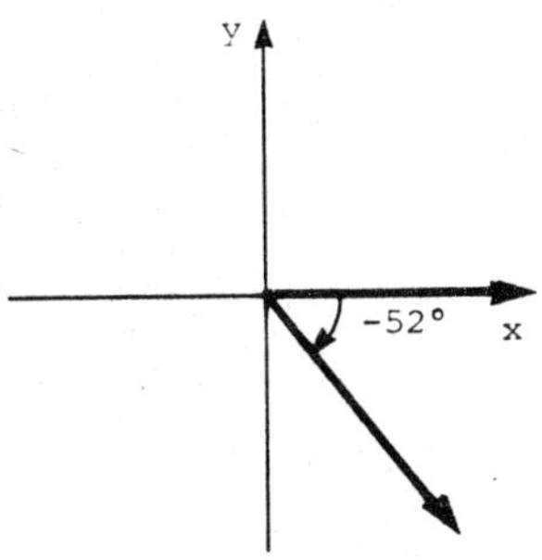

-52° is in quadrant IV.

A positive angle coterminal with -52° is

$$-52° + 360° = 308°.$$

A negative angle coterminal with -52° is

$$-52° - 360° = -412°.$$

71.

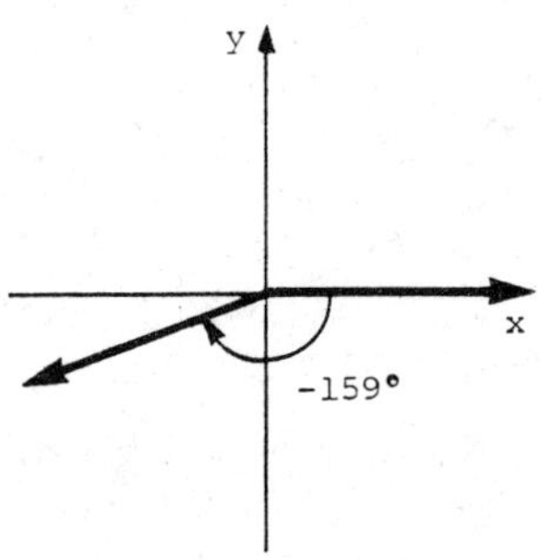

-159° is in quadrant III.

A positive angle coterminal with -159° is

$$-159° + 360° = 201°.$$

A negative angle coterminal with -159° is

$$-159° - 360° = -519°.$$

73.

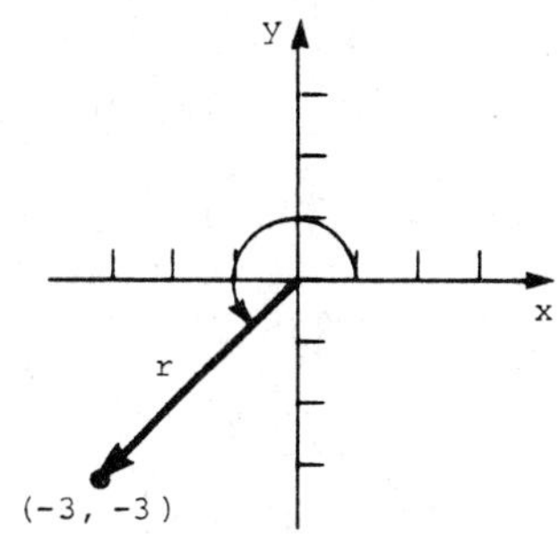

Points: (-3, -3), (0, 0)

$$\begin{aligned} r &= \sqrt{(-3 - 0)^2 + (-3 - 0)^2} \\ &= \sqrt{9 + 9} \\ &= \sqrt{18} \\ &= \sqrt{9 \cdot 2} \\ &= 3\sqrt{2} \end{aligned}$$

75.

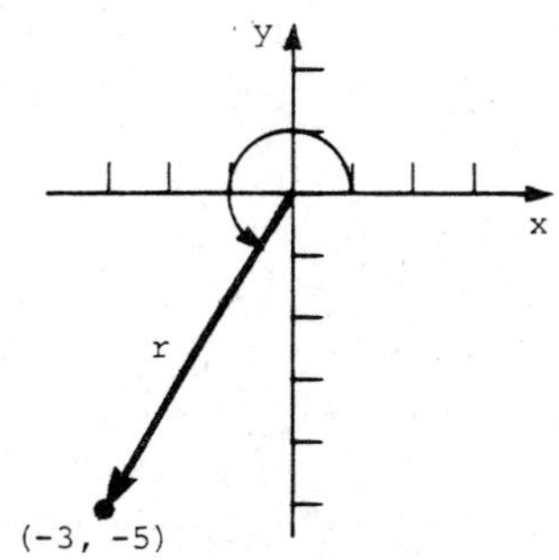

Points: (-3, -5), (0, 0)

$$r = \sqrt{(-3 - 0)^2 + (-5 - 0)^2}$$
$$= \sqrt{9 + 25}$$
$$= \sqrt{34}$$

77.

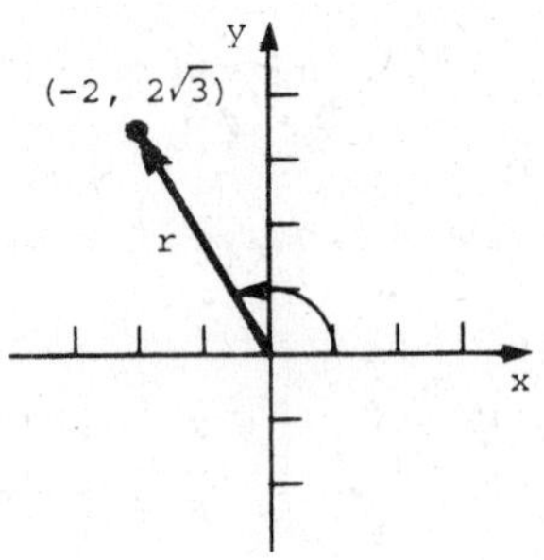

Points: $(-2, 2\sqrt{3})$, (0, 0)

$$r = \sqrt{(-2 - 0)^2 + (2\sqrt{3} - 0)^2}$$
$$= \sqrt{4 + 12}$$
$$= \sqrt{16}$$
$$= 4$$

79. 600 rotations per minute

= 600/60 rotations per second

= 10 rotations per second

= 5 rotations per 1/2 second

= 5(360°) = 1800° per 1/2 second

81. 75° per minute

= (75)(60°) per hour

= 4500° per hour

= $\frac{4500}{360}$ rotations per hour

= 12.5 rotations per hour

Section 1.3

1. $5x - 129 = 2x - 21$

since vertical angles are equal.

$$3x - 129 = -21$$
$$3x = 108$$
$$x = 36$$

$$5x - 129 = 5(36) - 129$$
$$= 180 - 129 = 51$$
$$2x - 21 = 2(36) - 21$$
$$= 72 - 21 = 51$$

Both angles are 51°.

3. $x + (x + 20) + (210 - 3x) = 180$

since the sum of the angles of a triangle is 180°.

$$-x + 230 = 180$$
$$-x = -50$$
$$x = 50$$
$$x + 20 = 50 + 20 = 70$$
$$210 - 3x = 210 - 3(50)$$
$$= 210 - 150 = 60$$

The three angles are 50°, 70°, and 60°.

5. $(x - 30) + (2x - 120) + \left(\frac{1}{2}x + 15\right) = 180$

since the sum of the angles of a triangle is 180°.

$$\frac{7}{2}x - 135 = 180$$
$$\frac{7}{2}x = 315$$
$$x = (315)\frac{2}{7} = 90$$
$$x - 30 = 90 - 30 = 60$$

$\frac{1}{2}x + 15 = \frac{1}{2}(90) + 15$

$= 45 + 15 = 60$

$2x - 120 = 2(90) - 120$

$= 180 - 120 = 60$

All three angles are 60°.

7. $(3x + 5) + (5x + 15) = 180$

since these two angles are supplementary.

$8x + 20 = 180$

$8x = 160$

$x = 20$

$3x + 5 = 3(20) + 5$

$= 60 + 5 = 65$

$5x + 15 = 5(20) + 15$

$= 100 + 15 = 115$

The two angles are 65° and 115°.

9. $2x - 5 = x + 22$

since alternate interior angles are equal.

$x - 5 = 22$

$x = 27$

$2x - 5 = 2(27) - 5$

$= 54 - 5 = 49$

$x + 22 = 27 + 22 = 49$

Both angles have a measure of 49°.

11. $(x + 1) + (4x - 56) = 180$

since the sum of interior angles on the same side of the transversal is 180°.

$5x - 55 = 180$

$5x = 235$

$x = 47$

$x + 1 = 47 + 1 = 48$

$4x - 56 = 4(47) - 56$

$= 188 - 56 = 132$

The angles are 48° and 132°.

13. Let x = the measure of the third angle.

$37° + 52° + x = 180°$

$89° + x = 180°$

$x = 91°$

15. Let x = the measure of the third angle.

$147°\ 12' + 30°\ 19' + x = 180°$

$177°\ 31' + x = 180°$

$x = 179°\ 60' - 177°\ 31'$

$x = 2°\ 29'$

17. Let x = the measure of the third angle.

$74°\ 12'\ 59'' + 80°\ 58'\ 05'' + x = 180°$

$154°\ 70'\ 64'' + x = 180°$

$154°\ 71'\ 04'' + x = 180°$

$155°\ 11'\ 04'' + x = 179°\ 59'\ 60''$

$x = 179°\ 59'\ 60'' - 155°\ 11'\ 04''$

$x = 24°\ 48'\ 56''$

19. A triangle cannot have two angles of measures 85° and 100° since the sum of the three angles of a triangle is 180° and 85° + 100° = 185°, 5° more than 180°.

21. angle 1 = 55° (vertical angle with 55°)

angle 2 = 65° (vertical angle with angle 4, see below)

angle 3 = 60° (vertical angle with angle 5, see below)

angle 4 = 180° − 60° − 55° = 65° (supplemental angle with angle 5 and 55°)

angle 5 = 180° − 120° = 60° (interior angles on same side of transversal)

angle 6 = 120° (vertical angles)

angle 7 = 180° − 120° = 60° (supplemental angle with angle 6)

angle 8 = 60° (vertical angle with angle 7)

angle 9 = 55° (vertical angle with angle 10)

angle 10 = 180° − 60° − 65° = 55° (sum of angles of triangle is 180°)

23. The triangle has a right angle, but each side has a different measure. The triangle is a right triangle and a scalene triangle.

25. The triangle has three acute angles and three equal sides, so it is acute and equilateral.

27. The triangle has a right angle and three unequal sides, so it is right and scalene.

29. The triangle has a right angle and two equal sides, so it is right and isosceles.

31. The triangle has one obtuse angle and three unequal sides, so it is obtuse and scalene.

33. The triangle has three acute angles and two equal sides, so it is acute and isosceles.

35. An isosceles right triangle is a triangle with two sides perpendicular (form a right angle) and the same length.

37. All equilateral triangles are similar because all angles have a measure of 60°, and since all sides have equal measure, the ratio of any side of triangle 1 to the corresponding side of triangle 2 will always be the same (proportional).

39. Corresponding angles are A and P, B and Q, C and R; corresponding sides are AB and PQ, CB and RQ, AC and PR.

41. Corresponding angles are H and F, K and E, HGK and FGE; corresponding sides are HK and FE, GK and GE, HG and FG.

43. A + B + C = 180° so

$$A = 180° - 46° - 78° = 56°.$$

M = B so M = 46°.

N = A so N = 56°.

P = C so P = 78°.

45. $Y = V$ so $Y = 28°$.

$X + Y + Z = 180°$ so

$$Z = 180° - 74° - 28° = 78°.$$

$T = X$ so $T = 74°$.

$W = Z$ so $W = 78°$.

47. $P + Q + R = 180°$ so

$$R = 180° - 64° - 20° = 96°.$$

$T = P$ so $T = 20°$.

$U = R$ so $U = 96°$.

$V = Q$ so $V = 64°$.

In Exercises 49–63, corresponding sides of similar triangles are proportional.

49. $\frac{25}{10} = \frac{a}{8}$

$200 = 10a$

$20 = a$

$\frac{25}{10} = \frac{b}{6}$

$150 = 10b$

$15 = b$

51. $\frac{6}{12} = \frac{a}{12}$

$72 = 12a$

$6 = a$

$\frac{b}{15} = \frac{6}{12}$

$\frac{b}{15} = \frac{1}{2}$

$2b = 15$

$b = \frac{15}{2}$, or $7\frac{1}{2}$

53. $\frac{x}{4} = \frac{9}{6}$

$\frac{x}{4} = \frac{3}{2}$

$2x = 12$

$x = 6$

55. Let T = the height of the tree.

The triangle formed by the tree and its shadow is similar to the triangle formed by the stick and its shadow.

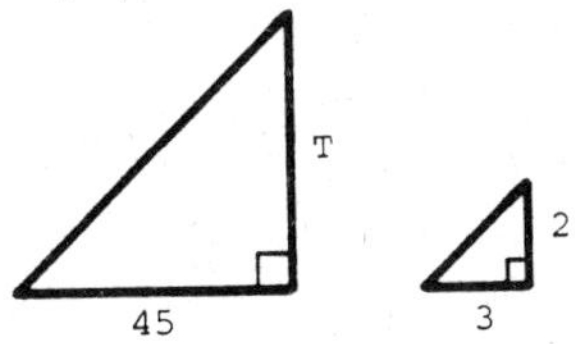

$\frac{T}{2} = \frac{45}{3}$

$\frac{T}{2} = 15$

$T = 30$

The tree is 30 m high.

57. Let x = the middle side of the actual triangle,

and y = the longest side of the actual triangle.

The triangles in the photograph and on the actual piece of land are similar. The shortest side on the land corresponds to the shortest side in the photograph.

$$\frac{400}{4} = \frac{x}{5}$$

$$100 = \frac{x}{5}$$

$$x = 500$$

$$\frac{400}{4} = \frac{y}{7}$$

$$100 = \frac{y}{7}$$

$$y = 700$$

The other two sides are 500 m and 700 m long.

59. Let h = the height of the building.

The triangle formed by the house and its shadow is similar to the triangle formed by the building and its shadow.

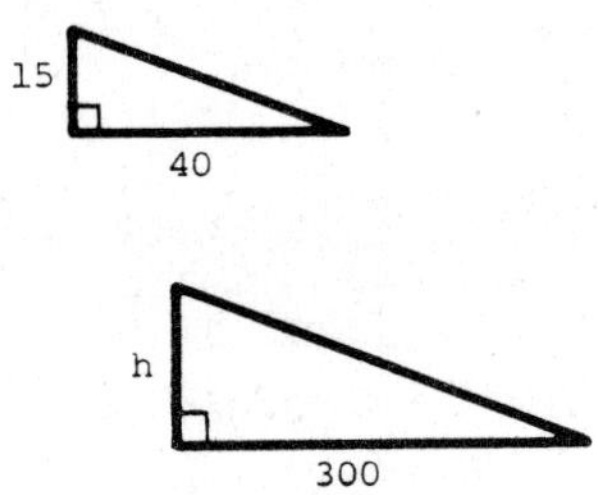

$$\frac{15}{40} = \frac{h}{300}$$

$$4500 = 40h$$

$$112.5 = h$$

The height of the building is 112.5 ft.

61. $\frac{x}{50} = \frac{100 + 120}{100}$

$$\frac{x}{50} = \frac{220}{100}$$

$$\frac{x}{50} = \frac{11}{5}$$

$$5x = 550$$

$$x = 110$$

63. $\frac{c}{100} = \frac{10 + 90}{90}$

$$\frac{c}{100} = \frac{100}{90}$$

$$\frac{c}{100} = \frac{10}{9}$$

$$c = \frac{1000}{9} \approx 111.1$$

65. Let a = the length of the longest side of the first quadrilateral,

x = the length of the shortest side of the second quadrilateral,

and y = the other unknown length.

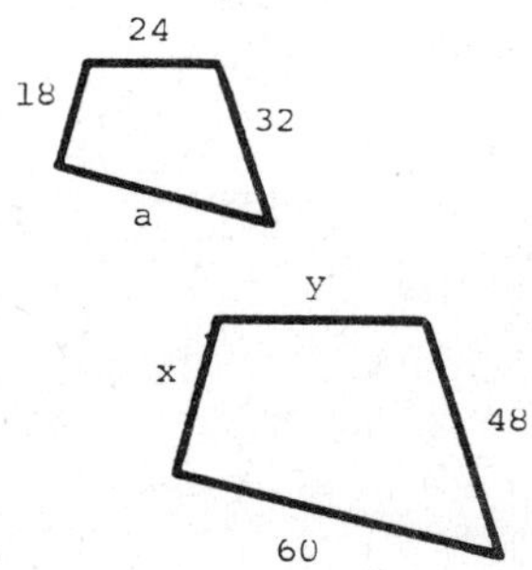

Corresponding sides are in proportion.

Since 32/48 = 2/3,

$$\frac{a}{60} = \frac{2}{3}$$

$$3a = 120$$

$$a = 40$$

$$\frac{24}{y} = \frac{2}{3}$$

$$2y = 72$$

$$y = 36$$

$$\frac{18}{x} = \frac{2}{3}$$

$$2x = 54$$

$$x = 27.$$

The unknown side in the first quadrilateral is 40 cm; the unknown sides in the second quadrilateral are 36 cm and 27 cm.

67. The two triangles are similar. Corresponding angles are equal.

$$10y + 8 = 58$$
$$10y = 50$$
$$y = 5$$

Corresponding sides are in proportion.

$$\frac{x - y}{x + y} = \frac{6}{18} = \frac{1}{3}$$
$$3(x - y) = 1(x + y)$$
$$3x - 3y = x + y$$
$$2x = 4y$$
$$x = 2y$$

Substitute 5 for y.

$$x = 2(5) = 10$$

Section 1.4

1.

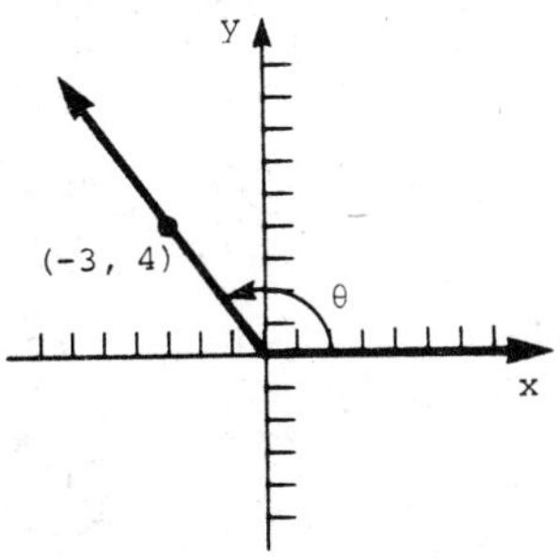

3.

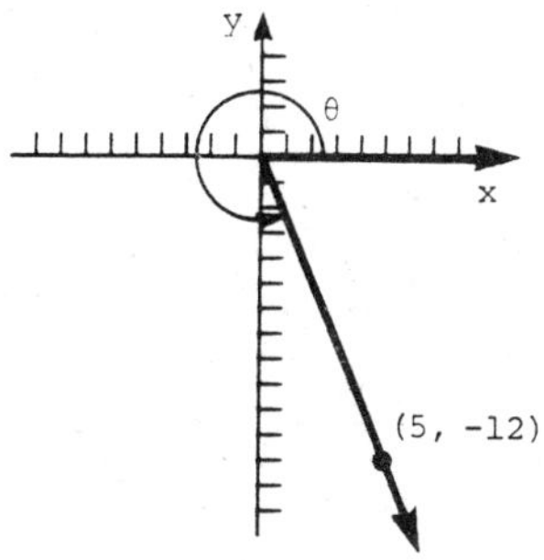

5. $(-3, 4)$

$x = -3,\ y = 4$

$$r = \sqrt{(-3)^2 + 4^2}$$
$$= \sqrt{9 + 16}$$
$$= \sqrt{25}$$
$$= 5$$

$$\sin\theta = \frac{y}{r} = \frac{4}{5}$$
$$\cos\theta = \frac{x}{r} = -\frac{3}{5}$$
$$\tan\theta = \frac{y}{x} = -\frac{4}{3}$$
$$\cot\theta = \frac{x}{y} = -\frac{3}{4}$$
$$\sec\theta = \frac{r}{x} = -\frac{5}{3}$$
$$\csc\theta = \frac{r}{y} = \frac{5}{4}$$

7. $(5, -12)$

$x = 5,\ y = -12$

$$r = \sqrt{5^2 + (-12)^2}$$
$$= \sqrt{25 + 144}$$
$$= \sqrt{169}$$
$$= 13$$

$$\sin\theta = -\frac{12}{13}$$
$$\cos\theta = \frac{5}{13}$$
$$\tan\theta = -\frac{12}{5}$$

$\cot \theta = -\frac{5}{12}$

$\sec \theta = \frac{13}{5}$

$\csc \theta = -\frac{13}{12}$

9. (6, 8)

$x = 6,\ y = 8$

$r = \sqrt{6^2 + 8^2}$
$= \sqrt{36 + 64}$
$= \sqrt{100}$
$= 10$

$\sin \theta = \frac{8}{10} = \frac{4}{5}$

$\cos \theta = \frac{6}{10} = \frac{3}{5}$

$\tan \theta = \frac{8}{6} = \frac{4}{3}$

$\cot \theta = \frac{3}{4}$

$\sec \theta = \frac{5}{3}$

$\csc \theta = \frac{5}{4}$

11. (-7, 24)

$x = -7,\ y = 24$

$r = \sqrt{(-7)^2 + 24^2}$
$= \sqrt{49 + 576}$
$= \sqrt{625}$
$= 25$

$\sin \theta = \frac{24}{25}$

$\cos \theta = -\frac{7}{25}$

$\tan \theta = -\frac{24}{7}$

$\cot \theta = -\frac{7}{24}$

$\sec \theta = -\frac{25}{7}$

$\csc \theta = \frac{25}{24}$

13. (0, 2)

$x = 0,\ y = 2$

$r = \sqrt{0^2 + 2^2}$
$= \sqrt{0 + 4}$
$= \sqrt{4}$
$= 2$

$\sin \theta = \frac{2}{2} = 1$

$\cos \theta = \frac{0}{2} = 0$

$\tan \theta = \frac{2}{0}$ Undefined

$\cot \theta = \frac{0}{2} = 0$

$\sec \theta = \frac{2}{0}$ Undefined

$\csc \theta = \frac{2}{2} = 1$

15. (8, 0)

$x = 8,\ y = 0$

$r = \sqrt{8^2 + 0^2}$
$= \sqrt{64 + 0}$
$= \sqrt{64} = 8$

$\sin \theta = \frac{0}{8} = 0$

$\cos \theta = \frac{8}{8} = 1$

$\tan \theta = \frac{0}{8} = 0$

$\cot \theta = \frac{8}{0}$ Undefined

$\sec \theta = \frac{8}{8} = 1$

$\csc \theta = \frac{8}{0}$ Undefined

17. $(1, \sqrt{3})$

$x = 1,\ y = \sqrt{3}$

$$r = \sqrt{1^2 + (\sqrt{3})^2} = \sqrt{1 + 3} = \sqrt{4} = 2$$

$\sin\theta = \frac{\sqrt{3}}{2}$

$\cos\theta = \frac{1}{2}$

$\tan\theta = \frac{\sqrt{3}}{1} = \sqrt{3}$

$\cot\theta = \frac{1}{\sqrt{3}} = \frac{\sqrt{3}}{3}$

$\sec\theta = \frac{2}{1} = 2$

$\csc\theta = \frac{2}{\sqrt{3}} = \frac{2\sqrt{3}}{3}$

19. $(5\sqrt{3}, -5)$

$x = 5\sqrt{3},\ y = -5$

$$r = \sqrt{(5\sqrt{3})^2 + (-5)^2} = \sqrt{75 + 25} = \sqrt{100} = 10$$

$\sin\theta = -\frac{5}{10} = -\frac{1}{2}$

$\cos\theta = \frac{5\sqrt{3}}{10} = \frac{\sqrt{3}}{2}$

$\tan\theta = -\frac{5}{5\sqrt{3}} = -\frac{1}{\sqrt{3}} = -\frac{\sqrt{3}}{3}$

$\cot\theta = -\frac{\sqrt{3}}{1} = -\sqrt{3}$

$\sec\theta = \frac{2}{\sqrt{3}} = \frac{2\sqrt{3}}{3}$

$\csc\theta = -\frac{2}{1} = -2$

21. $(2\sqrt{2}, -2\sqrt{2})$

$x = 2\sqrt{2},\ y = -2\sqrt{2}$

$$r = \sqrt{(2\sqrt{2})^2 + (-2\sqrt{2})^2} = \sqrt{8 + 8} = \sqrt{16} = 4$$

$\sin\theta = -\frac{2\sqrt{2}}{4} = -\frac{\sqrt{2}}{2}$

$\cos\theta = \frac{2\sqrt{2}}{4} = \frac{\sqrt{2}}{2}$

$\tan\theta = -\frac{2\sqrt{2}}{2\sqrt{2}} = -1$

$\cot\theta = \frac{2\sqrt{2}}{-2\sqrt{2}} = -1$

$\sec\theta = \frac{4}{2\sqrt{2}} = \sqrt{2}$

$\csc\theta = \frac{4}{-2\sqrt{2}} = -\sqrt{2}$

23. $(\sqrt{5}, -2)$

$x = \sqrt{5},\ y = -2$

$$r = \sqrt{(\sqrt{5})^2 + (-2)^2} = \sqrt{5 + 4} = \sqrt{9} = 3$$

$\sin\theta = -\frac{2}{3}$

$\cos\theta = \frac{\sqrt{5}}{3}$

$\tan\theta = -\frac{2}{\sqrt{5}} = -\frac{2\sqrt{5}}{5}$

$\cot\theta = -\frac{\sqrt{5}}{2}$

$\sec\theta = \frac{3}{\sqrt{5}} = \frac{3\sqrt{5}}{5}$

$\csc\theta = -\frac{3}{2}$

25. $(-\sqrt{13}, \sqrt{3})$

$x = -\sqrt{13}, \ y = \sqrt{3}$

$r = \sqrt{(-\sqrt{13})^2 + (\sqrt{3})^2}$

$= \sqrt{13 + 3}$

$= \sqrt{16}$

$= 4$

$\sin \theta = \frac{\sqrt{3}}{4}$

$\cos \theta = -\frac{\sqrt{13}}{4}$

$\tan \theta = \frac{\sqrt{3}}{-\sqrt{13}} = -\frac{\sqrt{39}}{13}$

$\cot \theta = \frac{-\sqrt{13}}{\sqrt{3}} = -\frac{\sqrt{39}}{3}$

$\sec \theta = \frac{4}{-\sqrt{13}} = -\frac{4\sqrt{13}}{13}$

$\csc \theta = \frac{4}{\sqrt{3}} = \frac{4\sqrt{3}}{3}$

27. $(\sqrt{15}, -\sqrt{10})$

$x = \sqrt{15}, \ y = -\sqrt{10}$

$r = \sqrt{(\sqrt{15})^2 + (-\sqrt{10})^2}$

$= \sqrt{15 + 10}$

$= \sqrt{25}$

$= 5$

$\sin \theta = \frac{-\sqrt{10}}{5} = -\frac{\sqrt{10}}{5}$

$\cos \theta = \frac{\sqrt{15}}{5}$

$\tan \theta = \frac{-\sqrt{10}}{\sqrt{15}} = -\frac{\sqrt{2}}{\sqrt{3}} = -\frac{\sqrt{6}}{3}$

$\cot \theta = \frac{\sqrt{15}}{-\sqrt{10}} = -\frac{\sqrt{3}}{\sqrt{2}} = -\frac{\sqrt{6}}{2}$

$\sec \theta = \frac{5}{\sqrt{15}} = \frac{5\sqrt{15}}{15} = \frac{\sqrt{15}}{3}$

$\csc \theta = -\frac{5}{\sqrt{10}} = \frac{-5\sqrt{10}}{10} = -\frac{\sqrt{10}}{2}$

29. $(8.7691, -3.2473)$

$x = 8.7691, \ y = -3.2473$

$r = \sqrt{(8.7691)^2 + (-3.2473)^2}$

$= \sqrt{87.442072}$

$= 9.3510466$

$\sin \theta = \frac{y}{r} = -.34727$

$\cos \theta = \frac{x}{r} = .93777$

$\tan \theta = \frac{y}{x} = -.37031$

$\cot \theta = \frac{x}{y} = -2.7004$

$\sec \theta = \frac{r}{x} = 1.0664$

$\csc \theta = \frac{r}{y} = -2.8796$

31. Since angle θ is a nonquadrantal angle, $\sin \theta = \frac{y}{r}$ and $\csc \theta = \frac{r}{y}$. Since $r = \sqrt{x^2 + y^2}$, r is always positive. Therefore, if y is positive, $\frac{y}{r}$ and $\frac{r}{y}$ will be positive. If y is negative, $\frac{y}{r}$ and $\frac{r}{y}$ will be negative. Therefore, $\sin \theta$ and $\csc \theta$ will have the same sign.

33. In the definitions of the sine, cosine, secant, and cosecant functions, the value of r is the hypotenuse of the triangle formed by the x-axis, the line connecting (x, y) and (0, 0) (assuming (x, y) is on the terminal side of the angle), and the vertical line through (x, y), perpendicular to the x-axis.

In Exercises 35-45, $r = \sqrt{x^2 + y^2}$, which is positive.

35. In quadrant II, y is positive so y/r is positive.

37. In quadrant III, y is negative so y/r is negative.

39. In quadrant III, both x and y are negative so y/x is positive.

41. In quadrant IV, x is positive so x/r is positive.

43. In quadrant IV, x is positive and y is negative so y/x is negative.

45. In quadrant III, x is negative so r/x is negative.

47. Since $x \geq 0$, the graph of the line $2x + y = 0$ is shown to the right of the y-axis. A point on this line is (1, -2) since $2(1) + (-2) = 0$.

$r = \sqrt{1^2 + (-2)^2} = \sqrt{1 + 4} = \sqrt{5}$

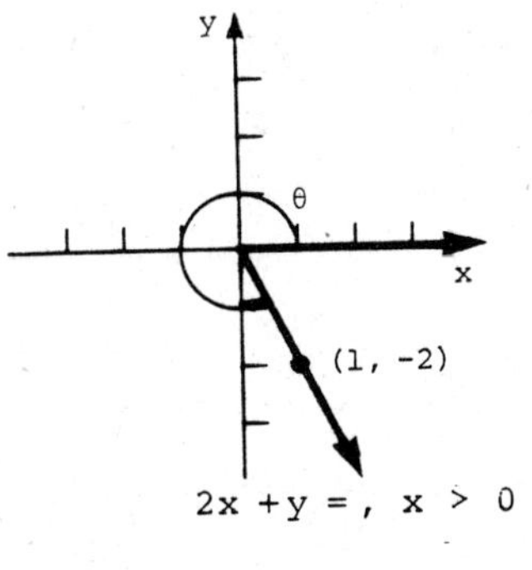

$\sin\theta = -\frac{2\sqrt{5}}{5}$

$\cos\theta = \frac{\sqrt{5}}{5}$

$\tan\theta = -2$

$\cot\theta = -\frac{1}{2}$

$\sec\theta = \sqrt{5}$

$\csc\theta = -\frac{\sqrt{5}}{2}$

49. Since $x \leq 0$, the graph of the line $-4x + 7y = 0$ is shown to the left of the y-axis. A point on this line is (-7, -4) since $-4(-7) + 7(-4) = 0$.

$r = \sqrt{(-7)^2 + (-4)^2} = \sqrt{49 + 16} = \sqrt{65}$

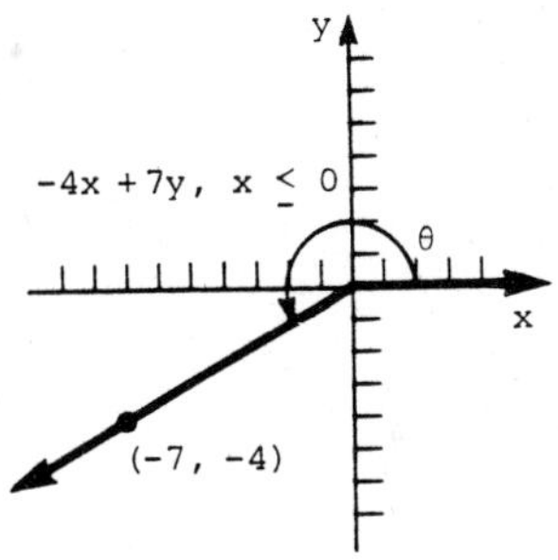

$\sin\theta = -\frac{4\sqrt{65}}{65}$

$\cos\theta = -\frac{7\sqrt{65}}{65}$

$\tan\theta = \frac{4}{7}$

$\cot\theta = \frac{7}{4}$

$\sec\theta = -\frac{\sqrt{65}}{7}$

$\csc\theta = -\frac{\sqrt{65}}{4}$

51. Since $x \le 0$, the graph of the line $-5x - 3y = 0$ is shown to the left of the y-axis. A point on this line is $(-3, 5)$ since $-5(-3) - 3(5) = 0$.

$r = \sqrt{(-3)^2 + (5)^2} = \sqrt{9 + 25} = \sqrt{34}$

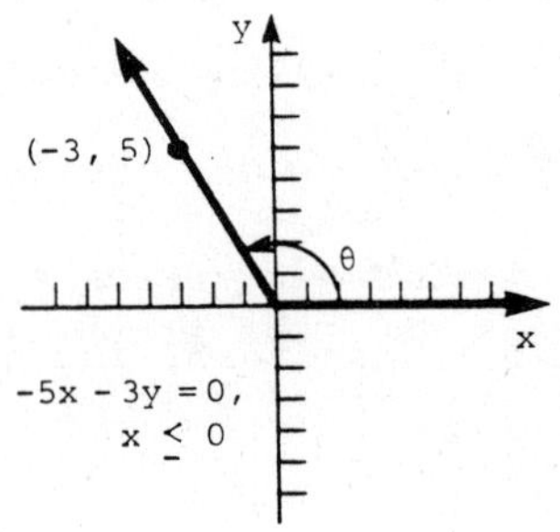

$\sin\theta = \frac{5\sqrt{34}}{34}$

$\cos\theta = -\frac{3\sqrt{34}}{34}$

$\tan\theta = -\frac{5}{3}$

$\cot\theta = -\frac{3}{5}$

$\sec\theta = -\frac{\sqrt{34}}{3}$

$\csc\theta = \frac{\sqrt{34}}{5}$

53. $\cos 90° + 3\sin 270°$
$= 0 + 3(-1) = -3$

55. $3\sec 180° - 5\tan 360°$
$= 3(-1) - 5(0) = -3$

57. $\tan 360° + 4\sin 180° + \cos^2 180°$
$= 0 + 4(0) + 5(-1)^2 = 5$

59. $\sin^2 180° + \cos^2 180°$
$= 0^2 + (-1)^2$
$= 1$

61. $\sec^2 180° - 3\sin^2 360° + 2\cos 180°$
$= (-1)^2 - 3(0)^2 + 2(-1)$
$= 1 - 0 - 2$
$= -1$

63. $\sin[n \cdot 180°]$

This angle is a quadrantal angle whose terminal side lies on either the positive part of the x-axis or the negative part of the x-axis. Any point on these terminal sides would have the form $(k, 0)$, k is any real number $k \neq 0$.

$$\sin[n \cdot 180°] = \frac{y}{r} = \frac{0}{\sqrt{k^2 + 0^2}} = 0$$

65. $\tan[(2n + 1) \cdot 90°]$

This angle is a quadrantal angle whose terminal side lies on either the positive part of the y-axis or the negative part of the y-axis. Any point on these terminal sides would have the form $(0, k)$, k is any real number, $k \neq 0$.

$$\tan[(2n + 1) \cdot 90°] = \frac{y}{x} = \frac{k}{0}$$

$\tan[(2n + 1) \cdot 90°]$ is undefined.

Section 1.5

1. $\csc\theta = 3$

$\sin\theta = \frac{1}{\csc\theta} = \frac{1}{3}$

3. $\tan \beta = -\frac{1}{5}$

$\cot \beta = \frac{1}{\tan \beta}$

$= \frac{1}{-\frac{1}{5}}$

$= -5$

5. $\sin \alpha = \frac{\sqrt{2}}{4}$

$\csc \alpha = \frac{1}{\sin \alpha}$

$= \frac{1}{\frac{\sqrt{2}}{4}}$

$= \frac{4}{\sqrt{2}} \cdot \frac{\sqrt{2}}{\sqrt{2}}$

$= \frac{4\sqrt{2}}{2} = 2\sqrt{2}$

7. $\cot \theta = -\frac{\sqrt{5}}{3}$

$\tan \theta = \frac{1}{\cot \theta}$

$= \frac{1}{-\frac{\sqrt{5}}{3}}$

$= -\frac{3}{\sqrt{5}} \cdot \frac{\sqrt{5}}{\sqrt{5}}$

$= -\frac{3\sqrt{5}}{5}$

9. $\csc \theta = 1.42716321$

$\sin \theta = \frac{1}{\csc \theta}$

$= \frac{1}{1.42716327}$

$= .70069071$

11. $\cot \alpha = .43900273$

$\tan \alpha = \frac{1}{\cot \alpha}$

$= \frac{1}{.43900273}$

$= 2.2778902$

13. It is impossible for $\sin \gamma > 0$ and $\csc \gamma < 0$. Suppose $\sin \gamma > 0$, since $\csc \gamma = \frac{1}{\sin \gamma}$, one divided by a positive number yields a positive number. Therefore, $\csc \gamma$ must be positive.

15. Since $\tan \theta = \frac{1}{\cot \theta}$, multiply both sides by $\cot \theta$.

$$\tan \theta \cot \theta = 1$$

Now divide both sides by $\tan \theta$.

$$\cot \theta = \frac{1}{\tan \theta}$$

17. $\cot \gamma = 2$

$\tan \gamma = \frac{1}{\cot \gamma}$

$= \frac{1}{2}$

19. $\cot \omega = \frac{\sqrt{3}}{3}$

$\tan \omega = \frac{1}{\cot \omega}$

$= \frac{1}{\frac{\sqrt{3}}{3}}$

$= \frac{3}{\sqrt{3}} = \sqrt{3}$

21. $\cot \alpha = -.01$

$$\tan \alpha = \frac{1}{\cot \alpha} = \frac{1}{-.01} = -100$$

23. $$\cos (6A + 5°) = \frac{1}{\sec (4A + 15°)}$$
$$= \cos (4A + 15°)$$
$$6A + 5° = 4A + 15°$$
$$2A = 10°$$
$$A = 5°$$

25. $$\sin (4\theta + 2°) \csc (3\theta + 5°) = 1$$
$$\sin (4\theta + 2°) = \frac{1}{\csc (30 + 5°)}$$
$$\sin (4\theta + 2°) = \sin (3\theta + 5°)$$
$$4\theta + 2° = 3\theta + 5°$$
$$\theta = 3°$$

27. $$\frac{1}{\sin (3\theta - 1°)} = \csc (2\theta + 3°)$$
$$\csc (3\theta - 1°) = \csc (2\theta + 3°)$$
$$3\theta - 1° = 2\theta + 3°$$
$$\theta = 4°$$

29. $\sin \alpha > 0$ implies α is in quadrant I or II. $\cos \alpha < 0$ implies α is in quadrant II or III. α is in quadrant II.

31. $\sec \theta < 0$ implies θ is in quadrant II or III. $\csc \theta < 0$ implies θ is in quadrant III or IV. θ is in quadrant III.

33. $\sin \beta < 0$ implies β is in quadrant III or IV. $\cos \beta > 0$ implies β is in quadrant I or IV. β is in quadrant IV.

35. $\tan \omega < 0$ implies ω is in quadrant II or IV. $\cot \omega < 0$ implies ω is in quadrant II or IV. ω is in quadrant II or quadrant IV.

37. $\sin \alpha > 0$ implies α is in quadrant I or II.

39. $\tan \theta > 0$ implies θ is in quadrant I or III.

In Exercises 41–51, the signs + for positive and – for negative are given in the following order: sine and cosecant, cosine and secant, tangent and cotangent.

41. 74° is in quadrant I.

+; +; +

43. 183° is in quadrant III.

–; –; +

45. 302° is in quadrant IV.

–; +; –

47. 406° is in quadrant I.

+; +; +

49. –82° is in quadrant IV.

–; +; –

51. –121° is in quadrant III.

–; –; +

53. $-1 \le \sin\theta \le 1$ for all θ and 2 is not in this range.

$\sin\theta = 2$ is impossible.

55. $\tan\beta$ can take on all values.

$\tan\beta = .92$ is possible.

57. $\csc\alpha \ge 1$ or $\csc\alpha \le -1$ for all α and 1/2 is not in this range.

$\csc\alpha = 1/2$ is impossible.

59. $\tan\theta$ can take on all values.

$\tan\theta = 1$ is possible.

61. $\sin\beta + 1 = .6$ means $\sin\beta = -.4$, which is possible because

$-1 \le \sin\beta \le 1.$

$\sin\beta + 1 = .6$ is possible.

63. $\csc\theta - 1 = -.2$ means $\csc\theta = .8$, which is not possible because

$\csc\theta \le -1$ or $\csc\theta \ge 1$.

$\csc\theta - 1 = -.2$ is impossible.

65. $\sin\alpha = 1/2$ is possible because $-1 \le \sin\alpha \le 1$. Furthermore, when

$$\sin\alpha = \frac{1}{2},$$

$$\csc\alpha = \frac{1}{\sin\alpha} = \frac{1}{1/2} = 2.$$

$\sin\alpha = 1/2$ and $\csc\alpha = 2$ is possible.

67. $\tan\beta = 2$ is possible, but when $\tan\beta = 2$,

$$\cot\beta = \frac{1}{\tan\beta} = \frac{1}{2}.$$

$\tan\beta = 2$ and $\cot\beta = -2$ is impossible.

69. Find $\cos\theta$ if $\sin\theta = 2/3$ and θ is in quadrant II.

$\sin^2\theta + \cos^2 = 1,$

$\cos^2\theta = 1 - \sin^2\theta$

$$\cos^2\theta = 1 - \left(\frac{2}{3}\right)^2 = 1 - \frac{4}{9} = \frac{5}{9}$$

$$\cos\theta = \pm\sqrt{\frac{5}{9}} = \pm\frac{\sqrt{5}}{\sqrt{9}} = \pm\frac{\sqrt{5}}{3}$$

θ is in quadrant II so $\cos\theta = -\frac{\sqrt{5}}{3}$.

71. Find $\csc\beta$ if $\cot\beta = -1/2$ and β is in quadrant IV.

$1 + \cot^2\beta = \csc^2\beta$

$$\csc^2\beta = 1 + \left(-\frac{1}{2}\right)^2 = 1 + \frac{1}{4} = \frac{5}{4}$$

$$\csc\beta = \pm\sqrt{\frac{5}{4}} = \pm\frac{\sqrt{5}}{\sqrt{4}} = \pm\frac{\sqrt{5}}{2}$$

β is in quadrant IV so $\csc\beta = -\frac{\sqrt{5}}{2}$.

73. Find $\sec\theta$ if $\tan\theta = \sqrt{7}/3$ and θ is in quadrant III.

$\tan^2\theta + 1 = \sec^2\theta$

$$\sec^2 \theta = \left(\frac{\sqrt{7}}{3}\right)^2 + 1$$
$$= \frac{7}{9} + \frac{9}{9} = \frac{16}{9}$$
$$\sec \theta = \pm\sqrt{\frac{16}{9}} = \pm\frac{4}{3}$$

θ is in quadrant III so $\sec \theta = -\frac{4}{3}$.

75. Find $\sin \theta$, if $\sec \theta = 2$, θ is in quadrant IV.

Since $\sec \theta = 2$ and $\cos \theta = \frac{1}{\sec \theta}$, $\cos \theta = \frac{1}{2}$.

Since $\sin^2 \theta + \cos^2 \theta = 1$, and $\cos \theta = 1/2$,

$$\sin^2 \theta + \left(\frac{1}{2}\right)^2 = 1$$
$$\sin^2 \theta + \frac{1}{4} = 1$$
$$\sin^2 \theta = \frac{3}{4}$$
$$\sin \theta = \pm\sqrt{\frac{3}{4}} = \pm\frac{\sqrt{3}}{2}$$

Since θ is in quadrant IV, and $\sin \theta$ is negative in quadrant IV, $\sin \theta = -\frac{\sqrt{3}}{2}$.

77. Find $\cos \alpha$ if $\sin \alpha = -\frac{\sqrt{5}}{7}$, with α in quadrant IV.

$$\sin^2 \alpha + \cos^2 \alpha = 1$$
$$\left(-\frac{\sqrt{5}}{7}\right)^2 + \cos^2 \alpha = 1$$
$$\frac{5}{49} + \cos^2 \alpha = 1$$
$$\cos^2 \alpha = \frac{44}{49}$$
$$\cos \alpha = \pm\sqrt{\frac{44}{49}} = \pm\frac{2\sqrt{11}}{7}$$

Since α is in quadrant IV, $\cos \alpha > 0$, so

$$\cos \alpha = \frac{2\sqrt{11}}{7}.$$

79. Find $\cot \alpha$ if $\csc \alpha = -3.5891420$, with α in quadrant III.

$$1 + \cot^2 \alpha = \csc^2$$
$$1 + \cot^2 \alpha = (-3.5891420)^2$$
$$1 + \cot^2 \alpha = 12.8819403$$
$$\cot^2 \alpha = 11.8819403$$
$$\cot \alpha = \pm 3.44701905$$

Since α is in quadrant III, $\cot \alpha > 0$, so

$$\cot \alpha = 3.44701905.$$

81. Find $\tan \beta$, if $\sin \beta = .49268329$, with β in quadrant II.

$$\sin^2 \beta + \cos^2 \beta = 1$$
$$\cos^2 \beta = 1 - \sin^2 \beta$$
$$\cos \beta = \pm\sqrt{1 - \sin^2 \beta}$$

$\cos \beta$ is in quadrant II, and the cosine is negative in quadrant II,

$$\cos \beta = -\sqrt{1 - \sin^2 \beta}$$
$$= -\sqrt{1 - (.49268329)^2}$$
$$= -\sqrt{1 - .24273682}$$
$$= -\sqrt{.75726318}$$
$$= -.87020870$$

$$\tan \beta = \frac{\sin \theta}{\cos \theta} = \frac{.49268329}{-.87020870}$$
$$= -.56616682$$

Hint for Exercises 83–89: Remember that r is always positive.

83. $\cos \alpha = -\frac{3}{5}$

$\cos \alpha = \frac{x}{r}$ so let

$x = -3,\ r = 5.$

$x^2 + y^2 = r^2$

$(-3)^2 + y^2 = 5^2$

$9 + y^2 = 25$

$y^2 = 25 - 9 = 16$

$y = \pm\sqrt{16}$

$y = \pm 4$

α is in quadrant III so $y = -4$.

$\sin \alpha = -\frac{4}{5}$

$\cos \alpha = -\frac{3}{5}$

$\tan \alpha = \frac{-4}{-3} = \frac{4}{3}$

$\cot \alpha = \frac{3}{4}$

$\sec \alpha = -\frac{5}{3}$

$\csc \alpha = -\frac{5}{4}$

85. $\sin \beta = \frac{7}{25}$, $\sin \beta = \frac{y}{r}$ so let $y = 7$,

$r = 25.$

$x^2 + y^2 = r^2,$

$x^2 + 7^2 = 25^2,$

$x^2 + 49 = 625$

$x^2 = 625 - 49 = 576,$

$x = \pm\sqrt{576} = \pm 24$

β is in quadrant II so $x = -24$.

$\sin \beta = \frac{7}{25}$

$\cos \beta = -\frac{24}{25}$

$\tan \beta = -\frac{7}{24}$

$\cot \beta = -\frac{24}{7}$

$\sec \beta = -\frac{25}{24}$

$\csc \beta = \frac{25}{7}$

87. $\csc \theta = 2 = \frac{2}{1} = \frac{r}{y}$, so

let $y = 1,\ r = 2.$

$x^2 + y^2 = r^2$

$x^2 + 1^2 = 2^2$

$x^2 + 1 = 4$

$x^2 = 4 - 1 = 3$

$x = \pm\sqrt{3}$

θ is in quadrant II, so $x = -\sqrt{3}$.

$\sin \theta = \frac{1}{2}$

$\cos \theta = -\frac{\sqrt{3}}{2}$

$\tan \theta = \frac{1}{-\sqrt{3}} = -\frac{\sqrt{3}}{3}$

$\cot \theta = -\frac{\sqrt{3}}{1} = -\sqrt{3}$

$\sec \theta = -\frac{2}{\sqrt{3}} = -\frac{2\sqrt{3}}{3}$

$\csc \theta = \frac{2}{1} = 2$

89. $\cot \alpha = \frac{\sqrt{3}}{8} = \frac{x}{y}$

$\sin \alpha > 0$ means $y > 0$, so let

$x = \sqrt{3},\ y = 8.$

$$x^2 + y^2 = r^2$$
$$(\sqrt{3})^2 + 8^2 = r^2$$
$$3 + 64 = r^2$$
$$r^2 = 67$$
$$r = \sqrt{67}$$

$$\sin \alpha = \frac{8}{\sqrt{67}} = \frac{8\sqrt{67}}{67}$$

$$\cos \alpha = \frac{\sqrt{3}}{\sqrt{67}} = \frac{\sqrt{201}}{67}$$

$$\tan \alpha = \frac{8}{\sqrt{3}} = \frac{8\sqrt{3}}{3}$$

$$\cos \alpha = \frac{\sqrt{3}}{8}$$

$$\sec \alpha = \frac{\sqrt{67}}{\sqrt{3}} = \frac{\sqrt{201}}{3}$$

$$\csc \alpha = \frac{\sqrt{67}}{8}$$

91. $\tan \theta = \frac{3}{2} = \frac{y}{x}$, $\csc \theta = \frac{\sqrt{13}}{3} = \frac{r}{y}$

Let $x = 2$, $y = 3$, $r = \sqrt{13}$.

$$\sin \theta = \frac{3}{\sqrt{13}} = \frac{3\sqrt{13}}{13}$$

$$\cos \theta = \frac{2}{\sqrt{13}} = \frac{2\sqrt{13}}{13}$$

$$\tan \theta = \frac{3}{2}$$

$$\cot \theta = \frac{2}{3}$$

$$\sec \theta = \frac{\sqrt{13}}{2}$$

$$\csc \theta = \frac{\sqrt{13}}{3}$$

93. $\sin \alpha = .164215 = \frac{.164215}{1}$

$$\sin \alpha = \frac{y}{r}$$

so let $y = .164215$, $r = 1$.

$$x^2 + y^2 = r^2$$
$$x^2 + (.164215)^2 = 1^2$$
$$x^2 + .026966 = 1$$
$$x^2 = 1 - .026966$$
$$= .973034$$
$$x = \pm\sqrt{.973034}$$
$$= \pm.986425$$

α is in quadrant II, so,

$x = -.986425$.

$$\sin \alpha = .164215$$

$$\cos \alpha = -.986425$$

$$\tan \alpha = -.166475$$

$$\cot \alpha = -\frac{.986425}{.164215}$$
$$= -6.00691$$

$$\sec \alpha = \frac{1}{-.986425}$$
$$= -1.01376$$

$$\csc \alpha = \frac{1}{.164215}$$
$$= 6.08958$$

95. $\sin \gamma = a = \frac{a}{1} = \frac{y}{r}$

so let $y = a$, $r = 1$.

$$x^2 + y^2 = r^2$$
$$x^2 + a^2 = 1$$
$$x^2 = 1 - a^2$$
$$x = \pm\sqrt{1 - a^2}$$

γ is in quadrant I, so $x = \sqrt{1 - a^2}$.

$$\sin \gamma = a$$

$$\cos \gamma = \frac{\sqrt{1 - a^2}}{1}$$
$$= \sqrt{1 - a^2}$$

$$\tan \gamma = \frac{1}{\sqrt{1 - a^2}}$$
$$= \frac{a\sqrt{1 - a^2}}{1 - a^2}$$

$$\cot \gamma = \frac{\sqrt{1 - a^2}}{a}$$

$$\sec \gamma = \frac{1}{\sqrt{1 - a^2}}$$

$$= \frac{\sqrt{1 - a^2}}{1 - a^2}$$

$$\csc \gamma = \frac{1}{a}$$

97. $x^2 + y^2 = r^2$

Divide by y^2.

$$\frac{x^2}{y^2} + \frac{y^2}{y^2} = \frac{r^2}{y^2}$$

$$\frac{x^2}{y^2} + 1 = \frac{r^2}{y^2}$$

$$\left(\frac{x}{y}\right)^2 + 1 = \left(\frac{r}{y}\right)^2$$

$$\cot^2 \theta + 1 = \csc^2 \theta$$

99. True or false: For all angles θ, $\sin \theta + \cos \theta = 1$.

The statement is false.

For example, let $\theta = 270°$.

$$\sin 270° + \cos 270° = -1 + 0$$
$$= -1 \neq 1$$

Chapter 1 Review Exercises

1. $\{x | x \leq -4\}$ This interval is a half-open interval with interval notation $(-\infty, -4]$.

3. $(4, -2)$ and $(1, -6)$

$$d = \sqrt{(1 - 4)^2 + [-6 - (-2)]^2}$$
$$= \sqrt{(-3)^2 + (-4)^2}$$
$$= \sqrt{9 + 16}$$
$$= \sqrt{25} = 5$$

5. Let $A = (-2, -2)$, $B = (8, 4)$ and $C = (2, 14)$.

Distance from A to B is

$$\sqrt{[8 - (-2)]^2 + [4 - (-2)]^2}$$
$$= \sqrt{100 + 36}$$
$$= \sqrt{136} = 2\sqrt{34}.$$

Distance A to C is

$$\sqrt{(-2 - 2)^2 + (-2 - 14)^2} = \sqrt{16 + 256}$$
$$= \sqrt{272}$$
$$= 4\sqrt{17}.$$

Distance from B to C is

$$\sqrt{(8 - 2)^2 + (4 - 14)^2} = \sqrt{36 + 100}$$
$$= \sqrt{136} = 2\sqrt{34}.$$

Does $(2\sqrt{34})^2 + (2\sqrt{34})^2 = (4\sqrt{17})^2$?

Yes, since $136 + 136 = 272$.

Therefore, these points are vertices of a right triangle.

In Exercises 7 and 9, $f(x) = -x^2 + 3x + 2$.

7. $f(0) = -(0)^2 + 3(0) + 2$
$= -0 + 0 + 2$
$= 2$

9. $f(a) = -a^2 + 3a + 2$

11. $y = 9x + 2$

Domain: x can take on any value so the domain is $(-\infty, \infty)$.

Range: y can take on any value, so the range is $(-\infty, \infty)$.

Each x-value gives one y-value so $y = 9x + 2$ is a function.

13. $y = |x|$

Domain: x can take on any value, so the domain is $(-\infty, \infty)$.

Range: y will be any nonnegative number since the absolute value of a number is non-negative. The range is $[0, \infty)$.

Each x-value gives one y-value, so $y = |x|$ is a function.

15. $x + 1 = y^2$

Domain: y^2 is always ≥ 0, so $x + 1 \geq 0$ and $x \geq -1$. The domain is $[-1, \infty)$.

Range: y can take on any value, so the range is $(-\infty, \infty)$.

Some x-values give more than one y-value. For example, if $x = 3$,

$y = 2$ or $y = -2$.

So $x + 1 = y^2$ is not a function.

17. Domain: Since every point on this graph has an x-value between −3 and 3 inclusive, the domain is $[-3, 3]$.

Range: Since every point on this graph has a y-value between −5 and 5 inclusive, the range is $[-5, 5]$.

Notice that the points (0, 5) and (0, −5) lie on the graph. Therefore, one x-value gives more than one y-value. So this is not the graph of a function.

19. −51° is coterminal with

$-51° + 360° = 309°$.

21. 792° is coterminal with

$792° - 2(360°) = 72°$.

23. 320 rotations per minute

$= 320(360°)$ per minute

$= 115{,}200°$ per minute

$= \frac{115{,}200°}{60}$ per second

$= 1920°$ per second

$= 1920° \left(\frac{2}{3}\right)$ per $\frac{2}{3}$ second

$= 1280°$ per $\frac{2}{3}$ second

25. 47° 25′ 11″

$= 47° + \frac{25}{60}° + \frac{11}{3600}°$

$= 47° + .417° + .003°$

$= 47.420°$

27. 74.2983°

$= 74° + (.2983)(60')$

$= 74° + 17.898'$

$= 74° + 17' + (.898)(60'')$

$= 74° + 17' + 53.88''$

$\approx 74°\ 17'\ 54''$

29. 183.0972°

$= 183° + (.0972)(60')$

$= 183° + 5.832'$

$= 183° + 5' + (.832)(60'')$

$= 183° + + 5' + 49.92''$

$= 183°\ 05'\ 50''$

31. The angles $(9x + 4)°$ and $(12x - 14)°$ are vertical angles, which have equal measure.

$$9x + 4 = 12x - 4$$
$$4 = 3x - 14$$
$$18 = 3x$$
$$6 = x$$

The angles would be 58° and 58°.

Hint for Exercises 33–41: In similar triangles, corresponding sides are in proportion and correspnding angles are equal.

33. $V = X = 41°$

$Z = T = 32°$

The sum of the angles in a triangle is 180° so

$$Y = U = 180° - (41° + 32°)$$
$$= 107°.$$

35. $\frac{75}{50} = \frac{3}{2}$

$$\frac{m}{30} = \frac{n}{40} = \frac{3}{2}$$

$$m = \frac{3(30)}{2} = 45$$

$$n = \frac{3(40)}{2} = 60$$

37. $\frac{6}{r} = \frac{7}{11 + 7}$

$$\frac{6}{r} = \frac{7}{18}$$

$$r = \frac{6(18)}{7} = \frac{108}{7}$$

39. It two triangles are similar, their corresponding sides are <u>proportional</u>, and their corresponding angles are <u>equal</u>.

41. $x = -3,\ y = -3$

$$r = \sqrt{x^2 + y^2}$$
$$= \sqrt{(-3)^2 + (-3)^2}$$
$$= \sqrt{9 + 9}$$
$$= \sqrt{18}$$
$$= 3\sqrt{2}$$

$$\sin \theta = \frac{y}{r} = -\frac{3}{3\sqrt{2}} = -\frac{1}{\sqrt{2}} = -\frac{\sqrt{2}}{2}$$

$$\cos \theta = \frac{x}{r} = -\frac{3}{3\sqrt{2}} = -\frac{1}{\sqrt{2}} = -\frac{\sqrt{2}}{2}$$

$$\tan \theta = \frac{y}{x} = \frac{-3}{-3} = 1$$

$$\cot \theta = \frac{x}{y} = \frac{-3}{-3} = 1$$

$$\sec \theta = \frac{r}{x} = \frac{3\sqrt{2}}{-3} = -\sqrt{2}$$

$$\csc \theta = \frac{r}{y} = \frac{3\sqrt{2}}{-3} = -\sqrt{2}$$

43. $\sin 180° = 0$

$\cos 180° = -1$

$\tan 180° = 0$

$\cot 180°$ is undefined.

$\sec 180° = -1$

$\csc 180°$ is undefined.

45. $(-8, 15),\ x = -8,\ y = 15,$

$$r = \sqrt{(-8)^2 + 15^2}$$
$$= \sqrt{64 + 225}$$
$$= \sqrt{289} = 17$$

$$\sin \theta = \frac{y}{r} = \frac{15}{17}$$

$$\cos \theta = \frac{x}{r} = -\frac{8}{17}$$

$$\tan \theta = \frac{y}{x} = \frac{15}{-8} = -\frac{15}{8}$$

$\cot\theta = \frac{x}{y} = -\frac{8}{15}$

$\sec\theta = \frac{r}{x} = \frac{17}{-8} = -\frac{17}{8}$

$\csc\theta = \frac{r}{y} = \frac{17}{15}$

47. $(1, -5)$, $x = 1$, $y = -5$,

$r = \sqrt{1^2 + (-5)^2}$
$= \sqrt{1 + 25}$
$= \sqrt{26}$

$\sin\theta = \frac{y}{r} = -\frac{5}{\sqrt{26}} = -\frac{5\sqrt{26}}{26}$

$\cos\theta = \frac{x}{r} = \frac{1}{\sqrt{26}} = \frac{\sqrt{26}}{26}$

$\tan\theta = \frac{y}{x} = -\frac{5}{1} = -5$

$\cot\theta = \frac{x}{y} = \frac{1}{-5} = -\frac{1}{5}$

$\sec\theta = \frac{r}{x} = \frac{\sqrt{26}}{1} = \sqrt{26}$

$\csc\theta = \frac{r}{y} = \frac{\sqrt{26}}{-5} = -\frac{\sqrt{26}}{5}$

49. $(6\sqrt{3}, -6)$, $x = 6\sqrt{3}$, $y = -6$

$r = \sqrt{(6\sqrt{3})^2 + (-6)^2}$
$= \sqrt{108 + 36}$
$= \sqrt{144} = 12$

$\sin\theta = \frac{y}{r} = -\frac{6}{12} = -\frac{1}{2}$

$\cos\theta = \frac{x}{r} = \frac{6\sqrt{3}}{12} = \frac{\sqrt{3}}{2}$

$\tan\theta = \frac{y}{x} = -\frac{6}{6\sqrt{3}} = -\frac{1}{\sqrt{3}}$
$= -\frac{\sqrt{3}}{3}$

$\cot\theta = \frac{x}{y} = \frac{6\sqrt{3}}{-6} = -\sqrt{3}$

$\sec\theta = \frac{r}{x} = \frac{12}{6\sqrt{3}}$
$= \frac{2}{\sqrt{3}}$
$= \frac{2\sqrt{3}}{3}$

$\csc\theta = \frac{r}{y} = \frac{12}{-6}$
$= -2$

51. Since the terminal side of the angle is defined by $5x - 3y = 0$, $x \geq 0$, a point on this terminal side would be $(3, 5)$ since $5(3) - 3(5) = 0$.

$r = \sqrt{3^2 + 5^2} = \sqrt{9 + 25} = \sqrt{34}$

$\sin\theta = \frac{y}{r} = \frac{5}{\sqrt{34}} = \frac{5\sqrt{34}}{34}$

$\cos\theta = \frac{x}{r} = \frac{3}{\sqrt{34}} = \frac{3\sqrt{34}}{34}$

$\tan\theta = \frac{y}{x} = \frac{5}{3}$

$\cot\theta = \frac{x}{y} = \frac{3}{5}$

$\sec\theta = \frac{r}{x} = \frac{\sqrt{34}}{3}$

$\csc\theta = \frac{r}{y} = \frac{\sqrt{34}}{5}$

53. $4\sec 180° - 2\sin^2 270°$
$= 4(-1) - 2(-1)^2$
$= -4 - 2(1)$
$= -4 - 2 = -6$

55. $\sin\theta = 3/4$ is possible because $-1 \leq \sin\theta \leq 1$ for all θ. When $\sin\theta = 3/4$ then $\csc = 4/3$ because $\csc\theta = 1/\sin\theta$.

$\sin\theta = \frac{3}{4}$ and $\csc\theta = \frac{4}{3}$ is possible.

57. $\tan\theta$ can take on all values.

$\tan\theta = 1.4$ is possible.

59. $\sin\theta = \sqrt{3}/5$, $\sin\theta = y/r$ so $y = \sqrt{3}$ and $r = 5$.

$$x^2 + y^2 = r^2$$
$$x^2 + (\sqrt{3})^2 = 5^2$$
$$x^2 + 3 = 25,\ x^2 = 25 - 3 = 22$$
$$x = \pm\sqrt{22}$$

$\cos\theta < 0$ means $x < 0$ so $x = -\sqrt{22}$.

$$\sin\theta = \frac{y}{r} = \frac{\sqrt{3}}{5}$$
$$\cos\theta = \frac{x}{r} = -\frac{\sqrt{22}}{5}$$
$$\tan\theta = \frac{y}{x} = \frac{\sqrt{3}}{-\sqrt{22}} = -\frac{\sqrt{66}}{22}$$
$$\cot\theta = \frac{x}{y} = -\frac{\sqrt{22}}{\sqrt{3}} = -\frac{\sqrt{66}}{3}$$
$$\sec\theta = \frac{r}{x} = \frac{5}{-\sqrt{22}} = \frac{5\sqrt{22}}{22}$$
$$\csc\theta = \frac{r}{y} = \frac{5}{\sqrt{3}} = \frac{5\sqrt{3}}{3}$$

61. $\tan\alpha = 2 = 2/1$, $\tan\alpha = y/x$ with α in quadrant III, so let $x = -1$ and $y = -2$.

$$x^2 + y^2 = r^2,$$
$$(-1)^2 + (-2)^2 = r^2$$
$$1 + 4 = r^2,\ r = \sqrt{5}$$

$$\sin\alpha = \frac{y}{r} = -\frac{2}{\sqrt{5}} = -\frac{2\sqrt{5}}{5}$$
$$\cos\alpha = \frac{x}{r} = -\frac{1}{\sqrt{5}} = -\frac{\sqrt{5}}{5}$$
$$\tan\alpha = \frac{y}{x} = \frac{-2}{-1} = 2$$
$$\cot\alpha = \frac{x}{y} = \frac{-1}{-2} = \frac{1}{2}$$
$$\sec\alpha = \frac{r}{x} = \frac{\sqrt{5}}{-1} = -\sqrt{5}$$
$$\csc\alpha = \frac{r}{y} = \frac{\sqrt{5}}{-2} = -\frac{\sqrt{5}}{2}$$

63. $\sin\theta = -\frac{2}{5}$, $\sin\theta = \frac{y}{r}$ with θ in quadrant III, so let $y = -2$ and $r = 5$.

$$x^2 + y^2 = r^2$$
$$x^2 + (-2)^2 = 5^2$$
$$x^2 + 4 = 25$$
$$x^2 = 21$$
$$x = \pm\sqrt{21}$$

Since x is negative in quadrant III, $x = -\sqrt{21}$.

$$\sin\theta = \frac{y}{r} = \frac{-2}{5} = -\frac{2}{5}$$
$$\cos\theta = \frac{x}{r} = \frac{-\sqrt{21}}{5} = -\frac{\sqrt{21}}{5}$$
$$\tan\theta = \frac{y}{x} = \frac{-2}{-\sqrt{21}} = \frac{2\sqrt{21}}{21}$$
$$\cot\theta = \frac{x}{y} = \frac{-\sqrt{21}}{-2} = \frac{\sqrt{21}}{2}$$
$$\sec\theta = \frac{r}{x} = \frac{5}{-\sqrt{21}} = -\frac{5\sqrt{21}}{21}$$
$$\csc\theta = \frac{r}{y} = \frac{5}{-2} = -\frac{5}{2}$$

65. The sine is negative in quadrant III and IV. The cosine is positive in quadrant I and IV. Since $\sin\theta < 0$ and $\cos\theta > 0$, θ lies in quadrant IV. Since $\tan\theta = \frac{\sin\theta}{\cos\theta}$ and

$\sin \theta < 0$ and $\cos \theta > 0$, the sign of $\tan \theta$ is negative since a negative divided by a positive is a negative.

CHAPTER 1 TEST

[1.1] Find the distance between each of the following pairs of points.

1. $(-3, -1)$ and $(5, -3)$

2. $(-6\sqrt{3}, 3\sqrt{6})$ and $(4\sqrt{3}, -\sqrt{6})$

[1.1] Write each of the following sets using interval notation.

3. $\{x|x \geq -4\}$

4. $\{t|-2 < t \leq 3\}$

[1.1] Let $f(x) = -2x^2 + x - 3$. Find each of the following.

5. $f(-2)$

6. $f(-b)$

[1.1] Find the domain and the range of each of the following.

7. $2x - 3y + 1 = 0$

8. $y = \sqrt{x^2 - 1}$

[1.2] Perform each calculation.

9. $58° 43' + 121° 31'$

10. $90° - 36° 15' 38''$

[1.2] **11.** Convert 307.2104° to degrees, minutes, seconds. Round to the nearest second.

[1.2] **12.** Convert 94° 15′ 27″ to decimal degrees. Round to the nearest thousandth of a degree.

[1.2] Find the angles of smallest possible positive measure coterminal with the following angles.

13. 980°

14. −47°

[1.3] **15.** The measures of two angles of a triangle are 38° 40′ and 51° 30′. Find the measure of the third angle.

[1.3] **16.** Find the unknown side lengths in the following pair of similar triangles.

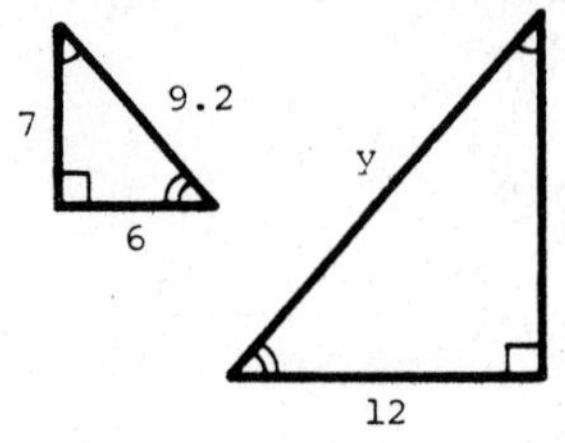

[1.4] **Find the values of the sine, cosine, and tangent functions for angle α in standard position having the given point on its terminal side.**

17. (-8, 6) **18.** (-3, -2)

[1.4] **Evaluate each of the following.**

19. $2 \tan 180° + 5 \sin 90°$ **20.** $\sec^2 180° + 2 \cot^2 90°$

[1.5] **21.** Find $\cot \theta$ and $\csc \theta$, given $\sin \theta = 3/5$ and $\cos \theta < 0$.

[1.5] **22.** Find the value of all the trigonometric functions, given $\cos \theta = -\frac{1}{2}$, where θ is in quadrant III.

[1.5] **Decide whether each of the following statements is *possible* or *impossible*.**

23. $\tan \alpha = \frac{3\sqrt{8}}{8}$ **24.** $\cos s = -\sqrt{3}$

[1.5] **25.** If for some particular angle θ, $\cot \theta$ is less in value than -1, what is the range in value of $\tan \theta$?

CHAPTER 1 TEST ANSWERS

1. $2\sqrt{17}$ **2.** $6\sqrt{11}$ **3.** $[-4, \infty)$ **4.** $(-2, 3]$ **5.** -13

6. $-2b^2 - b - 3$ **7.** $(-\infty, \infty)$; $(-\infty, \infty)$ **8.** $(-\infty, -1] \cup [1, \infty)$; $[0, \infty)$

9. $180° \, 14'$ **10.** $53° \, 44' \, 22''$ **11.** $307° \, 12' \, 37''$ **12.** $94.258°$

13. $260°$ **14.** $313°$ **15.** $89° \, 50'$ **16.** $x = 14$, $y = 18.4$

17. $\sin \alpha = 3/5$; $\cos \alpha = -4/5$; $\tan \alpha = -3/4$ **18.** $\sin \alpha = -2\sqrt{13}/13$; $\cos \alpha = -3\sqrt{13}/13$; $\tan \alpha = 2/3$ **19.** 5 **20.** 1

21. $\cot \theta = -4/3$; $\csc \theta = 5/3$ **22.** $\sin \theta = -\sqrt{3}/2$; $\cos \theta = -1/2$; $\tan \theta = \sqrt{3}$; $\cot \theta = \sqrt{3}/3$; $\sec \theta = -2$; $\csc \theta = -2\sqrt{3}/3$ **23.** Possible

24. Impossible **25.** $(-1, 0)$

$\sin \theta < 0$ and $\cos \theta > 0$, the sign of $\tan \theta$ is negative since a negative divided by a positive is a negative.

CHAPTER 2 ACUTE ANGLES AND RIGHT TRIANGLES

Section 2.1

1. $\sin A = \dfrac{\text{side opposite}}{\text{hypotenuse}} = \dfrac{3}{5}$

$\cos A = \dfrac{\text{side adjacent}}{\text{hypotenuse}} = \dfrac{4}{5}$

$\tan A = \dfrac{\text{side opposite}}{\text{side adjacent}} = \dfrac{3}{4}$

$\cot A = \dfrac{\text{side adjacent}}{\text{side opposite}} = \dfrac{4}{3}$

$\sec A = \dfrac{\text{hypotenuse}}{\text{side adjacent}} = \dfrac{5}{4}$

$\csc A = \dfrac{\text{hypotenuse}}{\text{side opposite}} = \dfrac{5}{3}$

3. $\sin A = \dfrac{\text{side opposite}}{\text{hypotenuse}} = \dfrac{21}{29}$

$\cos A = \dfrac{\text{side adjacent}}{\text{hypotenuse}} = \dfrac{20}{29}$

$\tan A = \dfrac{\text{side opposite}}{\text{side adjacent}} = \dfrac{21}{20}$

$\cot A = \dfrac{\text{side adjacent}}{\text{side opposite}} = \dfrac{20}{21}$

$\sec A = \dfrac{\text{hypotenuse}}{\text{side adjacent}} = \dfrac{29}{20}$

$\csc A = \dfrac{\text{hypotenuse}}{\text{side opposite}} = \dfrac{29}{21}$

5. $\sin A = \dfrac{\text{side opposite}}{\text{hypotenuse}} = \dfrac{n}{p}$

$\cos A = \dfrac{\text{side adjacent}}{\text{hypotenuse}} = \dfrac{m}{p}$

$\tan A = \dfrac{\text{side opposite}}{\text{side adjacent}} = \dfrac{n}{m}$

$\cot A = \dfrac{\text{side adjacent}}{\text{side opposite}} = \dfrac{m}{n}$

$\sec A = \dfrac{\text{hypotenuse}}{\text{side adjacent}} = \dfrac{p}{m}$

$\csc A = \dfrac{\text{hypotenuse}}{\text{side opposite}} = \dfrac{p}{n}$

7. $\sin A = \dfrac{\text{side opposite}}{\text{hypotenuse}} = \dfrac{.8974}{1.1819}$

$= .7593$

$\cos A = \dfrac{\text{side adjacent}}{\text{hypotenuse}} = \dfrac{.7692}{1.1819}$

$= .6508$

$\tan A = \dfrac{\text{side opposite}}{\text{side adjacent}} = \dfrac{.8974}{.7692}$

$= 1.1667$

$\cot A = \dfrac{\text{side adjacent}}{\text{side opposite}} = \dfrac{.7692}{.8974}$

$= .8571$

$\sec A = \dfrac{\text{hypotenuse}}{\text{side adjacent}} = \dfrac{1.1819}{.7692}$

$= 1.5365$

$\csc A = \dfrac{\text{hypotenuse}}{\text{side opposite}} = \dfrac{1.1819}{.8974}$

$= 1.3170$

9. $\sin A = \dfrac{\text{side opposite}}{\text{hypotenuse}} = \dfrac{8.5342}{9.7908}$

$= .8717$

$\cos A = \dfrac{\text{side adjacent}}{\text{hypotenuse}} = \dfrac{4.7986}{9.7908}$

$= .4901$

$\tan A = \dfrac{\text{side opposite}}{\text{side adjacent}} = \dfrac{8.5342}{4.7986}$

$= 1.7785$

$\cot A = \dfrac{\text{side adjacent}}{\text{side opposite}} = \dfrac{4.7986}{8.5342}$

$= .5623$

$\sec A = \dfrac{\text{hypotenuse}}{\text{side adjacent}} = \dfrac{9.7908}{4.7986}$

$= 2.0403$

$\csc A = \dfrac{\text{hypotenuse}}{\text{side opposite}} = \dfrac{9.7908}{8.5342}$

$= 1.1472$

11. $b = 7,\ c = 12$

$$c^2 = a^2 + b^2$$
$$12^2 = a^2 + 7^2$$
$$144 = a^2 + 49$$
$$a^2 = 95$$
$$a = \sqrt{95}$$

$$\sin B = \frac{b}{c} = \frac{7}{12}$$
$$\cos B = \frac{a}{c} = \frac{\sqrt{95}}{12}$$
$$\tan B = \frac{b}{a} = \frac{7}{\sqrt{95}} = \frac{7\sqrt{95}}{95}$$
$$\cot B = \frac{a}{b} = \frac{\sqrt{95}}{7}$$
$$\sec B = \frac{c}{a} = \frac{12}{\sqrt{95}} = \frac{12\sqrt{95}}{95}$$
$$\csc B = \frac{c}{b} = \frac{12}{7}$$

15. $\tan 50° = \cot (90° - 50°)$
$= \cot 40°$

17. $\csc 47° = \sec (90° - 47°)$
$= \sec 43°$

19. $\sin 38°\ 29'$
$= \cos (89°\ 60' - 38°\ 29')$
$= \cos 51°\ 31'$

21. $\sin \gamma = \cos (90° - \gamma)$

23. $\cos (\alpha + 20°)$
$= \sin [90° - (\alpha + 20°)]$
$= \sin (70° - \alpha)$

25. $\cos \theta = \sin 2\theta$

Since sine and cosine are cofunctions, this equation is true if the sum of the angles is 90°, so

$$\theta + 2\theta = 90°$$
$$3\theta = 90°$$
$$\theta = 30°.$$

27. $\sec (\beta + 10°) = \csc (2\beta + 20°)$

Since secant and cosecant are cofunctions,

$$(\beta + 10°) + (2\beta + 20°) = 90°$$
$$3\beta + 30° = 90°$$
$$3\beta = 60°$$
$$\beta = 20°.$$

29. $\cot (5\theta + 2°) = \tan (2\theta + 4°)$

Since cotangent and tangent are cofunctions,

$$(5\theta + 2°) + (2\theta + 4°) = 90°$$
$$7\theta + 6° = 90°$$
$$7\theta = 84°$$
$$\theta = 12°.$$

31. $\sec \left(\frac{\beta}{2} + 5°\right) = \csc \left(\frac{\beta}{2} + 15°\right)$

Since secant and cosecant are cofunctions,

$$\left(\frac{\beta}{2} + 5°\right) + \left(\frac{\beta}{2} + 15°\right) = 90°$$
$$\beta + 20° = 90°$$
$$\beta = 70°.$$

33. $\tan 28° \le \tan 40°$

$\tan \theta$ increases as θ increases from 0° to 90°. Since

$$40° > 28°$$

$$\tan 40° > \tan 28°.$$

Therefore, the given statement is true.

35. $\sin 46° < \cos 46°$

$\sin \theta$ increases as θ increases from 0° to 90°. Since

$$46° > 44°$$

$$\sin 46° > \sin 44°.$$

But

$$\sin 44° = \cos 46°,$$

so

$$\sin 46° > \cos 46°.$$

False

37. $\tan 41° < \cot 41°$

$\tan \theta$ increases as θ increases from θ to 90°. Since

$$49° > 41°$$

$$\tan 49° > \tan 41°.$$

But

$$\tan 49° = \cot 41°,$$

so

$$\cot 41° > \tan 41°.$$

True

39. $\sin 60° \le \cos 30°$

$$\sin 60° = \cos (90° - 60°)$$
$$= \cos 30°,$$

so

$\sin 60° \le \cos 30°$ is true.

41.

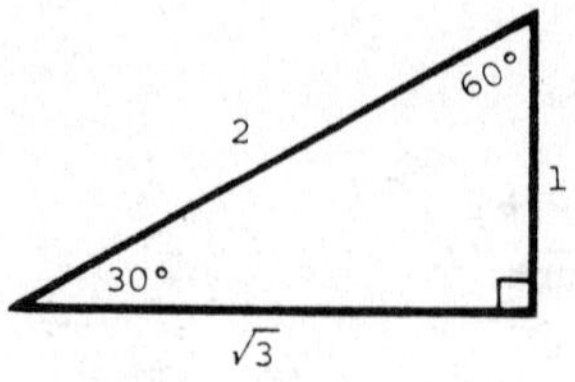

$$\tan 30° = \frac{\text{side opposite}}{\text{side adjacent}} = \frac{1}{\sqrt{3}} = \frac{\sqrt{3}}{3}$$

43. See figure in Exercise 41.

$$\sin 30° = \frac{\text{side opposite}}{\text{hypotenuse}} = \frac{1}{2}$$

45. See figure in Exercise 41.

$$\sec 30° = \frac{\text{hypotenuse}}{\text{side adjacent}} = \frac{2}{\sqrt{3}} = \frac{2\sqrt{3}}{3}$$

47.

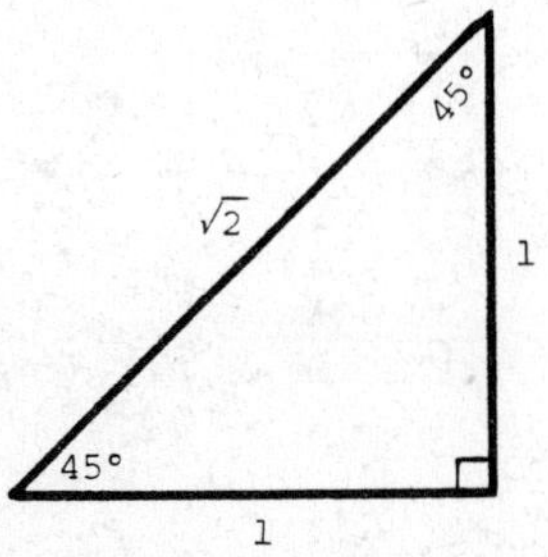

$$\csc 45° = \frac{\text{hypotenuse}}{\text{side opposite}} = \frac{\sqrt{2}}{1} = \sqrt{2}$$

49. See figure in Exercise 47.

$$\cos 45° = \frac{\text{side adjacent}}{\text{hypotenuse}} = \frac{1}{\sqrt{2}} = \frac{\sqrt{2}}{2}$$

51. See figure in Exercise 47.

$$\cot 45° = \frac{\text{side adjacent}}{\text{side opposite}} = \frac{1}{1} = 1$$

53. See figure in Exercise 41.

$$\sin 60° = \frac{\text{side opposite}}{\text{hypotenuse}} = \frac{\sqrt{3}}{2}$$

55. See figure in Exercise 41.

$$\tan 60° = \frac{\text{side opposite}}{\text{side adjacent}} = \frac{\sqrt{3}}{1} = \sqrt{3}$$

57. See figure in Exercise 41.

$$\sec 60° = \frac{\text{hypotenuse}}{\text{side adjacent}} = \frac{2}{1} = 2$$

59. Provided both functions are defined,

$\cot \theta = \frac{1}{\tan \theta}$.

Also, $\tan \theta = \cot (90 - \theta)$.

Thus, the tangent and cotangent functions are both reciprocals and cofunctions.

61.

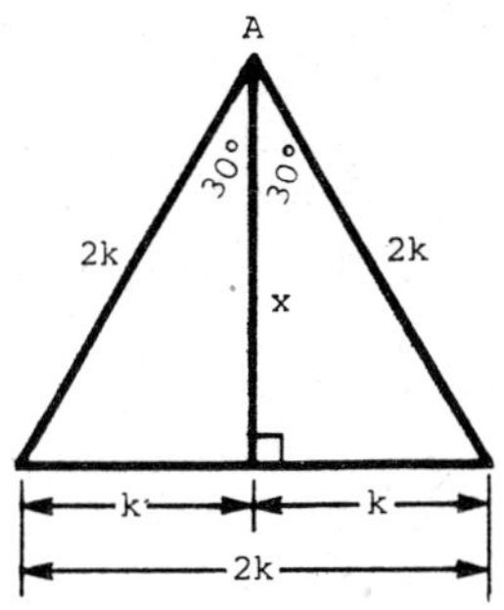

(a) Each angle of every equilateral triangle has a measure of 60°.

(b) The perpendicular bisects the opposite side, so the length of each side opposite each 30° angle is k.

(c) Let x equal the length of the perpendicular and apply the Pythagorean theorem.

$$x^2 + k^2 = (2k)^2$$
$$x^2 + k^2 = 4k^2$$
$$x^2 = 3k^2$$
$$x = k\sqrt{3}$$

The length of the perpendicular is $k\sqrt{3}$.

(d) In a 30°-60° right triangle, the the hypotenuse is always 2 times as long as the shorter leg, and the longer leg has a length that is $\sqrt{3}$ times as long as that of the shorter leg. Also, the shorter leg is opposite the 30° angle, and the longer leg is opposite the 60° angle.

63. $a = \frac{1}{2}(24)$, so $a = 12$.

$b = a\sqrt{3}$, so $b = 12\sqrt{3}$.

$d = b$, so $d = 12\sqrt{3}$.

$c = \sqrt{2}$, so $d = (12\sqrt{3})(\sqrt{2}) = 12\sqrt{6}$.

65. $7 = m\sqrt{3}$, so $m = \frac{7}{\sqrt{3}} = \frac{7\sqrt{3}}{3}$.

$a = 2m$, so $a = 2\left(\frac{7\sqrt{3}}{3}\right) = \frac{14\sqrt{3}}{3}$.

$n = a$, so $n = \frac{14\sqrt{3}}{3}$.

$q = n\sqrt{2}$, so $q = \left(\frac{14\sqrt{3}}{3}\right)\sqrt{2}, = \frac{14\sqrt{6}}{3}$.

67. Let h be the height of the equilateral triangle. By geometry, h bisects the base, s, and forms two 30°-60° right triangles.

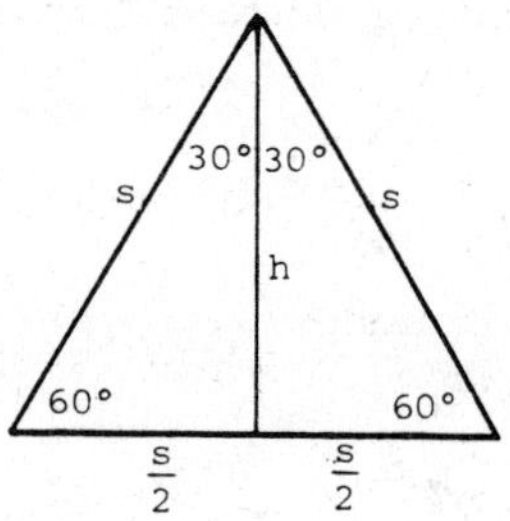

The formula for the area of a triangle is

$$\text{Area} = \frac{1}{2}bh.$$

In this triangle, b = s. The height h of the triangle is the side opposite the 60° angle in either 30°-60° right triangle. The side opposite the 30° angle is s/2. The height is $\sqrt{3}/(s/2)$. So the area of the entire triangle is

$$\frac{1}{2}s\left(\sqrt{3}\frac{s}{2}\right) = \frac{s^2\sqrt{3}}{4}.$$

69. If the pattern were to continue past 90°, we would have $\sin\theta = \frac{\sqrt{5}}{2}$. However, this is impossible because $\frac{\sqrt{5}}{2} > 1$ and $\sin\theta$ cannot be greater than 1.

Section 2.2

1. $\theta = 120°$

θ is in quadrant II.

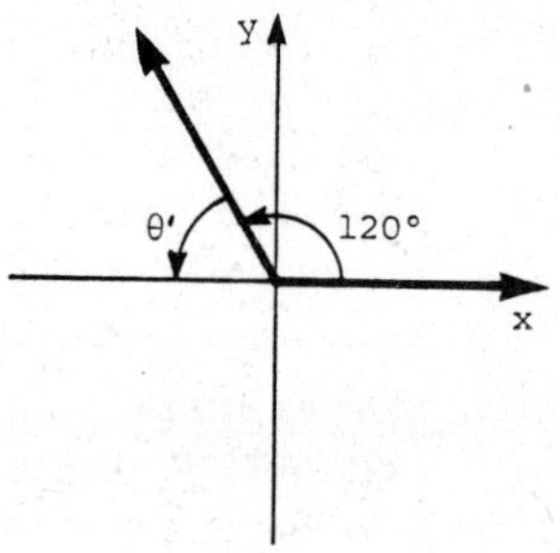

$\theta' = 180° - 120° = 60°$

3. $\theta = 225°$

θ is in quadrant III.

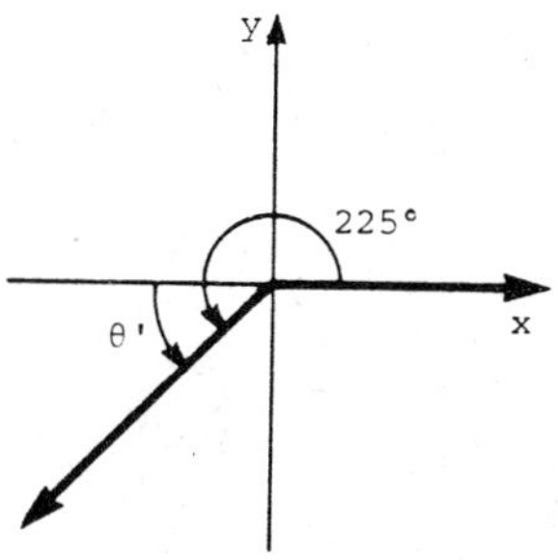

$\theta' = 225° - 180° = 45°$

5. $\theta = 300°$

θ is in quadrant IV.

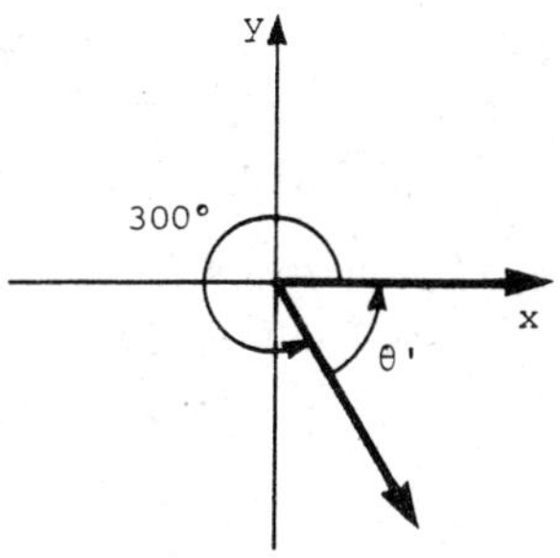

$\theta' = 360° - 300° = 60°$

7. $\theta = -135°$

Since $\theta < 0°$, add a multiple of 360° to find a coterminal angle between 0° and 360°.

$\theta + 360° = -135° + 360° = 225°$

The coterminal angle is in quadrant III. Find θ'.

$$\theta' = 225° - 180° = 45°$$

9. $\theta = 98°$

θ is in quadrant II.

$$\theta' = 180° - 98° = 82°$$

11. $\theta = 285°\ 30'$

θ is in quadrant IV.

$$\theta' = 360° - 285°\ 30' = 74°\ 30'$$

13. $\theta = 480°$

Since $\theta > 360°$, subtract a multiple of 360° to find a coterminal angle.

$$\theta - 360° = 480° - 360° = 120°$$

The coterminal angle is in quadrant II.

$$\theta' = 180° - 120° = 60°$$

15. $\theta = -600°$

Since $\theta < 0$, add a multiple of 360° to find a coterminal angle.

$$\theta + 2(360°) = -600° + 720° = 120°$$

The coterminal angle is in quadrant II.

$$\theta = 180° - 120° = 60°$$

17. By definition, a reference angle is a positive acute angle. Since 120° is not the measure of an acute angle (120° > 90°), 120° cannot be the measure of a reference angle.

19. 615° divided by 360° is approximately 1.71.

$$615° - 1 \cdot (360°) = 255°$$

21. 458° 20′ divided by 360° is approximately 1.27.

$$458°\ 20' - 1 \cdot (360°) = 98°\ 20'$$

23. −321° 50′ divided by 360° is approximately −.89. (As before we round down to obtain −1.)

$$-321°\ 50' - (-1)(360°) = 38°\ 10'$$

25. 120°

Use the method illustrated in Example 2.

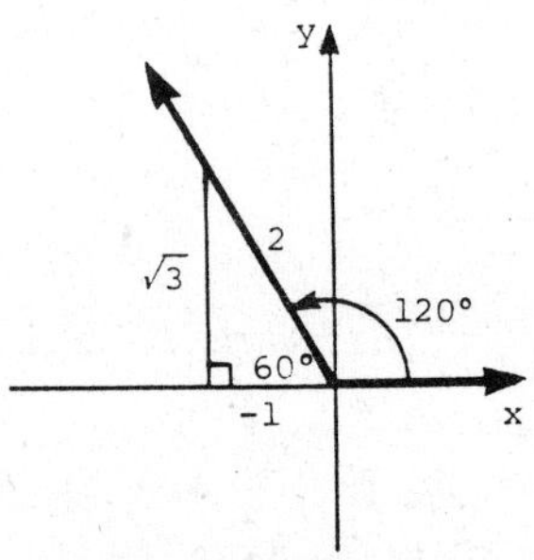

$x = -1,\ y = \sqrt{3},\ r = 2$

$\sin 120° = \frac{y}{r} = \frac{\sqrt{3}}{2}$

$\cos 120° = \frac{x}{r} = \frac{-1}{2} = -\frac{1}{2}$

$\tan 120° = \frac{y}{x} = \frac{\sqrt{3}}{-1} = -\sqrt{3}$

$\cot 120° = \frac{x}{y} = -\frac{1}{\sqrt{3}} = -\frac{\sqrt{3}}{3}$

$\sec 120° = \frac{r}{x} = \frac{2}{-1} = -2$

$\csc 120° = \frac{r}{y} = \frac{2}{\sqrt{3}} = \frac{2\sqrt{3}}{3}$

27. 150°

Use the method illustrated in Example 2.

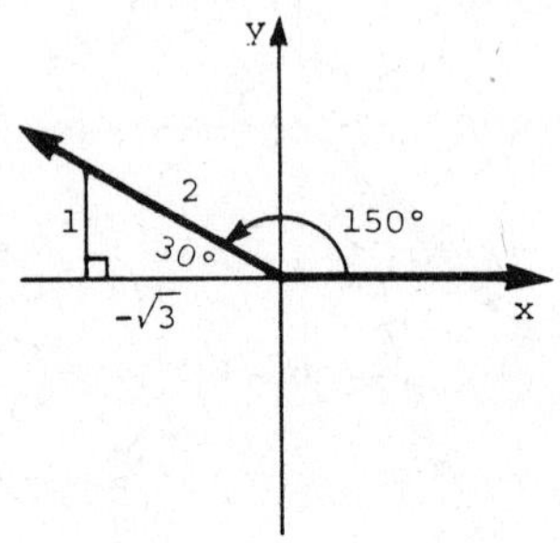

$x = -\sqrt{3},\ y = 1,\ r = 2$

$\sin 150° = \frac{y}{r} = \frac{1}{2}$

$\cos 150° = \frac{x}{r} = \frac{-\sqrt{3}}{2} = -\frac{\sqrt{3}}{2}$

$\tan 150° = \frac{y}{x} = \frac{1}{-\sqrt{3}} = -\frac{\sqrt{3}}{3}$

$\cot 150° = \frac{x}{y} = -\frac{\sqrt{3}}{1} = -\sqrt{3}$

$\sec 150° = \frac{r}{x} = \frac{2}{-\sqrt{3}} = -\frac{2\sqrt{3}}{3}$

$\csc 150° = \frac{r}{y} = \frac{2}{1} = 2$

29. 240°

Use the method illustrated in Example 3.

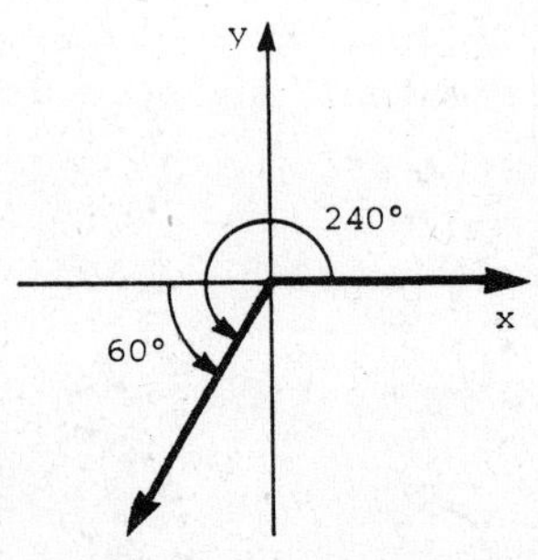

The reference angle is 60°. Since 240° is in quadrant III, the sine, cosine, secant, and cosecant are negative.

$\sin 240° = -\sin 60° = -\frac{\sqrt{3}}{2}$

$\cos 240° = -\cos 60° = -\frac{1}{2}$

$\tan 240° = \tan 60° = \sqrt{3}$

$\cot 240° = \cot 60° = \frac{\sqrt{3}}{3}$

$\sec 240° = -\sec 60° = -2$

$\csc 240° = -\csc 60° = -\frac{2\sqrt{3}}{3}$

31. 330°

Use the method illustrated in Example 3.

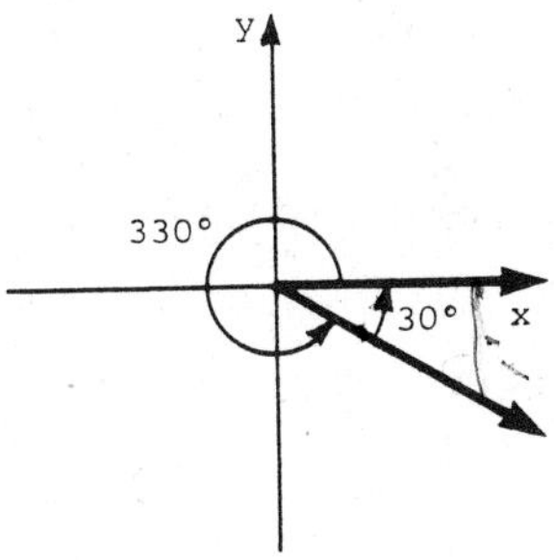

The reference angle is 30°. Since 330° is in quadrant IV, the sine, tangent, cotangent, and cosecant are negative.

$\sin 330° = -\sin 30° = -\frac{1}{2}$

$\cos 330° = \cos 30° = \frac{\sqrt{3}}{2}$

$\tan 330° = -\tan 30° = -\frac{\sqrt{3}}{3}$

$\cot 330° = -\cot 30° = -\sqrt{3}$

$\sec 330° = \sec 30° = \frac{2\sqrt{3}}{3}$

$\csc 330° = -\csc 30° = -2$

33. 420°

Use the method illustrated in Example 3.

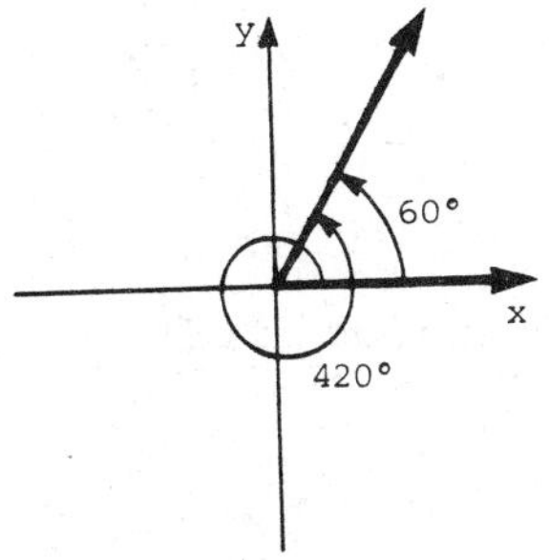

The reference angle is 60°. Since 420° is in quadrant I, all functions are positive.

$\sin 420° = \sin 60° = \frac{\sqrt{3}}{2}$

$\cos 420° = \cos 60° = \frac{1}{2}$

$\tan 420° = \tan 60° = \sqrt{3}$

$\cot 420° = \cot 60° = \frac{\sqrt{3}}{3}$

$\sec 420° = \sec 60° = 2$

$\csc 420° = \csc 60° = \frac{2\sqrt{3}}{3}$

35. 510°

Use the method illustrated in Example 3.

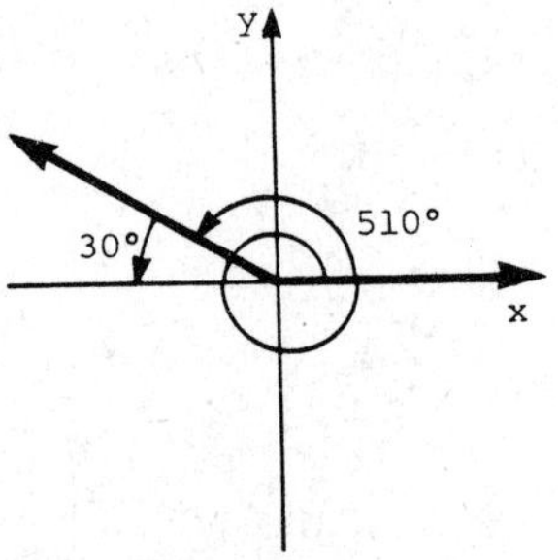

The reference angle is 30°. Since 510° is in quadrant II, the cosine, tangent, cotangent, and secant are negative.

$\sin 510° = \sin 30° = \frac{1}{2}$

$\cos 510° = -\cos 30° = -\frac{\sqrt{3}}{2}$

$\tan 510° = -\tan 30° = -\frac{\sqrt{3}}{3}$

$\cot 510° = -\cot 30° = -\sqrt{3}$

$\sec 510° = -\sec 30° = -\frac{2\sqrt{3}}{3}$

$\csc 540° = \csc 30° = 2$

37. 750° is coterminal with

$$750° - 2(360°) = 750° - 720° = 30°.$$

$\sin 750° = \sin 30° = \frac{1}{2}$

$\cos 750° = \cos 30° = \frac{\sqrt{3}}{2}$

$\tan 750° = \tan 30° = \frac{\sqrt{3}}{3}$

$\cot 750° = \cot 30° = \sqrt{3}$

$\sec 750° = \sec 30° = \frac{2\sqrt{3}}{3}$

$\csc 750° = \csc 30° = 2$

39. 1500° is coterminal with

$$1500° - 4(360°) = 1500° - 1440° = 60°.$$

$\sin 1500° = \sin 60° = \frac{\sqrt{3}}{2}$

$\cos 1500° = \cos 60° = \frac{1}{2}$

$\tan 1500° = \tan 60° = \sqrt{3}$

$\cot 1500° = \cot 60° = \frac{\sqrt{3}}{3}$

$\sec 1500° = \sec 60° = 2$

$\csc 1500° = \csc 60° = \frac{2\sqrt{3}}{3}$

41. −390° is coterminal with

$$-390° - (-2)(360°) = 330°.$$

The reference angle is 30°. Since −390° is in quadrant IV, the sine, tangent, cotangent, and cosecant are negative.

$\sin(-390°) = -\sin 30° = -\frac{1}{2}$

$\cos(-390°) = \cos 30° = \frac{\sqrt{3}}{2}$

$\tan(-390°) = -\tan 30° = -\frac{\sqrt{3}}{3}$

$\cot(-390°) = -\cot 30° = -\sqrt{3}$

$\sec(-390°) = \sec 30° = \frac{2\sqrt{3}}{3}$

$\csc(-390°) = -\csc 30° = -2$

43. $-1020°$ is coterminal with

$$-1020° - (-3)(360°) = -1020° + 1080° = 60°.$$

$$\sin(-1020°) = \sin 60° = \frac{\sqrt{3}}{2}$$

$$\cos(-1020°) = \cos 60° = \frac{1}{2}$$

$$\tan(-1020°) = \tan 60° = \sqrt{3}$$

$$\cot(-1020°) = \cot 60° = \frac{\sqrt{3}}{3}$$

$$\sec(-1020°) = \sec 60° = 2$$

$$\csc(-1020°) = \csc 60° = \frac{2\sqrt{3}}{3}$$

45. $30°$:

$$\tan 30° = \frac{\sqrt{3}}{3}$$

$$\cot 30° = \sqrt{3}$$

47. $60°$:

$$\sin 60° = \frac{\sqrt{3}}{2}$$

$$\cot 60° = \frac{\sqrt{3}}{3}$$

$$\csc 60° = \frac{2\sqrt{3}}{3}$$

49. $135°$:

$$\tan 135° = -\tan 45° = -1$$

$$\cot 135° = -\cot 45° = -1$$

51. $210°$:

$$\cos 210° = -\cos 30° = -\frac{\sqrt{3}}{2}$$

$$\sec 210° = -\sec 30° = -\frac{2\sqrt{3}}{3}$$

53. $\sin\theta = \frac{1}{2}$

$\sin\theta > 0$ for θ in quadrant I or II. Since $\sin\theta = \frac{1}{2}$, the reference angle θ' must be $30°$. Draw the two possible angles θ.

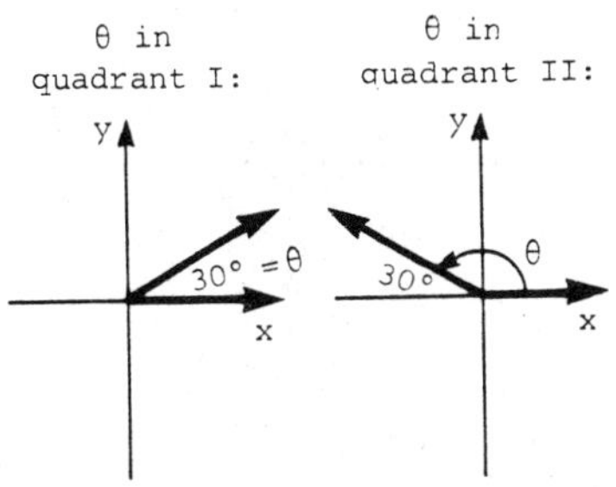

In quadrant I,

$$\theta = 30°.$$

In quadrant II,

$$\theta = 180° - 30° = 150°.$$

55. $\tan\theta = \sqrt{3}$

$\tan\theta > 0$ for θ in quadrant I or III. Since $\tan\theta = \sqrt{3}$, the reference angle θ' must be $60°$. Draw the two possible angles θ.

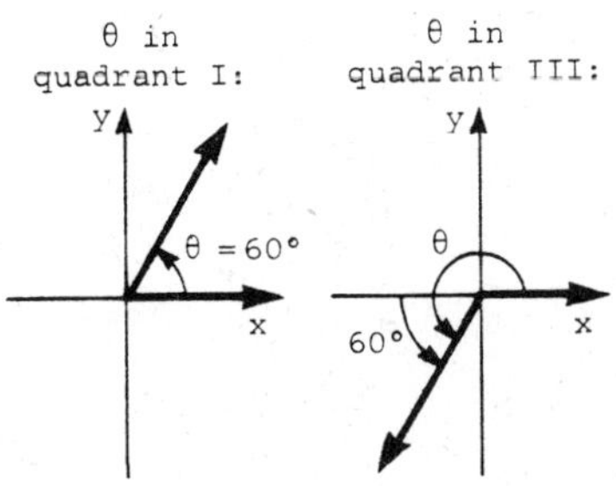

In quadrant I,

$$\theta = 60°.$$

In quadrant III,

$$\theta = 180° + 60° = 240°.$$

57. $\cos \theta = -\frac{1}{2}$

$\cos \theta < 0$ for θ in quadrant II or III. Since $\cos \theta = -\frac{1}{2}$, the reference angle must be 60°. Draw the two possible angles θ.

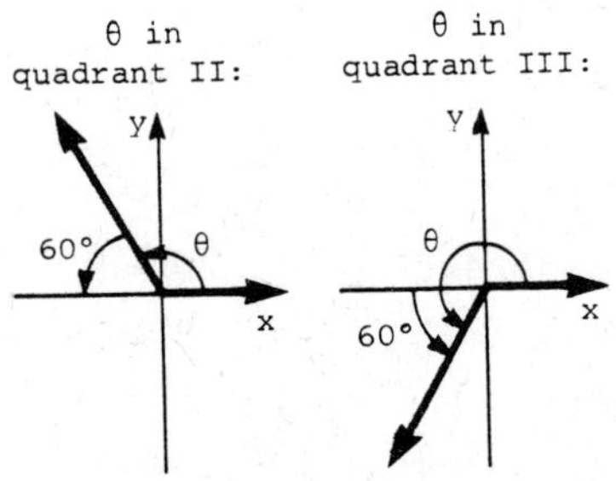

In quadrant II,

$$\theta = 180° - 60° = 120°.$$

In quadrant III,

$$\theta = 180° + 60° = 240°.$$

59. $\sin \theta = -\frac{\sqrt{3}}{2}$

$\sin \theta < 0$ for θ in quadrant III or IV. Since $\sin \theta = -\frac{\sqrt{3}}{2}$, the reference angle is 60°. Draw the two possible angles θ.

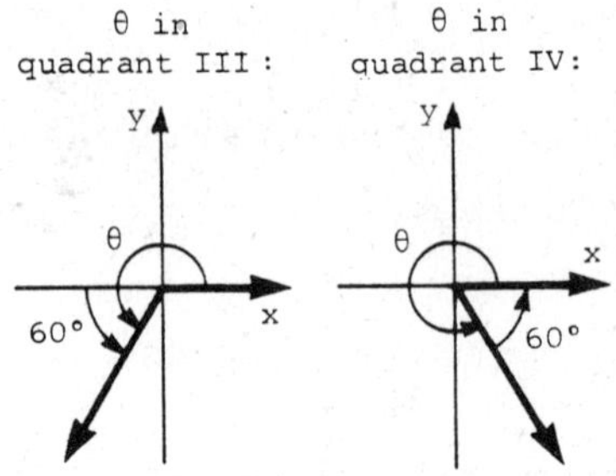

In quadrant III,

$$\theta = 180° + 60° = 240°.$$

In quadrant IV,

$$\theta = 360° - 60° = 300°.$$

61. $\tan \theta = -1$

$\tan \theta < 0$ for θ in quadrant II or IV. Since $\tan \theta = -1$, the reference angle is 45°. Draw the two possible angles.

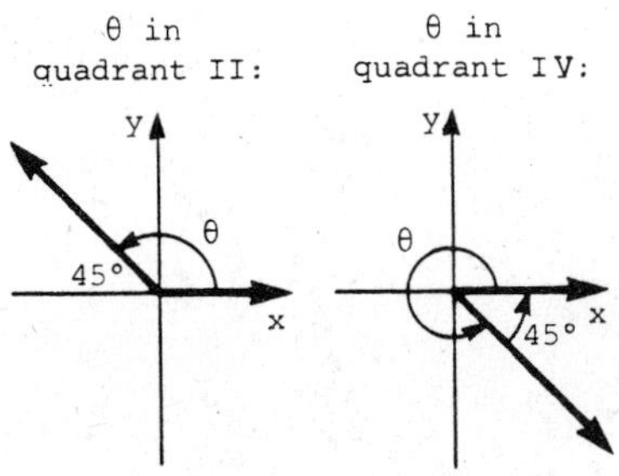

In quadrant II,

$$\theta = 180° - 45° = 135°.$$

In quadrant IV,

$$\theta = 360° - 45° = 315°.$$

63. $\cos \theta = 0$

when $\theta = 90°$ or $\theta = 270°$.

65. cot θ is undefined.

Since $\cot \theta = \frac{1}{\tan \theta}$, unless $\tan \theta = 0$, cot θ is undefined when $\tan \theta = 0$. $\tan \theta = 0$ when

$$\theta = 0° \quad \text{or} \quad \theta = 180°.$$

67. For every angle θ, $\sin \theta = \frac{y}{r}$ and $\cos \theta = \frac{x}{r}$. Thus, for every angle θ,

$$(\sin \theta)^2 + (\cos \theta)^2 = \left(\frac{y}{r}\right)^2 + \left(\frac{x}{r}\right)^2 = \frac{y^2 + x^2}{r^2} = \frac{r^2}{r^2} = 1.$$

Since $(.6)^2 + (-.8)^2 = .36 + .64 = 1$, there is an angle θ for which $\cos \theta = .6$ and $\sin \theta = -.8$. Since $\cos \theta > 0$ and $\sin \theta < 0$, it is an angle in quadrant IV.

69. If θ is in interval (90°, 180°),

$$90° < \theta < 180°.$$

Thus, dividing by 2, we get

$$45° < \frac{\theta}{2} < 90°.$$

Thus, $\frac{\theta}{2}$ is a quadrant I angle and $\sin \frac{\theta}{2}$ is positive.

71. If θ is in the interval (90°, 180°),

$$90° < \theta < 180°.$$

Adding 180° gives

$$270° < \theta + 180° < 360°.$$

Thus, $\theta + 180°$ is a quadrant IV angle and $\cot(\theta + 180°)$ is negative.

73. If θ is in the interval (90°, 180°),

$$90° < \theta < 180°.$$

Multiplying by -1 gives

$$-90° > -\theta > -180°.$$

Thus, $-\theta$ is a quadrant III angle and $\cos(-\theta)$ is negative.

75. $\sin^2 120° + \cos^2 120°$

$$= \left(\frac{\sqrt{3}}{2}\right)^2 + \left(-\frac{1}{2}\right)^2 = \frac{3}{4} + \frac{1}{4} = 1$$

77. $2 \tan^2 120° + 3 \sin^2 150° - \cos^2 180°$

$$= 2(-\sqrt{3})^2 + 3\left(\frac{1}{2}\right)^2 - (-1)^2 = 2 \cdot 3 + 3 \cdot \left(\frac{1}{4}\right) - 1 = 6 + \frac{3}{4} - 1 = \frac{23}{4}$$

79. $\sin^2 225° - \cos^2 270° + \tan 60°$

$$= \left(-\frac{\sqrt{2}}{2}\right)^2 - (0)^2 + \sqrt{3} = \frac{2}{4} - 0 + \sqrt{3} = \frac{1}{2} + \sqrt{3}$$

81. $\cos^2 60° + \sec^2 150° - \csc^2 210°$

$= \left(\frac{1}{2}\right)^2 + \left(-\frac{2\sqrt{3}}{3}\right)^2 - (-2)^2$

$= \frac{1}{4} + \frac{12}{9} - 4$

$= \frac{1}{4} + \frac{4}{3} - 4$

$= \frac{3}{12} + \frac{16}{12} - \frac{48}{12} = -\frac{29}{12}$

83. $\sin 30° + \sin 60° = \sin (30° + 60°)$

Evaluate each side to determine whether this statement is true or false.

$\sin 30° + \sin 60° = \frac{1}{2} + \frac{\sqrt{3}}{2}$

$= \frac{1 + \sqrt{3}}{2}$

$\sin (30° + 60°) = \sin 90°$

$= 1$

Since $\frac{1 + \sqrt{3}}{2} \neq 1$, the given statement is false.

85. $\cos 60° = 2 \cos^2 30° - 1$

$\cos 60° = \frac{1}{2}$

$2 \cos^2 30° - 1 = 2\left(\frac{\sqrt{3}}{2}\right)^2 - 1$

$= 2\left(\frac{3}{4}\right) - 1$

$= \frac{3}{2} - 1$

$= \frac{1}{2}$

Thus, $\frac{1}{2} = \frac{1}{2}$.

True

87. $\sin 120° = \sin 150° - \sin 30°$

$\sin 120° = \frac{\sqrt{3}}{2}$

$\sin 150° - \sin 30° = \frac{1}{2} - \frac{1}{2} = 0$

Thus, $\frac{\sqrt{3}}{2} \neq 0$.

False

89. $\sin 120° = \sin 180° \cdot \cos 60° - \sin 60° \cdot \cos 180°$

$\sin 120° = \frac{\sqrt{3}}{2}$

$\sin 180° \cdot \cos 60° - \sin 60° \cdot \cos 180°$

$= 0\left(\frac{1}{2}\right) - \left(\frac{\sqrt{3}}{2}\right)(-1)$

$= 0 - \left(-\frac{\sqrt{3}}{2}\right)$

$= \frac{\sqrt{3}}{2}$

Thus, $\frac{\sqrt{3}}{2} = \frac{\sqrt{3}}{2}$.

True

Section 2.3

For the following exercises be sure your calculator is in degree mode. If your calculator accepts angles in degrees, minutes and seconds, it is not necessary to change angles to decimal degrees.

1. $\sin 38° 40'$

$38° 40' = \left(38 + \frac{40}{60}\right)°$

$= 38.66666667°$

Enter: 38.66666667 [SIN]

Display: .62478851

3. cot 41° 20′

$41°\ 20' = \left(41 + \frac{20}{60}\right)^{\circ}$

$= 41.33333333°$

Enter: 41.33333333 [TAN] [1/x]

Display: 1.1369414

5. sin 58° 30′

$58°\ 30' = \left(58 + \frac{30}{60}\right)^{\circ}$

$= 58.5°$

Enter: 58.5 [SIN]

Display: .85264016

7. tan 17° 20′

$17°\ 20' = \left(17 + \frac{20}{60}\right)^{\circ}$

$= 17.33333333°$

Enter: 17.33333333 [TAN]

Display: .31210365

9. cot 128° 30′

$128°\ 30' = \left(128 + \frac{30}{60}\right)^{\circ}$

$= 128.5°$

Enter: 128.5 [TAN] [1/x]

Display: −.79543592

11. sin 179° 20′

$179°\ 20' = \left(179 + \frac{20}{60}\right)^{\circ}$

$= 179.33333333°$

Enter: 179.33333333 [SIN]

Display: .01163527

13. sin 204° 20′

$204°\ 20' = \left(204 + \frac{20}{60}\right)^{\circ}$

$= 204.3333333°$

Enter: 204.3333333 [SIN]

Display: −.41204452

15. cos 251° 10′

$251°\ 10' = \left(251 + \frac{10}{60}\right)^{\circ}$

$= 251.1666667°$

Enter: 251.1666667 [COS]

Display: −.32281638

17. sin 274° 30′

$274°\ 30' = \left(274 + \frac{30}{60}\right)^{\circ}$

$= 274.5°$

Enter: 274.5 [SIN]

Display: −.99691733

19. csc 421° 10′

$421°\ 10' = \left(421 + \frac{10}{60}\right)^{\circ}$

$= 421.1666667°$

Enter: 421.1666667 [SIN] [1/x]

Display: 1.1415183

21. sec (−108° 20′)

$108°\ 20' = -\left(108 + \frac{20}{60}\right)^{\circ}$

$= -108.3333333°$

Enter: 108.3333333 [+/−] [COS] [1/x]

Display: −3.1791978

23. tan (−197° 50′)

$-197°\ 50' = -\left(197 + \frac{50}{60}\right)^{\circ}$

$= -197.833333$

Enter: 197.833333 [+/−] [TAN]

Display: −.32170674

25. sin 59.642°

Enter: 59.642 [SIN]

Display: .86288438

27. tan (−80.612°)

Enter: 80.612 [+/−] [TAN]

Display: −6.0483717

29. cos 258.409°

Enter: 258.409 [COS]

Display: −.20092405

31. cot 109.713°

Enter: 109.713 [TAN] [1/x]

Display: −.35830784

33. cos 74° 11′

$74°\ 11' = \left(74 + \frac{11}{60}\right)^{\circ}$

$= 74.18333333$

Enter: 74.1333333 [COS]

Display: .27256013

35. cot 125° 52′ 10″

$125°\ 52'\ 10'' = \left(125 + \frac{52}{60} + \frac{10}{3600}\right)^{\circ}$

$= 125.8694444$

Enter: 125.864444 [TAN] [1/x]

Display: −.72306685

37. sin (−15° 28′ 17″)

−15° 28′ 17″

$= -\left(15 + \frac{28}{60} + \frac{17}{3600}\right)^{\circ}$

$= -15.471389°$

Enter: 15.471389 [+/−] [SIN]

Display: −.26675715

39. sec (−121° 53′ 42″)

$-121°\ 53'\ 42'' = -\left(121 + \frac{53}{60} + \frac{42}{3600}\right)^{\circ}$

$= -121.895°$

Enter: 121.895 [+/−] [COS] [1/x]

Display: −1.8926337

41. The student should check the mode of the calculator. The calculator was not set in degree mode.

43. $\sin \theta = .84802194$

Enter: .84802194 { [INV] [SIN] or [ARCSIN] or [SIN⁻¹] }

Display: 57.997172

$\theta = 57.997172°$

45. $\sec \theta = 1.1606249$

Enter: 1.1606249 [1/x] { [INV] [COS] or [ARCCOS] or [COS^{-1}] }

Display: 30.502748

$\theta = 30.502748°$

47. $\sin \theta = .72144101$

Enter: .72144101 { [INV] [SIN] or [ARCSIN] or [SIN^{-1}] }

Display: 46.173581

$\theta = 46.173581°$

49. $\tan \theta = 6.4358841$

Enter: 6.4358841 { [INV] [TAN] or [ARCTAN] or [TAN^{-1}] }

Display: = 81.168073

$\theta = 81.168073°$

51. $\cos \theta = .68716510$

Since $\cos \theta$ is positive, there will be one angle in quadrant I and one angle in quadrant IV. If θ' is the reference angle, the two angles will be θ' and $360° - \theta'$.

Enter: .68716510 { [INV] [COS] or [ARCCOS] or [COS^{-1}] }

Display: 46.593881

$\theta' = 46.593881°$

$360° - \theta' = 313.40612°$

53. $\sin \theta = .41298643$

Since $\sin \theta$ is positive, there will be one angle in quadrant I and one angle in quadrant II. If θ' is the reference angle, then the two angles with be θ' and $180° - \theta'$.

Enter: .41298643 [SIN^{-1}]

Display: 24.392576

$\theta' = 24.392576°$

$360° - \theta' = 155.60742°$

55. $\tan \theta = .87692035$

Since $\tan \theta$ is positive, there will be one angle in quadrant I and one angle in quadrant III. If θ' is the reference angle, then the two angles will be θ' and $180° + \theta'$.

Enter: .87692035 [TAN^{-1}]

Display: 41.248183

$\theta' = 41.248183°$

$180° + \theta' = 221.24818°$

57. $\cos 100° \cos 80° - \sin 100° \sin 80°$

Enter: 100 [COS] [×] 80 [COS] [−]

100 [SIN] [×] 80 [SIN] [=]

Display: -1

$\cos 100° \cos 80° - \sin 100° \sin 80° = -1$

59. $\sin 28° 14' \cos 61° 46'$
$+ \cos 28° 14' \sin 61° 46'$

Enter: [(] 28 [+] 14 [÷] 60 [)]

[SIN] [×] [(] 61 [+] 46 [÷]

60 [)] COS [+] [(] 28 [+]

14 [÷] 60 [)] [COS] [×] [(] 61

[+] 46 [÷] 60 [)] [SIN] [=]

Display: 1

$\sin 28° 14' \cos 61° 46'$
$+ \cos 28° 14' \sin 61° 46' = 1$

61. $\theta_1 = 46°$, $\theta_2 = 31°$, $c_1 = 3 \times 10^8$ m per sec

$$\frac{c_1}{c_2} = \frac{\sin \theta_1}{\sin \theta_2}$$

$$c_2 = \frac{c_1 \sin \theta_2}{\sin \theta_1}$$

$$c_2 = \frac{(3 \times 10^8)(\sin 31°)}{\sin 46°}$$

$$c_2 = 2 \times 10^8 \text{ m per sec}$$

63. $\theta_1 = 40°$, $c_2 = 1.5 \times 10^8$ m per sec, $c_1 = 3 \times 10^8$ m per sec

$$\frac{c_1}{c_2} = \frac{\sin \theta_1}{\sin \theta_2}$$

$$\sin \theta_2 = \frac{c_2 \sin \theta_1}{c_1}$$

$$\sin \theta_2 = \frac{1.5 \times 10^8(\sin 40°)}{3 \times 10^8}$$

$$\theta_2 = 19°$$

65. $\theta_1 = 90°$, $c_1 = 3 \times 10^8$ m per sec, $c_2 = 2.254 \times 10^8$

$$\frac{c_1}{c_2} = \frac{\sin \theta_1}{\sin \theta_2}$$

$$\sin \theta_2 = \frac{c_2 \sin \theta_1}{c_1}$$

$$\sin \theta_2 = \frac{2.254 \times 10^8(\sin 90°)}{3 \times 10^8}$$

$$= \frac{2.254 \times 10^8(1)}{3 \times 10^8}$$

$$= \frac{2.254}{3} = .7513$$

$$\theta_2 = 48.7°$$

67. $\sin 10° + \sin 10° = \sin 20°$

$\sin 10° = .17364818$

$\sin 20° = .34202014$

$.17364818 + .17364818 \neq .34202014$

False

69. $\sin 50° = 2 \sin 25° \cdot \cos 25°$

$\sin 50° = .76604444$

$\sin 25° = .42261826$

$\cos 25° = .90630779$

$2 \sin 25° \cdot \cos 25°$

$= 2(.42261826)(.90630779)$

$= .76604444$

$.76604444 = .76604444$

True

71. $\cos 40° = 1 - 2 \sin^2 80°$

$\cos 40° = .76604444$

$\sin 80° = .98480775$

$1 - 2 \sin^2 80°$

$= 1 - 2(.98480775)^2$

$= -.93969262$

$.76604444 \neq -.93969262$

False

73. $\sin 39° 48' + \cos 39° 48' = 1$

$\sin 39° 48' = \sin 39.8°$

$= .64010970$

$\cos 39° 48' = \cos 39.8°$

$= .76828352$

$\sin 39° 48' + \cos 39° 48'$

$= .64010970 + .76828352$

$= 1.40839322$

$1.40839322 \neq 1$

False

Trigonometric Function Values: Exact and Approximate

1. $\sin 240°$

$\theta' = 60°$ so we can find an exact value.

$$\sin 240° = -\sin 60° = -\frac{\sqrt{3}}{2}$$

3. $\cos 130°$

$\theta' = 50°$ so we must find a calculator approximation.

$$\cos 130° \approx -.64278761$$

5. $\tan (-60°)$

$\theta' = 60°$ so we can find an exact value.

$$\tan (-60°) = -\tan 60° = -\sqrt{3}$$

7. $\sin (-60°)$

$\theta' = 60°$ so we can find an exact value.

$$\sin (-60°) = -\sin 60° = -\frac{\sqrt{3}}{2}$$

9. $\sin 190°$

$\theta' = 10°$ so we must find a calculator approximation.

$$\sin 190° \approx -.17364818$$

11. $\tan 210°$

$\theta' = 30°$ so we can find an exact value.

$$\tan 210° = \tan 30° = \frac{\sqrt{3}}{3}$$

13. $\sec 150°$

θ' so we can find and exact value.

$$\sec 150° = -\sec 30° = -\frac{2\sqrt{3}}{3}$$

15. $\csc 140°$

$\theta = 40°$ so we must find a calculator approximation.

$$\csc 140° \approx 1.5557238$$

17. $\tan 495°$

$\theta' = 45°$ so we can find an exact value.

$$\tan 495° = -\tan 45° = -1$$

19. $\cot (-80°)$

$\theta' = 10°$ so we must find a calculator approximation.

$$\cot (-80°) \approx -.17632698$$

Section 2.4

1. 29,000 ft

This measurement is to the nearest foot. It represents a range of 28,999.5 ft to 29,000.5 ft.

3. If 1650 feet and 160 feet represent accuracy to the nearest foot, then the ranges represented by these two numbers are 1649.5 to 1650.5 and 159.5 to 160.5, respectively.

5. The error is that both 2 and 65 are counting numbers, not measurements. They are exact numbers so all the digits in their product are significant.

7. $A = 36°\ 20', c = 964$ m

$A + B = 90°$

$B = 90° - A$

$B = 90° - 36°\ 20'$

$= 89°\ 60' - 36°\ 20'$

$B = 53°\ 40'$

$\sin A = \frac{a}{c}$

$a = c \sin A$

$a = 964 \sin 36°\ 20'$

Use a calculator and round answer to three significant digits.

$a = 571$ m

$\cos A = \frac{b}{c}$

$b = c(\cos A)$

$b = 964 \cos 36°\ 20'$

Use a calculator and round answer to three significant digits.

$b = 777$ m

9. $N = 51.2°, m = 124$ m

$M + N = 90°$

$M = 90° - N$

$M = 90° - 51.2°$

$M = 38.8°$

$\tan N = \frac{n}{m}$

$n = m \tan N$

$n = 124 \cdot \tan 51.2°$

$n = 154$ m

$\cos N = \frac{m}{p}$

$p = \frac{m}{\cos N}$

$p = \frac{124}{\cos 51.2°}$

$p = 198$ m

11. $B = 42.0892°, b = 56.851$ cm

$A + B = 90°$

$A = 90° - B$

$A = 90° - 42.0892°$

$= 47.9108°$

$\sin B = \frac{b}{c}$

$c = \frac{b}{\sin B}$

$c = \frac{56.851}{\sin 42.0892°}$

$c = 84.816$ cm

$$\tan B = \frac{b}{a}$$

$$a = \frac{b}{\tan B}$$

$$a = \frac{56.851}{\tan 42.0892°}$$

$$a = 62.942 \text{ cm}$$

13. There are infinitely many right triangles with the same two acute angles. They are similar to each other, but not congruent, so it is impossible to solve such a triangle.

15. $A = 28°\ 00'$, $c = 17.4$ ft

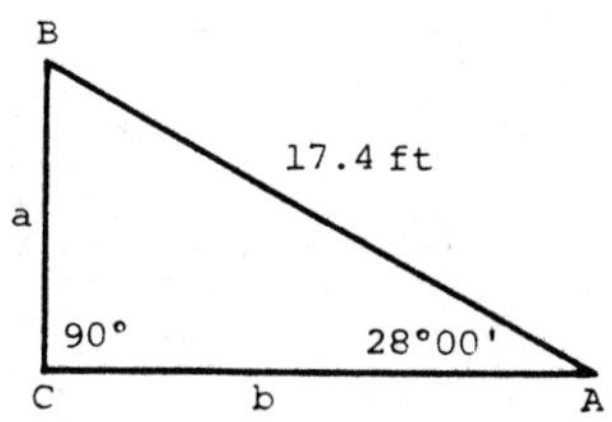

$$A + B = 90°$$
$$B = 90° - A$$
$$B = 90° - 28°\ 00'$$
$$B = 62°\ 00'$$

$$\sin A = \frac{a}{c}$$
$$a = c \sin A$$
$$a = 17.4 \sin 28°\ 00'$$
$$a = 8.17 \text{ ft}$$

$$\cos A = \frac{b}{c}$$
$$b = c \cos A$$
$$b = 17.4 \cos 28°\ 00'$$
$$b = 15.4 \text{ ft}$$

17. $B = 73°\ 00'$, $b = 128$ in

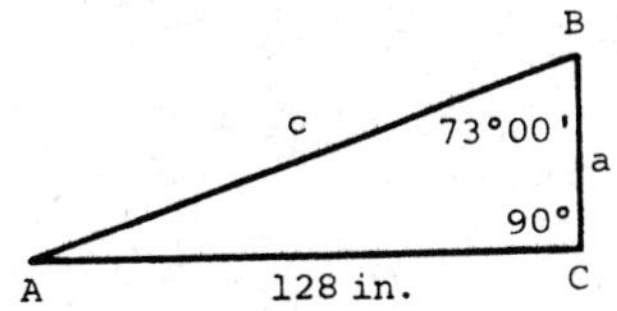

$$A = 90° - 73°\ 00'$$
$$= 17°\ 00'$$

$$\tan 73°\ 00' = \frac{128}{a}$$
$$a = \frac{128}{\tan 73°\ 00'}$$
$$a = 39.1 \text{ in}$$

$$\sin 73°\ 00' = \frac{128}{c}$$
$$c = \frac{128}{\sin 73°\ 00'}$$
$$c = 134 \text{ in}$$

19. $a = 76.4$ yd, $b = 39.3$ yd

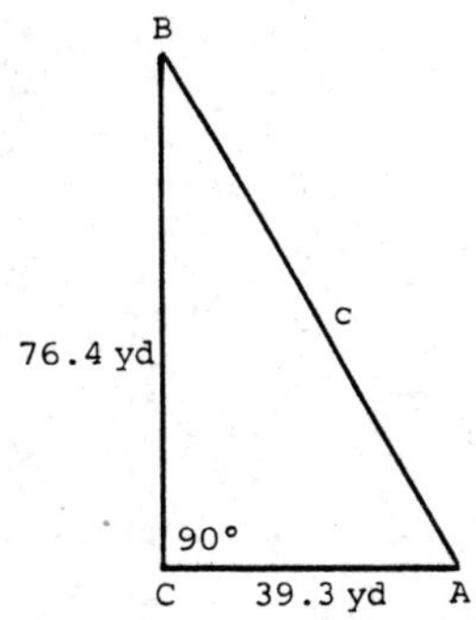

$$c^2 = a^2 + b^2$$
$$c = \sqrt{a^2 + b^2}$$
$$c = \sqrt{(76.4)^2 + (39.3)^2}$$
$$= 85.9 \text{ yd}$$

$$\tan A = \frac{76.4}{39.3}$$

$$A = 62° \ 50'$$

$$\tan B = \frac{39.3}{76.4}$$

$$B = 27° \ 10'$$

21. $a = 18.9$ cm, $c = 46.3$ cm

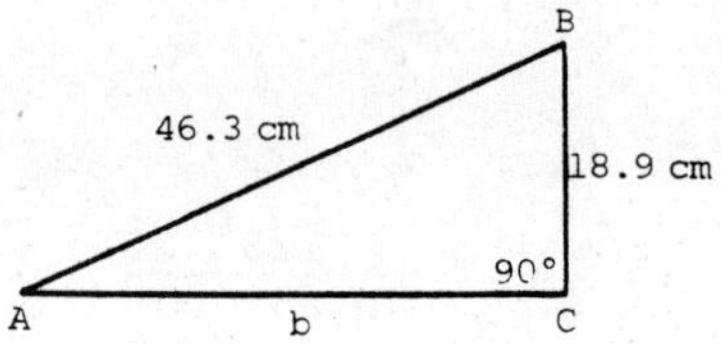

$$b = \sqrt{c^2 - a^2}$$

$$b = \sqrt{(46.3)^2 - (18.9)^2}$$

$$= 42.3 \text{ cm}$$

$$\sin A = \frac{18.9}{46.3}$$

$$A = 24° \ 10'$$

$$\cos B = \frac{18.9}{46.3}$$

$$B = 65° \ 50'$$

23. $A = 53° \ 24'$, $c = 387.1$ ft

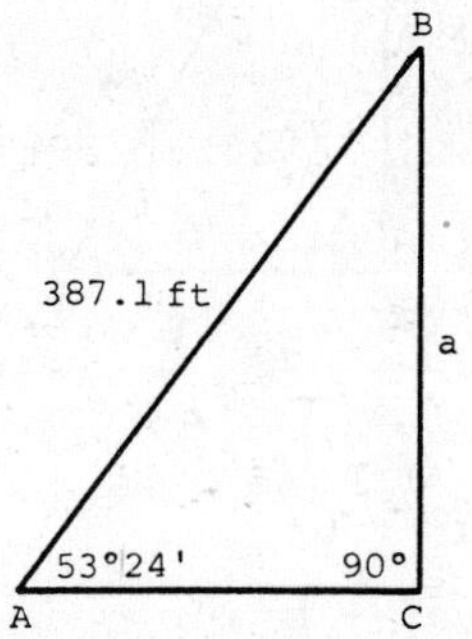

$$B = 90° - 53° \ 24'$$

$$= 89° \ 60' - 53° \ 24'$$

$$B = 36° \ 36'$$

$$\sin 53° \ 24' = \frac{a}{387.1}$$

$$a = 387.1 \sin 53° \ 24'$$

$$a = 310.8 \text{ ft}$$

$$\cos 53° \ 24' = \frac{b}{387.1}$$

$$b = 387.1 \cos 53° \ 24'$$

$$b = 230.8 \text{ ft}$$

25. $B = 39° \ 09'$, $c = .6231$ m

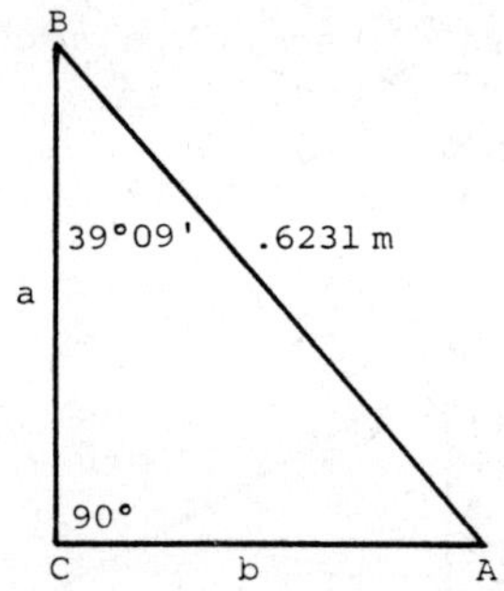

$$A = 90° - 39° \ 09'$$

$$A = 50° \ 51'$$

$$\cos 39° \ 09' = \frac{a}{.6231}$$

$$a = (.6231) \cos 39° \ 09'$$

$$a = .4832 \text{ m}$$

$$\sin 39° \ 09' = \frac{b}{.6231}$$

$$b = (.6231) \sin 39° \ 09'$$

$$b = .3934 \text{ m}$$

27. $c = 7.813$ m, $b = 2.467$ m

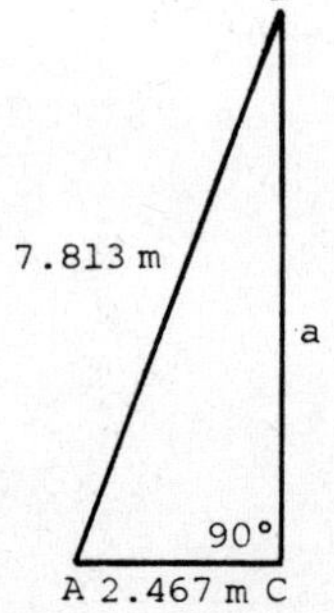

$$\cos A = \frac{2.467}{7.813}$$
$$A = 71°\ 36'$$

$$\sin B = \frac{2.467}{7.813}$$
$$B = 18°\ 24'$$

$$a = \sqrt{c^2 - b^2}$$
$$a = \sqrt{(7.813)^2 - (2.467)^2}$$
$$= 7.413 \text{ m}$$

29. $B = 42.432°$, $a = 157.49$ m

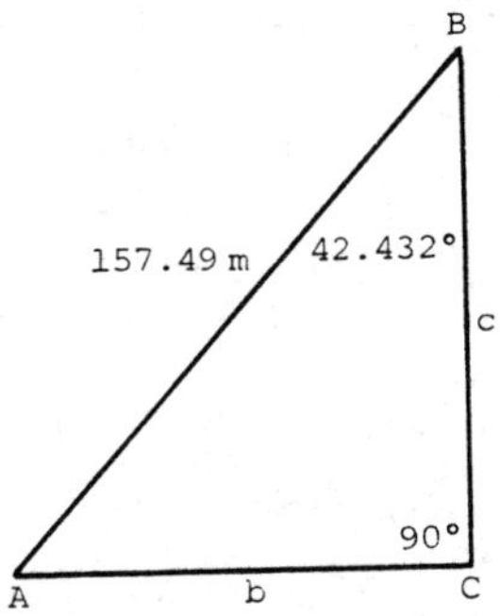

$$A = 90° - 42.432°$$
$$= 47.568°$$

$$\tan 42.432° = \frac{b}{157.49}$$
$$b = 157.49 \tan 42.432°$$
$$b = 143.97 \text{ m}$$

$$\cos 42.432° = \frac{157.49}{c}$$
$$c = \frac{157.49}{\cos 42.432°}$$
$$c = 213.38 \text{ m}$$

31. $A = 57.209°$, $c = 186.49$ cm

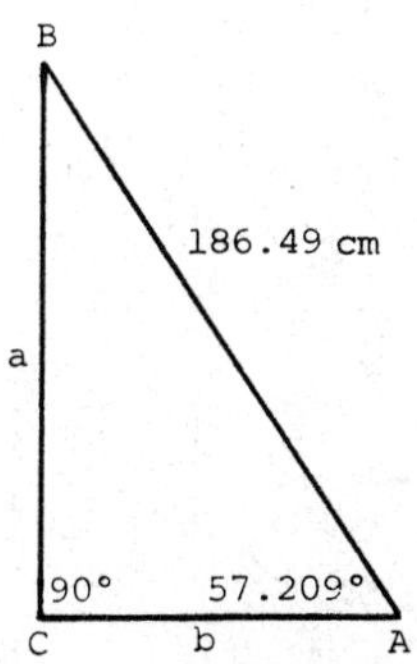

$$B = 90° - 57.209°$$
$$= 32.791°$$

$$\sin 57.209° = \frac{a}{186.49}$$
$$a = 186.49 \sin 57.209°$$
$$a = 156.77 \text{ cm}$$

$$\cos 57.209° = \frac{b}{186.49}$$
$$b = 186.49 \cos 57.209°$$
$$b = 101.00 \text{ cm}$$

33. $b = 173.921$ m, $c = 208.543$ m

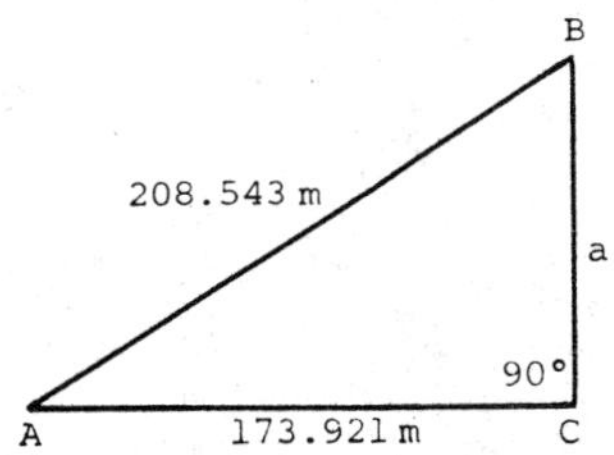

$$a = \sqrt{208.543^2 - 173.921^2}$$
$$= 115.072 \text{ m}$$

$$\cos A = \frac{173.921}{208.543}$$
$$A = 33.4901° = 33°\ 29.40'$$

$$\sin B = \frac{173.921}{208.543}$$
$$B = 56.5099° = 56°\ 30.60'$$

35. If two parallel lines are intersected by a transversal, then alternate interior angles are congruent.
In the figure in the textbook angle DAB and angle ABC are alternate interior angles and are congruent. Therefore, they have the same measure.

37. Let h = the distance the ladder goes up the wall.

$$\sin 43° \ 50' = \frac{h}{13.5}$$

$$h = 13.5 \sin 43° \ 50'$$

$$h \approx 9.3496000$$

The ladder goes up the wall 9.3 m.

39. Let θ = the angle the guy wire makes with the ground.

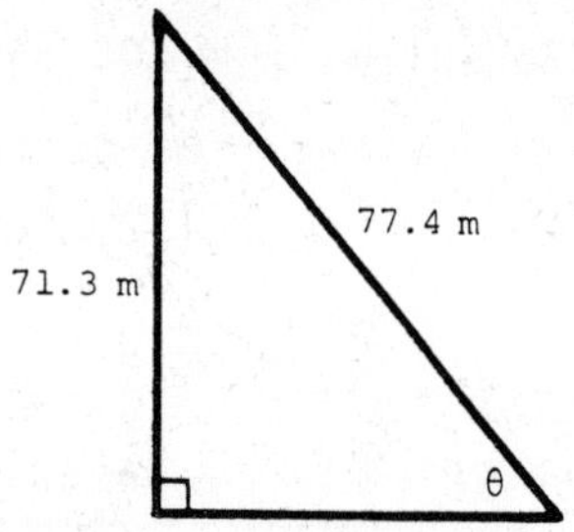

$$\sin \theta = \frac{71.3}{77.4}$$

$$\theta \approx 67.100475°$$

$$\theta \approx 67° + (.100475)(60')$$

$$\theta \approx 67° \ 10'$$

The guy wire makes an angle of 67° 10′ with the ground.

41. Let h = the height of the flagpole.

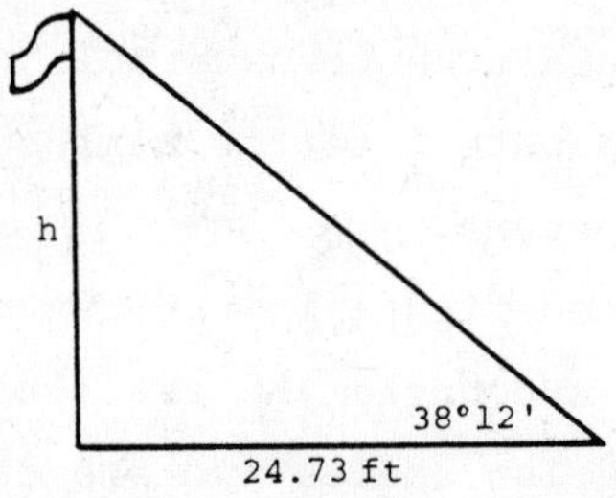

$$\tan 38° \ 12' = \frac{h}{24.73}$$

$$h = 24.73 \tan 38° \ 12'$$

$$h = 19.46$$

The height of the flagpole is 19.46 ft.

43. Since angle T = 32° 10′ and angle S = 57° 50′, S + T = 90°, and triangle RST is a right triangle.

$$\tan 32° \ 10' = \frac{RS}{53.1}$$

$$RS = 53.1 \tan 32° \ 10'$$

$$RS \approx 33.395727$$

The distance across the lake is 33.4 m.

45. The two right triangles are congruent.
So corresponding sides are congruent. Since the sides that compose the base of the isosceles triangle are congruent, each side is 42.36/2, or 21.18 in.

Let x = the length of each of the two equal sides of the isosceles triangle.

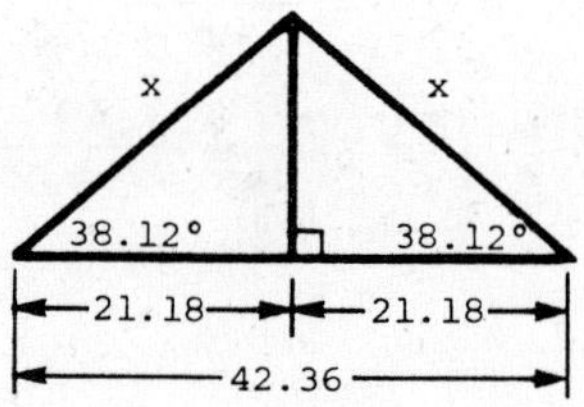

$$\cos 38.12° = \frac{21.18}{x}$$

$$x \cos 38.12° = 21.18$$

$$x = \frac{21.18}{\cos 38.12°}$$

$$x \approx 26.921918$$

The length of each of the two equal sides of the triangle is 26.92 in.

47. Let x = the length of the shadow.

$$\tan 23.4° = \frac{5.75}{x}$$

$$x \tan 23.4° = 5.75$$

$$x = \frac{5.75}{\tan 23.4°}$$

$$x \approx 13.287466$$

The length of the shadow is 13.3 ft.

49. Let θ = the angle of elevation of the sun.

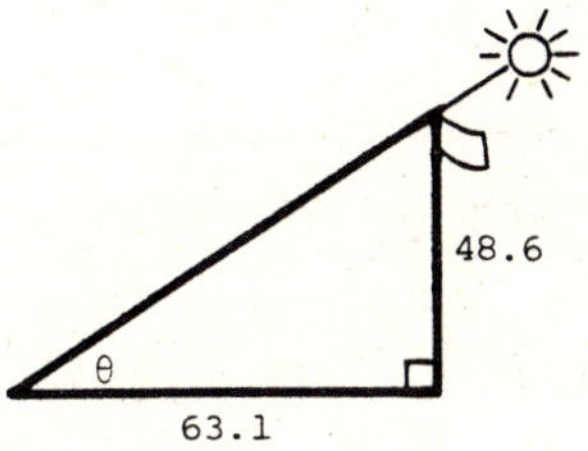

$$\tan \theta = \frac{48.6}{63.1}$$

$$\theta \approx 37.603681$$

$$\approx 37° + (.603681)(60)$$

$$\theta \approx 37° 40'$$

The angle of elevation of the sun is 37° 40′.

51. Let x = the horizontal distance that the plane must fly to be directly over the tree.

$$\tan 13° 50' = \frac{10,500}{x}$$

$$x \tan 13° 50' = 10,500$$

$$x = \frac{10,500}{\tan 13° 50'}$$

$$x \approx 42,600$$

The horizontal distance that the plane must fly to be directly over the tree is 42,600 ft.

53. Let α = the angle of depression that the lens make with the horizontal

and θ = the angle that the lens make with the vertical.

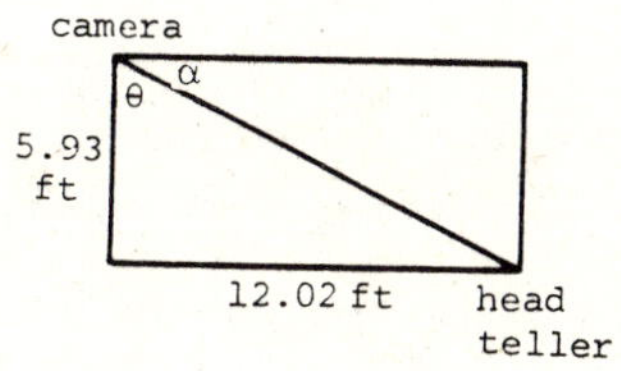

$$\tan \theta = \frac{12.02}{5.93}$$

$$\theta = 63° 40'$$

$$\alpha = 90° - \theta$$

$$= 90° - 63° 40'$$

$$\alpha = 26° 20'$$

The lens should make an angle of 26° 20′ with the horizontal.

55. Let θ = the angle between the cell edge and the cube diagonal.

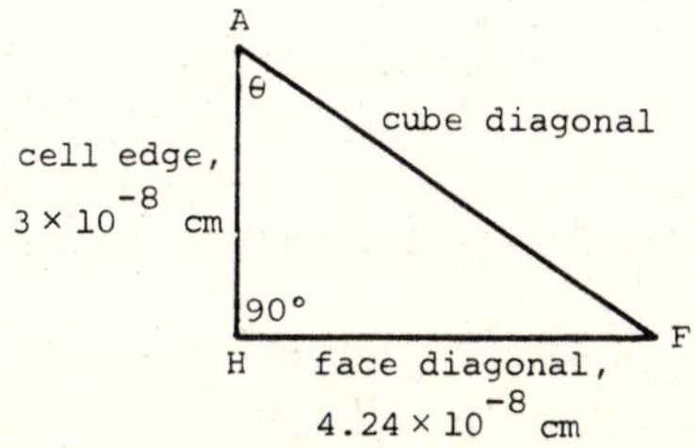

$$\tan \theta = \frac{4.24 \times 10^{-8}}{3 \cdot 10^{-8}}$$

$$\theta \approx 54° \ 40'$$

57. Let point S represent the star.

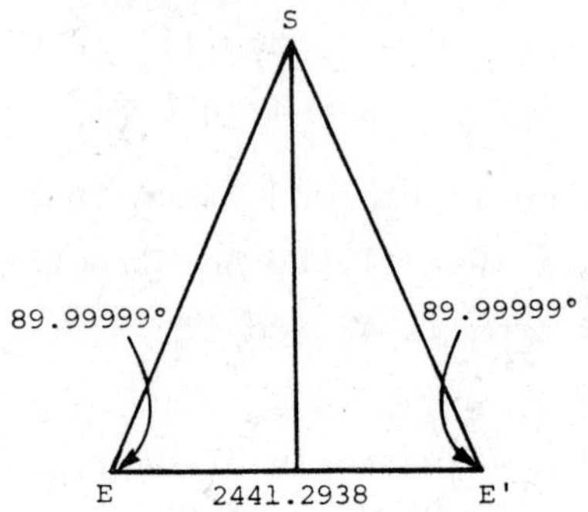

Triangle SEE′ has equal base angles each 89.99999° so the triangle is isosceles. The altitude of the triangle bisects the base of the triangle and forms two right triangles. In one of the right triangles, the side adjacent to the 89.99999° angle is 2441.2938/2, or 1220.6469 mi.

$$\cos E' = \frac{1220.6469}{SE'}$$

$$SE' = \frac{1220.6469}{\cos 89.99999°}$$

$$= 6.993792 \times 10^9 \text{ mi}$$

The distance from the Haystack Observatory to the star is

$$6.993792 \times 10^9 \text{ mi}.$$

59. Let θ = the angle at which the rafters are to be cut.

$$\tan \theta = \frac{8}{1.5}$$

$$\theta \approx 79.380345$$

The rafters should be cut at a 79° angle.

Section 2.5

1. Let θ = the angle of the sun above the horizon.

$$\tan \theta = \frac{5.25}{8.65}$$

$$\theta = 31° \ 20'$$

The angle of the sun above the horizon is 31° 20′.

3. Let α = the angle the panel makes with the horizontal.

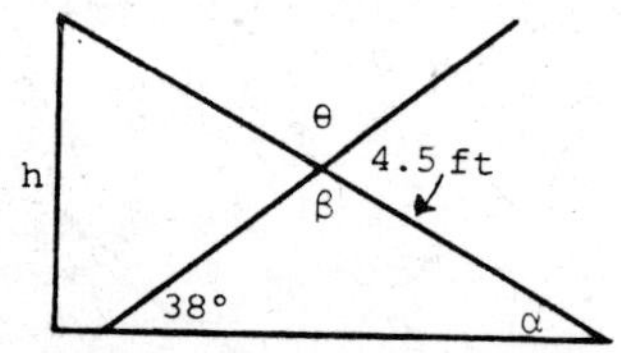

In the sketch,

$$\beta = \theta$$

since they are vertical angles.

So $\beta = 94°$

since the sum of the measures of the angles of a triangle is 180°,

$$38° + 94° + \alpha = 180°$$

$$\alpha = 180° - 38° - 94°$$

$$= 48°.$$

$$\sin \alpha = \frac{h}{4.5}$$

$$h = 4.5 \sin \alpha$$

$$= 4.5 \sin 48°$$

$$h = 3.3 \text{ ft}$$

5. Let x = the minimum distance that a plant needing full sun can be placed from the fence.

$$\tan 23° \; 20' = \frac{4.65}{x}$$

$$x \tan 23° \; 20' = 4.65$$

$$x = \frac{4.65}{\tan 23° \; 20'}$$

$$x \approx 10.8$$

The minimum distance is 10.8 ft.

7. Let y = the common hypotenuse of the two right triangles.

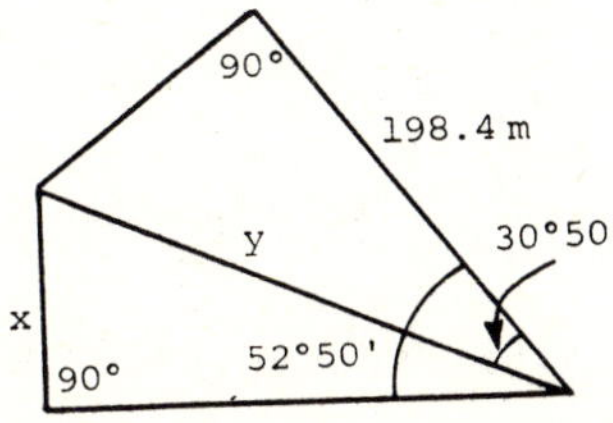

$$\cos 30° \; 50' = \frac{198.4}{y}$$

$$y = \frac{198.4}{\cos 30° \; 50'}$$

$$y = 231.06$$

To find x, first find the angle opposite x in the right triangle by subtracting.

$$52° \; 20' - 30° \; 50' = 51° \; 80' - 30° \; 50'$$
$$= 21° \; 30'$$

$$\sin 21° \; 30' = \frac{x}{y} = \frac{x}{231.06}$$

$$x = 231.06(\sin 21° \; 30')$$
$$= 84.7 \text{ m}$$

9. Let θ = the angle at which the tower deviates from the vertical.

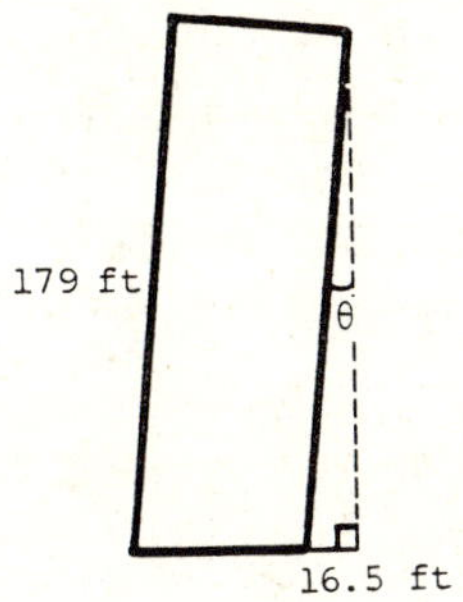

$$\sin \theta = \frac{16.5}{179}$$

$$\theta = 5.29° \text{ or } 5° \; 20'$$

The tower deviates from the vertical 5° 20′.

11. Let x be the side adjacent to 49° 10′ in the smaller triangle.

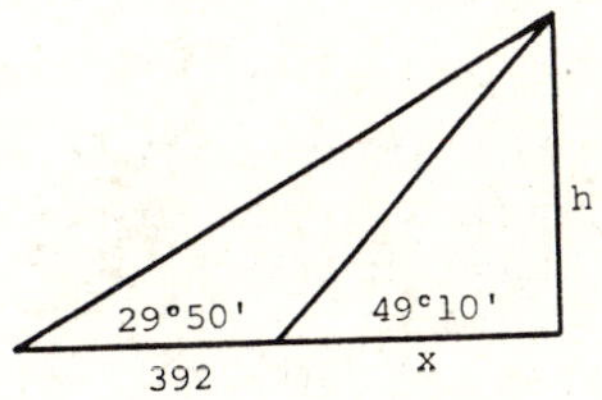

In the larger right triangle:

$$\tan 29° \; 50' = \frac{h}{392 + x}$$

$$h = (392 + x) \tan 29° \; 50'.$$

In the smaller right triangle:

$$\tan 49° \; 10' = \frac{h}{x}$$

$$h = x \tan 49° \; 10'$$

Substitute for h in this equation and solve for x.

$$(392 + x) \tan 29^\circ 50' = x \tan 49^\circ 10'$$

$$392 \tan 29^\circ 50' + x \tan 20^\circ 50'$$
$$= x \tan 49^\circ 10'$$

$$392 \tan 29^\circ 50'$$
$$= x \tan 49^\circ 10' - x \tan 29^\circ 50'$$

$$392 \tan 29^\circ 50'$$
$$= x(\tan 49^\circ 10' - \tan 29^\circ 50')$$

$$\frac{392 \tan 29^\circ 50'}{\tan 49^\circ 10' - \tan 29^\circ 50'} = x$$

Then, substitute for x in the equation for the small triangle, $h = x \tan 49^\circ 10'$. Thus,

$$h = \frac{392 \tan 29^\circ 50' (\tan 49^\circ 10')}{\tan 49^\circ 10' - \tan 29^\circ 50'}.$$

Use a calculator to find h.

$h = 446$

13. Let x = the distance from the closer point on the ground to the base of height h of the pyramid.

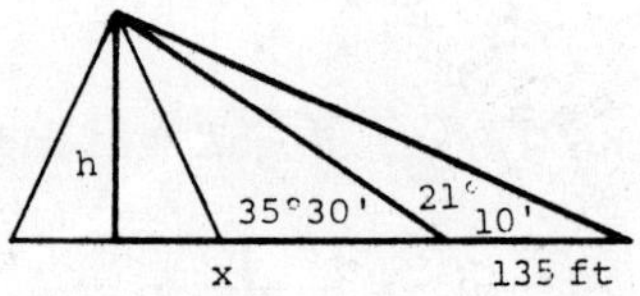

In the larger right triangle:

$$\tan 21^\circ 10' = \frac{h}{135 + x}$$

$$h = (135 + x) \tan 21^\circ 10'$$

In the smaller right triangle:

$$\tan 35^\circ 30' = \frac{h}{x}$$

$$h = x \tan 35^\circ 30'$$

Substitute for h in this equation, and solve for x.

$$(135 + x) \tan 21^\circ 10'$$
$$= x \tan 35^\circ 30'$$

$$135 \tan 21^\circ 10' + x \tan 21^\circ 10'$$
$$= x \tan 35^\circ 30'$$

$$135 \tan 21^\circ 10'$$
$$= x \tan 35^\circ 30' - x \tan 21^\circ 10'$$

$$135 \tan 21^\circ 10'$$
$$= x(\tan 35^\circ 30' - \tan 21^\circ 10')$$

$$\frac{135 \tan 21^\circ 10'}{(\tan 35^\circ 30' - \tan 21^\circ 10')} = x$$

Then substitute for x in the equation for the smaller triangle.

$$h = \frac{135 \tan 21^\circ 10' (\tan 35^\circ 30')}{(\tan 35^\circ 30' - \tan 21^\circ 10')}$$

$h = 114$

The height of the pyramid is 114 ft.

15. Let x = the height of the antenna and h = the height of the house.

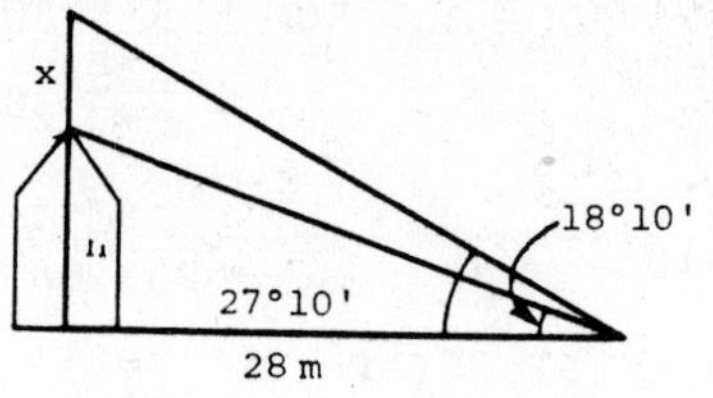

In the smaller right triangle:

$$\tan 18° \; 10' = \frac{h}{28}$$

$$h = 28 \; (\tan 18° \; 10')$$

In the larger right triangle:

$$\tan 27° \; 10' = \frac{x + h}{28}$$

$$x + h = 28 \; (\tan 27° \; 10')$$

$$x = 28(\tan 27° \; 10') - 28(\tan 18° \; 10')$$

$$x = 5.18$$

The height of the antenna is 5.18 m.

17. Let x = the distance the plane flies in 15 sec.

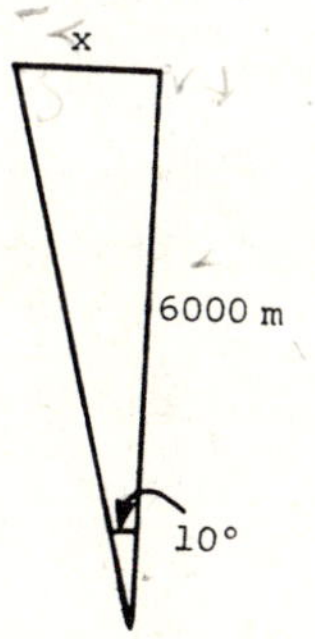

$$\tan 10° = \frac{x}{6000}$$

$$x = 6000 \tan 10°$$

$$x = 1058$$

The plane flies 1058 m in 15 sec so it flies

$$\frac{1058}{15} = 70 \text{ m per sec.}$$

19. Let x = the distance the plane is from its starting point.

In the figure, the measure of angle ACB is 40° + (180° - 130°) or 90°. Therefore, triangle ACB is a right triangle.

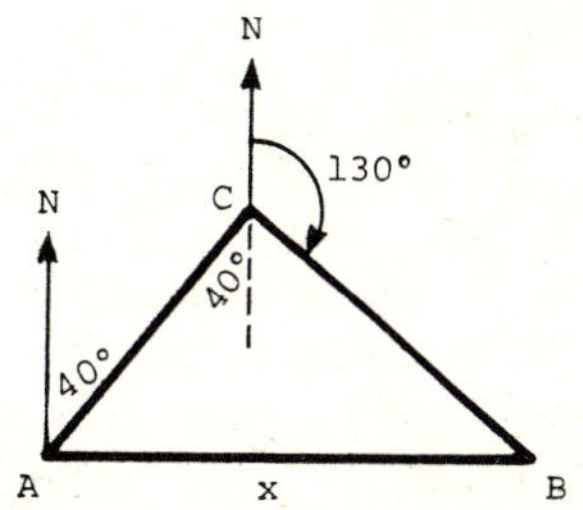

Distance traveled in 1.5 hr:

$$(1.5 \text{ hr})(110 \text{ mph}) = 165 \text{ mi}$$

Distance traveled in 1.3 hr:

$$(1.3 \text{ hr})(110 \text{ mph}) = 143 \text{ mi}$$

Using the Pythagorean theorem, we have

$$x^2 = 165^2 + 143^2$$

$$x^2 = 47{,}674$$

$$x \approx 220.$$

The plane is 220 mi from its starting point.

21. Let x = distance the ships are apart.

$$(1.5 \text{ hours})(18 \text{ knots}) = 27 \text{ nautical mi}$$

$$(1.5 \text{ hours})(26 \text{ knots}) = 39 \text{ nautical mi}$$

Since 130° - 40° = 90°, we have a right triangle. Applying the Pythagorean theorem, we have

$$x^2 = 27^2 + 39^2$$

$$x^2 = 2250$$

$$x \approx 47$$

The ships are 47 nautical mi apart.

23. Let x = distance between the two ships.

The angle between the bearings of the ships is 180° − (28° 10′ + 61° 50′) or 90°. The triangle formed is a right triangle.

Distance traveled at 24.0 mph:

(4 hrs)(24.0 mph) = 96 mi

Distance traveled at 28.0 mph:

(4 hrs)(28.0 mph) = 112 mi

Applying the Pythagorean theorem gives

$$x^2 = 96^2 + 112^2$$
$$x^2 = 21{,}760$$
$$x = 148$$

The ships are 148 miles apart.

25. Draw triangle WDG with W representing Winston-Salem, D representing Danville, and G representing Goldsboro. Name any point X on the line due south from D.

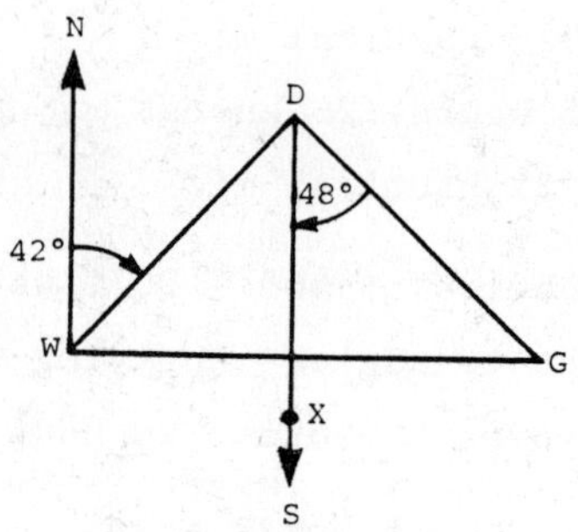

Since the bearing from W to D is 42°, angle WDX = 42°.

Angle XDG = 48°

So angle D = 42° + 48° = 90°, and triangle WDG is a right triangle.

Using d = rt and the Pythagorean theorem, we have

$$WG = \sqrt{(WD)^2 + (DG)^2}$$
$$= \sqrt{[60(1)]^2 + [60(1.05)]^2}$$
$$= \sqrt{15{,}300}$$
$$= 120$$

The distance from Winston-Salem to Goldsboro is 120 mi.

Chapter 2 Review Exercises

1. $\sin A = \dfrac{\text{side opposite}}{\text{hypotenuse}} = \dfrac{60}{61}$

$\cos A = \dfrac{\text{side adjacent}}{\text{hypotenuse}} = \dfrac{11}{61}$

$\tan A = \dfrac{\text{side opposite}}{\text{side adjacent}} = \dfrac{60}{11}$

$\cot A = \dfrac{\text{side adjacent}}{\text{side opposite}} = \dfrac{11}{60}$

$\sec A = \dfrac{\text{hypotenuse}}{\text{side adjacent}} = \dfrac{61}{11}$

$\csc A = \dfrac{\text{hypotenuse}}{\text{side opposite}} = \dfrac{61}{60}$

3. $\sin 4\beta = \cos 5\beta$

Since sine and cosine are cofunctions,

$$4\beta + 5\beta = 90°$$
$$9\beta = 90°$$
$$\beta = 10°.$$

5. $\tan (5x + 11°) = \cot (6x + 2°)$

Since tangent and cotangent are cofunctions,

$$5x + 11° + 6x + 2° = 90°$$
$$11x = 90° - 11° - 2°$$
$$11x = 77°$$
$$x = 7°.$$

7. $\sin 46° < \sin 58°$

$\sin\theta$ increases as θ increases from $0°$ to $90°$. Since

$$58° > 46°$$
$$\sin 58° > \sin 46°$$

True

9. $\sec 48° \geq \cos 42°$

$\sec\theta \geq 1$ for all θ and $\cos\theta \leq 1$ for all θ.
Therefore,

$$\cos\theta \leq 1 \leq \sec\theta.$$

True

11. $\theta = 219°$

Since θ is in quadrant III,

$$\theta' = 219° - 180° = 39°.$$

13. $\theta = 578.94°$

$578.94°$ is coterminal with $578.94° - 360° = 218.94°$, which is in quadrant III.

$$\theta = 218.94° - 180°$$
$$= 38.94°$$

The reference angle θ' is $38.94°$.

17. $120°$

This angle is in quadrant II, so the reference angle is $180° - 120° = 60°$.
Since $120°$ is in quadrant II, the cosine, tangent, cotangent, and secant are negative.

$$\sin 120° = \sin 60° = \frac{\sqrt{3}}{2}$$

$$\cos 120° = -\cos 60° = -\frac{1}{2}$$

$$\tan 120° = -\tan 60° = -\sqrt{3}$$

$$\cot 120° = -\cot 60° = -\frac{\sqrt{3}}{3}$$

$$\sec 120° = -\sec 60° = -2$$

$$\csc 120° = \csc 60° = \frac{2\sqrt{3}}{3}$$

19. $300°$

This angle is in quadrant IV, so the reference angle is $360° - 300° = 60°$.
Since $300°$ is in quadrant IV, the sine, tangent, cotangent, and cosecant are negative.

$$\sin 300° = -\sin 60° = -\frac{\sqrt{3}}{2}$$

$$\cos 300° = \cos 60° = \frac{1}{2}$$

$$\tan 300° = -\tan 60° = -\sqrt{3}$$

$$\cot 300° = -\cot 60° = -\frac{\sqrt{3}}{3}$$

$$\sec 300° = \sec 60° = 2$$

$$\csc 300° = -\csc 60° = -\frac{2\sqrt{3}}{3}$$

21. $-225°$

$-225°$ is coterminal with

$$-225° + 360° = 135°.$$

The reference angle is $45°$. Since $-225°$ is in quadrant II, the cosine, tangent, cotangent, and secant are negative.

$\sin(-225°) = \sin 45° = \frac{\sqrt{2}}{2}$

$\cos(-225°) = -\cos 45° = -\frac{\sqrt{2}}{2}$

$\tan(-225°) = -\tan 45° = -1$

$\cot(-225°) = -\cot 45° = -1$

$\sec(-225°) = -\sec 45° = -\sqrt{2}$

$\csc(-225°) = \csc 45° = \sqrt{2}$

23. $\sin\theta = -\frac{1}{2}$

Since $\sin\theta < 0$, θ is in quadrant III or IV. Since $\sin\theta = -\frac{1}{2}$, the reference angle is 30°.

In quadrant III,

$$\theta = 180° + 30° = 210°.$$

In quadrant IV,

$$\theta = 360° - 30° = 330°.$$

25. $\cot\theta = -1$

Since $\cot\theta < 0$, θ is in quadrant II or IV. Since $\cot\theta = -1$, the refer-erence angle is 45°.

In quadrant II,

$$\theta = 180° - 45° = 135°.$$

In quadrant IV,

$$\theta = 360° - 45° = 315°.$$

27. $\cos 60° + 2\sin^2 30°$

$= \frac{1}{2} + 2\left(\frac{1}{2}\right)^2$

$= \frac{1}{2} + 2\left(\frac{1}{4}\right)$

$= \frac{1}{2} + \frac{1}{2}$

$= 1$

29. $\cot^2 300° + \cos^2 120° - 3\sin^2 240°$

$= \left(-\frac{\sqrt{3}}{3}\right)^2 + \left(-\frac{1}{2}\right)^2 - 3\left(-\frac{\sqrt{3}}{2}\right)^2$

$= \frac{3}{9} + \frac{1}{4} - 3\left(\frac{3}{4}\right)$

$= \frac{1}{3} + \frac{1}{4} - \frac{9}{4}$

$= \frac{4}{12} + \frac{3}{12} - \frac{27}{12}$

$= -\frac{20}{12}$

$= -\frac{5}{3}$

In Exercises 31–39, be sure that your calculator is in degree mode.

31. $\sin 72°\ 30'$

$72°\ 30' = \left(72 + \frac{30}{60}\right)^\circ = 72.5$

Enter: 72.5 [SIN]

Display: .95371695

33. $\cot 305.6°$

Enter: 305.6 [TAN] [1/x]

Display: -.71592968

35. $\cot 32°\ 42'$

$32°\ 42' = \left(32 + \frac{42}{60}\right)^\circ = 32.7°$

Enter: 32.7 [TAN] [1/x]

Display: 1.5576601

37. $\sec 58.9041°$

Enter: 58.9041 [COS] [1/x]

Display: 1.9362132

39. sin 89.0043°

Enter: 89.0043 [SIN]

Display: .99984900

41. The exact answer is $\frac{\sqrt{2}}{2}$. The value .7071067812 is a ten-place decimal approximation; it is not exact.

43. $\sin \theta = .82584121$

Enter: .82584121 [SIN^{-1}]

Display: 55.673870

$$\theta \approx 55.673870°$$

45. $\cos \theta = .97540415$

Enter: .97540415 [COS^{-1}]

Display: 12.733938

$$\theta \approx 12.733938°$$

47. $\tan \theta = 1.9633124$

Enter: 1.9633124 [TAN^{-1}]

Display: 63.008286

$$\theta \approx 63.008286°$$

49. $\sin \theta = .73254290$

Since $\sin \theta$ is positive, there will be one angle in quadrant I and one angle in quadrant II. If θ' is the reference angle, then the two angles are θ' and $180° - \theta'$.

Enter: .73254290 [INV] [SIN]

Display: 47.100000

$$\theta' = 47.1°$$

$$180° - \theta' = 180° - 47.1° = 132.4°$$

51. $\sin 50° + \sin 40° = \sin 90°$

$\sin 50° = .76604444$

$\sin 40° = .64278761$, so

$\sin 50° + \sin 40°$

$= .76604444 + .64278761$

$= 1.40883205$

$\sin 90° = 1$, so

$$1.40883205 \neq 1$$

False

53. $\sin 240° = 2 \sin 120° \cdot \cos 120°$

$\sin 240° = -\sin 60° = -\frac{\sqrt{3}}{2}$

$2 \sin 120° \cdot \cos 120°$

$= 2(\sin 60°)(-\cos 60°)$

$= 2\left(\frac{\sqrt{3}}{2}\right)\left(-\frac{1}{2}\right)$

$= -\frac{\sqrt{3}}{2}$

True

55. $A = 58° 30'$, $c = 748$

$A + B = 90°$

$B = 90° - A$

$B = 90° - 58° 30'$

$B = 31° 30'$

$\sin A = \frac{a}{c}$

$a = c \sin A$

$a = 748 \sin 58° 30'$

$a = 638$

$\cos A = \frac{b}{c}$

$b = c \cos A$

$b = 748 \cos 58° \, 30'$

$b = 391$

57. $A = 39.72°$, $b = 38.97$ m

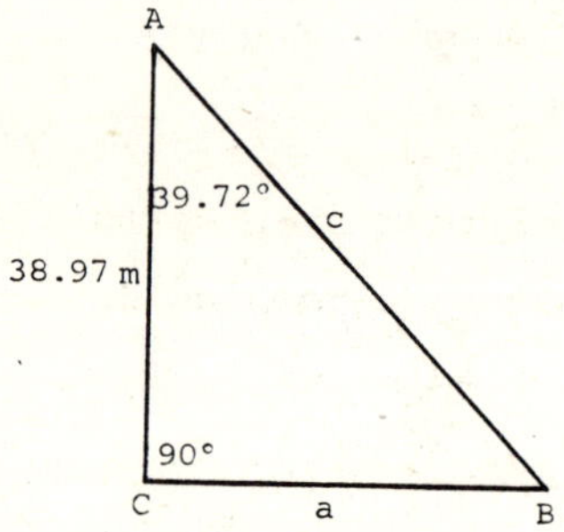

$A + B = 90° = C$

$B = 90° - A$

$B = 90° - 39.72°$

$= 50.28°$

$\tan A = \frac{a}{b}$

$a = b \tan A$

$a = 38.97 \tan 39.72°$

$a = 32.38$ m

$\cos A = \frac{b}{c}$

$c = \frac{b}{\cos A}$

$c = \frac{38.97}{\cos 39.72°}$

$c = 50.66$ m

59. Let x = height of the tower

$\tan 38° \, 20' = \frac{x}{93.2}$

$x = 93.2 \tan 38° \, 20'$

$x \approx 73.693005$

The height of the tower is 73.7 ft.

61. The diagonal d of the rectangle is the hypotenuse of a right triangle whose legs are the adjacent sides of the rectangle.

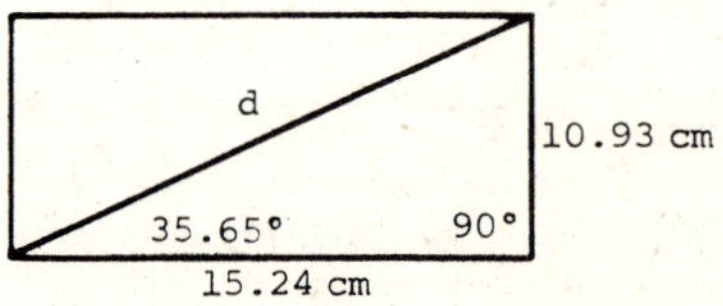

By the Pythagorean theorem, the length d of the diagonal is

$d = \sqrt{(15.24)^2 + (10.93)^2} = 18.75$ cm.

An alternative method gives the same answer.

$\sin 35.65° = \frac{10.93}{d}$

$d = \frac{10.93}{\sin 35.65°}$

$= 18.75$ cm

63. Draw triangle ABC and extend the north-south line to a point X south of A. Extend the north-south line to a point Y, north of C.

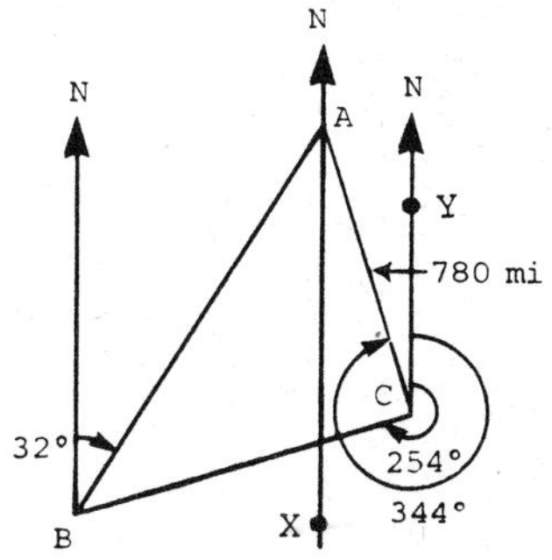

Angle ACB = 344° − 254° = 90°, so ABC is a right triangle.

Angle BAX = 32° since it is an alternate interior angle to 32°.

Angle YCA = 360° − 344° = 16°

Angle XAC = 16° since it is an alternate interior angle to angle YCA.

Angle BAC = 32° + 16° = 48°

In triangle ABC,

$$\cos A = \frac{AC}{AB}$$

$$AB = \frac{AC}{\cos A}$$

$$AB = \frac{780}{\cos 48°}$$

$$= 1200$$

The distance from A to B is 1200 m.

65. Suppose A is the car heading south at 55 mph, B is the car heading west at 40 mph, and point C is the intersection from which they start. After two hours, by d = rt,

$$AC = 55(2) = 110$$
$$BC = 40(2) = 80.$$

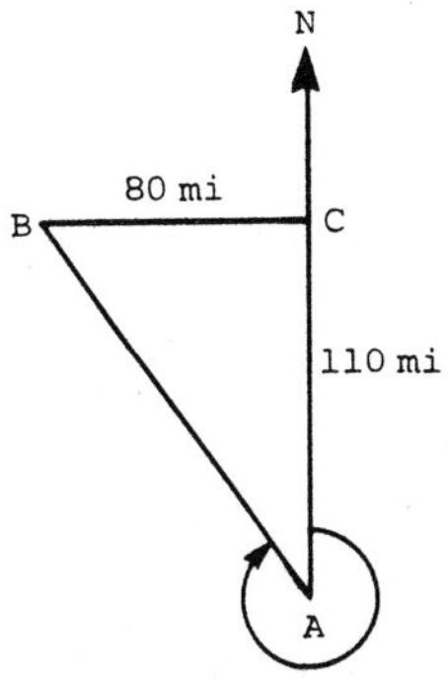

Since angle ACB is a right angle, use the Pythagorean formula to find the distance from A to B.

$$(AB)^2 = 80^2 + 110^2$$
$$(AB)^2 = 6400 + 12{,}100$$
$$AB = \sqrt{18{,}500} \approx 140 \text{ mi}$$

An alternative method follows. Since the bearing of A from B is 324°,

angle CAB = 360° − 324° = 36°.

$$\sin CAB = \frac{BC}{AB}$$

$$AB = \frac{BC}{\sin CAB} = \frac{80}{\sin 36°}$$

$$\approx 140 \text{ mi}$$

CHAPTER 2 TEST

[2.1] Find a solution for each of the following equations. Assume that all angles are cute angles.

1. $\sin 6\theta = \cos 9\theta$

2. $\tan (11\alpha + 8°) = \cot (12\alpha - 10°)$

[2.2] Find the reference angle θ' for each of the following.

3. 201° 14′

4. 517° 21′

[2.2] 5. Can 90° be the measure of a reference angle? Explain.

[2.2] In each exercise, find the exact values of the sine, cosine, and tangent for θ. Do not use a calculator. Rationalize denominators.

6. $\theta = 240°$

7. $\theta = 675°$

[2.2] Evaluate each of the following. Do not use a calculator.

8. $\tan^2 45° + \sin^2 150°$

9. $2 \cos^2 210° + \csc^2 300°$

10. $\sec^2 180° + 3 \cot^2 120°$

[2.3] Use a calculator to find approximations for each of the following.

11. sin 108° 20′

12. sec 41° 30′

13. cot 128.8°

[2.3] Find two angles θ in the interval [0°, 360°) that satisfy each of the following. Give each answer to the nearest tenth of a degree.

14. $\sin \theta = .91382416$

15. $\tan \theta = 1.3865127$

16. $\cos \theta = -.10875886$

[2.4] Solve each right triangle. The right angle is at C.

17. a = 28.4 cm, b = 34.7 cm

18. B = 38°, a = 17 ft

[2.5] **Solve the following problems.**

19. The angle of elevation from a park bench to the top of a tower is 9.8°. The tower is 31.2 ft high. How far is the bench from the top of the tower?

20. The bearing from A to B is 35°. The bearing from B to C is 152°. The bearing from A to C is 62°. If the distance from B to C is 18 km, find the distance from A to C.

CHAPTER 2 TEST ANSWERS

1. 6° **2.** 4° **3.** 21° 14′ **4.** 22° 39′ **5.** No, a reference angle is an acute angle. **6.** $-\sqrt{3}/2$; $-1/2$; $\sqrt{3}$ **7.** $-\sqrt{2}/2$; $\sqrt{2}/2$; -1

8. 5/4 **9.** 17/6 **10.** 2 **11.** .94924264 **12.** 1.3351924

13. −.80402064 **14.** 66.0°; 114.0° **15.** 54.2°; 234.2°

16. 96.2°; 263.8° **17.** c = 44.8 cm; A = 39.3°; B = 50.7°

18. A = 52°; b = 13 ft; c = 22 ft **19.** 183 ft **20.** 35 km

CHAPTER 3 RADIAN MEASURE AND THE CIRCULAR FUNCTIONS

Section 3.1

1. $60° = 60°\left(\frac{\pi}{180°}\right) = \frac{\pi}{3}$

This method is the formula method. The proportion method may also be used as follows.

$$\frac{\text{Radian measure}}{\pi} = \frac{\text{Degree measure}}{180}$$

$$\frac{\text{Radian measure}}{\pi} = \frac{60}{180}$$

$$\text{Radian measure} = \frac{60\pi}{180} = \frac{\pi}{3}$$

3. $90° = 90°\left(\frac{\pi}{180°}\right) = \frac{\pi}{2}$

5. $150° = 150°\left(\frac{\pi}{180°}\right) = \frac{5\pi}{6}$

7. $210° = 210°\left(\frac{\pi}{180°}\right) = \frac{7\pi}{6}$

9. $300° = 300°\left(\frac{\pi}{180°}\right) = \frac{5\pi}{3}$

11. $450° = 450°\left(\frac{\pi}{180°}\right) = \frac{5\pi}{2}$

13. $20° = 20°\left(\frac{\pi}{180°}\right) = \frac{\pi}{9}$

17. $\frac{\pi}{3} = \frac{\pi}{3}\left(\frac{180°}{\pi}\right) = 60°$

This method is the formula method. The proportion method may also be used.

$$\frac{\text{Radian measure}}{\pi} = \frac{\text{Degree measure}}{180}$$

$$\frac{\frac{\pi}{3}}{\pi} = \frac{\text{Degree measure}}{180}$$

$$\frac{1}{3} = \frac{\text{Degree measure}}{180}$$

$$\text{Degree measure} = \frac{180}{3} = 60$$

In $\frac{\pi}{3}$ radians, there are 60°.

19. $\frac{7\pi}{4} = \frac{7\pi}{4}\left(\frac{180°}{\pi}\right) = 315°$

21. $\frac{11\pi}{6} = \frac{11\pi}{6}\left(\frac{180°}{\pi}\right) = 330°$

23. $-\frac{\pi}{6} = -\frac{\pi}{6}\left(\frac{180°}{\pi}\right) = -30°$

25. $\frac{8\pi}{5} = \frac{8\pi}{5}\left(\frac{180°}{\pi}\right) = 288°$

27. $\frac{11\pi}{15} = \frac{11\pi}{15}\left(\frac{180°}{\pi}\right) = 132°$

29. $\frac{7\pi}{20} = \frac{7\pi}{20}\left(\frac{180°}{\pi}\right) = 63°$

31. $\frac{11\pi}{30} = \frac{11\pi}{30}\left(\frac{180°}{\pi}\right) = 66°$

33. $39° = 39°\left(\frac{\pi}{180°}\right) = .68$

35. $42°\ 30' = 42.5°\left(\frac{\pi}{180°}\right) = .742$

37. $139°\ 10' = 139\frac{1}{6}°\left(\frac{\pi}{180°}\right) = 2.43$

39. $64.29° = 64.29°\left(\frac{\pi}{180°}\right)$

$= 1.122$

41. $56°\ 25' = 56\frac{25}{60}°\left(\frac{\pi}{180°}\right)$

$= .9847$

43. $47.6925° = 47.6925°\left(\frac{\pi}{180°}\right)$

$= .832391$

45. $-29°\ 42'\ 36''$

$= -\left(29° + \frac{42}{60}° + \frac{36}{3600}°\right)$

$= -29.71°\left(\frac{\pi}{180°}\right)$

$= -.518537$

47. $-209°\ 46'\ 15''$

$= -\left(209° + \frac{46}{60}° + \frac{15}{3600}°\right)$

$= -209.7708°\left(\frac{\pi}{180°}\right)$

$= -3.66119$

49. 2 radians $= 2\left(\frac{180°}{\pi}\right) = 114.5916°$

$= 114° + (.5916)(60')$

$= 114°\ 35'$

51. 1.74 radians $= 1.74\left(\frac{180°}{\pi}\right)$

$= 99.6947°$

$= 99° + (.6947)(60')$

$= 99°\ 42'$

53. .0912 radian $= .0912\left(\frac{180°}{\pi}\right)$

$= 5.2254°$

$= 5° + (.2254)(60')$

$= 5°\ 14'$

55. 9.84763 radians $= 9.84763\left(\frac{180°}{\pi}\right)$

$= 564.228°$

$= 564° + (.228)(60')$

$= 564°\ 14'$

57. -3.47189 radians

$= -3.47189\left(\frac{180°}{\pi}\right)$

$= -198.925°$

$= -[198° + (.925)(60')]$

$= -198°\ 55'$

59. The value of sin 30 is not $\frac{1}{2}$, because 30, without a degree symbol, is seen as a radian measure. If you want to have an answer of $\frac{1}{2}$, you must say sin 30°.

61. $\sin\frac{\pi}{3} = \sin\frac{\pi}{3}\cdot\frac{180°}{\pi}$

$= \sin 60°$

$= \frac{\sqrt{3}}{2}$

63. $\tan\frac{\pi}{4} = \sin\frac{\pi}{4}\cdot\frac{180°}{\pi}$

$= \tan 45°$

$= 1$

65. $\sec\frac{\pi}{6} = \sec\frac{\pi}{6}\cdot\frac{180°}{\pi}$

$= \sec 30°$

$= \frac{2\sqrt{3}}{3}$

67. $\sin \frac{\pi}{2} = \sin \frac{\pi}{2} \cdot \frac{180°}{\pi}$

$= \sin 90°$

$= 1$

69. $\tan \frac{2\pi}{3} = \tan \frac{2\pi}{3} \cdot \frac{180°}{\pi}$

$= \tan 120°$

$= -\sqrt{3}$

71. $\sin \frac{5\pi}{6} = \sin \frac{5\pi}{6} \cdot \frac{180°}{\pi}$

$= \sin 150°$

$= \frac{1}{2}$

73. $\cos 3\pi = \cos 3\pi \cdot \frac{180°}{\pi}$

$= \cos 540°$

$= -1$

75. $\sin \frac{4\pi}{3} = \sin \frac{4\pi}{3} \cdot \frac{180°}{\pi}$

$= \sin 240°$

$= -\frac{\sqrt{3}}{2}$

77. $\tan \frac{5\pi}{4} = \tan \frac{5\pi}{4} \cdot \frac{180°}{\pi}$

$= \tan 225°$

$= 1$

79. $\sin 3\pi = \sin 3\pi \cdot \frac{180°}{\pi}$

$= \sin 540°$

$= 0$

81. $\tan \left(-\frac{\pi}{3}\right) = \tan \left(-\frac{\pi}{3}\right) \cdot \frac{180°}{\pi}$

$= \tan (-60°)$

$= -\sqrt{3}$

83. $\sin \left(-\frac{7\pi}{6}\right) = \sin \left(-\frac{7\pi}{6}\right) \cdot \frac{180°}{\pi}$

$= \sin (-210°)$

$= \frac{1}{2}$

85. Answers start at 30° and proceed counterclockwise around circle.

$30° = 30°\left(\frac{\pi}{180°}\right) = \frac{\pi}{6}$

$\frac{\pi}{4}$ radians $= \frac{\pi}{4}\left(\frac{180°}{\pi}\right) = 45°$

$60° = 60°\left(\frac{\pi}{180°}\right) = \frac{\pi}{3}$

$\frac{2\pi}{3}$ radians $= \frac{2\pi}{3}\left(\frac{180°}{\pi}\right) = 120°$

$\frac{3\pi}{4}$ radians $= \frac{3\pi}{4}\left(\frac{180°}{\pi}\right) = 135°$

$150° = 150°\left(\frac{\pi}{180°}\right) = \frac{5\pi}{6}$

$180° = 180°\left(\frac{\pi}{180°}\right) = \pi$

$210° = 210°\left(\frac{\pi}{180°}\right) = \frac{7\pi}{6}$

$225° = 225°\left(\frac{\pi}{180°}\right) = \frac{5\pi}{4}$

$\frac{4\pi}{3}$ radians $= \frac{4\pi}{3}\left(\frac{180°}{\pi}\right) = 240°$

$\frac{5\pi}{3}$ radians $= \frac{5\pi}{3}\left(\frac{180°}{\pi}\right) = 300°$

$315° = 315°\left(\frac{\pi}{180°}\right) = \frac{7\pi}{4}$

$330° = 330°\left(\frac{\pi}{180°}\right) = \frac{11\pi}{6}$

87. (a) The hour hand completes 2 revolutions in 24 hours. One revolution is 2π.

$$\theta = 2(2\pi) = 4\pi$$

(b) In 4 hours it rotates through

$$\theta = \left(\frac{4}{12}\right)(2\pi) = \frac{2\pi}{3}$$

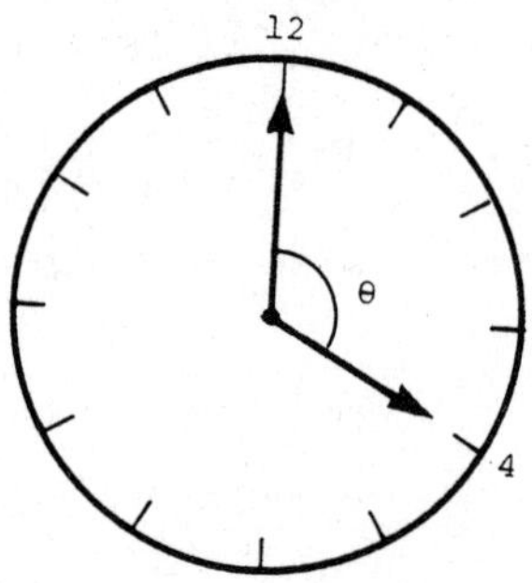

Section 3.2

1. $r = 8.00$ in, $\theta = \pi$ radians

$s = r\theta = 8.00(\pi) = 8\pi$ in

≈ 25.1 in

3. $r = 12.2$ cm, $\theta = \frac{2\pi}{3}$ radians

$s = r\theta = 12.3\left(\frac{2\pi}{3}\right)$

$= 8.2\pi$ cm

≈ 25.8 cm

5. $r = 253$ m, $\theta = \frac{2\pi}{5}$ radians

$s = r\theta = 253\left(\frac{2\pi}{5}\right) = 101.2\pi$ m

≈ 318 m

7. $r = 4.82$ m, $\theta = 60°$

Convert θ to radians.

$\theta = 60° = \frac{\pi}{3}$

$s = r\theta = 4.82\left(\frac{\pi}{3}\right)$

$= 1.61\pi$

≈ 5.05 m

9. $r = 58.402$ m, $\theta = 52.417°$

$\theta = 52.417° = 52.417\left(\frac{\pi}{180°}\right)$

$= .91485$

$s = r\theta = 58.402(.91485) = 53.429$ m

11. The formula for arc length is $s = r\theta$ where θ is in radians. Since γ is in degrees, to change this angle measure to radians, we multiply γ by $\frac{\pi}{180°}$. Therefore, $\theta = \gamma\frac{\pi}{180°}$. Substituting this value for θ into $s = r\theta$, we obtain

$$s = r\left(\gamma\frac{\pi}{180°}\right) = \frac{r\gamma\pi}{180°}.$$

For Exercises 13-17, note that since 6400 has two significant digits and the angles are given to the nearest degree, we can have only two significant digits in the answers.

13. 44° N, 33° N

$r = 6400$ km

$\theta = 44° - 33° = 11° = 11°\left(\frac{\pi}{180°}\right)$

$= \frac{11\pi}{180}$

The distance between the two cities is

$s = r\theta = 6400\left(\frac{11\pi}{180}\right)$

$= 1228.7$ km ≈ 1200 km.

15. 9° N, 40° N

$\theta = 40° - 9° = 31°$

$= 31°\left(\frac{\pi}{180°}\right)$

$= \frac{31\pi}{180}$

$s = r\theta$

$= 6400\left(\frac{31\pi}{180}\right)$

$= 3462.7$ km

≈ 3500 km

17. 41° N, 12° S

12° S = −12° N

$\theta = 41° - (-12°) = 53°$

$= 53°\left(\frac{\pi}{180°}\right)$

$= \frac{53\pi}{180}$

$s = r\theta = 6400\left(\frac{53\pi}{180}\right)$

$= 5920.2$ km

≈ 5900 km

19. 40° E, 105° E

$\theta = 105° - 40° = 65°$

$= 65°\left(\frac{\pi}{180°}\right)$

$= \frac{13\pi}{36}$

$s = r\theta = 6400\left(\frac{13\pi}{36}\right) = 7260.6$ km

≈ 7300 km

21. (a) The number of inches lifted is the arc length in a circle with $r = 9.27$ in and $\theta = 71°\ 50'$.

$71°\ 50' = \left(71° + \frac{50°}{60}\right)\left(\frac{\pi}{180°}\right)$

$s = r\theta$

$= 9.27\left(71° + \frac{50°}{60}\right)\left(\frac{\pi}{180°}\right)$

$= 11.6$

The weight will rise 11.6 in.

(b) When the weight is raised 6 in,

$s = r\theta$

$\theta = \frac{s}{r}$

$= \frac{6}{9.27}\left(\frac{180°}{\pi}\right)$

$= 37.085° = 37° + (.085)(60')$

$= 37°\ 05'$.

The pulley must be rotated through 37° 05′.

23. A rotation of

$\theta = 60.0° = \frac{\pi}{3}$

on the smaller wheel moves through an arc length of

$s = r\theta = 5.23\left(\frac{\pi}{3}\right) = 5.48$ cm.

Since both wheels move together, the larger wheel moves 5.48 cm, which rotates it through an angle θ, where

$5.48 = 8.16\theta$

$\theta = \frac{5.48}{8.16} = .671$ radians

$= .671\left(\frac{180°}{\pi}\right) = 38.5°$.

The larger wheel rotates through 38.5°.

25. The chain moves a distance equal to the arc length on the larger gear. So, for the large gear and pedal,

$$s = r\theta$$
$$= 4.72(180°)\frac{\pi}{180°}$$
$$= 4.72\pi.$$

Thus, the chain moves 4.72π in. The small gear rotates through an angle

$$\theta = \frac{s}{r}$$
$$= \frac{4.72\pi}{1.38}$$
$$= 3.42\pi.$$

θ for the wheel and θ for the small gear are the same, or 3.42π. So, for the wheel,

$$s = r\theta$$
$$= 13.6(3.42\pi)$$
$$= 146 \text{ in.}$$

The bicycle will move 146 in.

27. Let t = the height of the tree. t is approximately the arc length subtended by 2°. Convert 2° to radians.

$$\theta = 2°\left(\frac{\pi}{180°}\right)$$
$$= \frac{\pi}{90}$$
$$t \approx r\theta = 600\left(\frac{\pi}{90}\right)$$
$$\approx 21 \text{ m}$$

The height of the tree is about 21 m.

29. Let t = the length of the train. t is approximately the arc length subtended by 3° 20′.

$$\theta = 3°\ 20' = 3\frac{1}{3}°$$
$$\theta = \left(3\frac{1}{3}°\right)\left(\frac{\pi}{180°}\right)$$
$$= \frac{\pi}{54}$$
$$t \approx r\theta = 3.5\left(\frac{\pi}{54}\right)$$
$$\approx .20 \text{ km}$$

The trains is about .20 km long.

31. Let d = the diameter of the moon. d is approximately the arc length subtended by 1/2°.

$$\theta = \left(\frac{1}{2}°\right)\left(\frac{\pi}{180°}\right)$$
$$= \frac{\pi}{360}$$
$$d \approx r\theta = 240{,}000\left(\frac{\pi}{360}\right)$$
$$\approx 2100 \text{ mi}$$

The diameter of the moon is about 2100 mi.

33. Let r = the distance of the boat. The height of the mast, 32 ft, is approximately the arc length subtended by 2° 10′.

$$\theta = 2°\ 10' = 2\frac{1}{6}°\left(\frac{\pi}{180°}\right)$$
$$= \frac{13\pi}{1080}$$

$$r = \frac{s}{\theta}$$

$$r \approx \frac{32}{\frac{13\pi}{1080}}$$

$$\approx 850 \text{ ft}$$

The boat is about 850 ft away.

35. $r = 29.2$ m, $\theta = \frac{5\pi}{6}$

$$A = \frac{1}{2}r^2\theta = \frac{1}{2}(29.2)^2\left(\frac{5\pi}{6}\right)$$

$$= 355.27\pi$$

$$= 1116.112679$$

$$\approx 1120 \text{ m}^2$$

37. $r = 52$ cm, $\theta = \frac{3\pi}{10}$

$$A = \frac{1}{2}(52)^2\left(\frac{3\pi}{10}\right)$$

$$= 405.6\pi$$

$$= 1274.228904$$

$$\approx 1300 \text{ cm}^2$$

39. $r = 12.7$ cm, $\theta = 81°$

Convert θ to radians.

$$\theta = 81°\left(\frac{\pi}{180°}\right) = \frac{9\pi}{20}$$

$$A = \frac{1}{2}(12.7)^2\left(\frac{9\pi}{20}\right)$$

$$= 36.29\pi$$

$$\approx 114 \text{ cm}^2$$

41. $r = 32.6$ m, $\theta = 38°\ 40'$

Change $38°\ 40'$ to $38\frac{2}{3}°$ before converting to radians.

$$A = \frac{1}{2}(32.6)^2\left(38\frac{2}{3}°\right)\left(\frac{\pi}{180°}\right)$$

$$= 114.1\pi$$

$$\approx 359 \text{ m}^2$$

43. $r = 86.243$ m, $\theta = 11.7142°$

$$A = \frac{1}{2}(86.243)^2(11.7142)°\left(\frac{\pi}{180°}\right)$$

$$= 242.02\pi$$

$$\approx 760.34 \text{ m}^2$$

45. The formula for the area of a sector is $A = \frac{1}{2}r^2\theta$, where θ is an angle measured in radians. Since γ is measured in degrees, we must change it to radians by multiplying by $\frac{\pi}{180°}$. Therefore $\theta = \gamma \cdot \frac{\pi}{180°}$. Substitute this into the equation

$$A = \frac{1}{2}r^2\theta$$

$$= \frac{1}{2}r^2\left(\gamma\frac{\pi}{180°}\right)$$

$$= \frac{r^2\gamma\pi}{2(180°)}$$

$$= \frac{r^2\gamma\pi}{360°}.$$

47. (a) Central angle in degrees:

$$\frac{360°}{32} = 11.25°$$

Central angle in radians:

$$\frac{2\pi}{32} = \frac{\pi}{16}$$

(b) Circumference $= 2\pi r$;

radius = 76 ft, so

circumference $= 2(76)\pi$

≈ 480 ft

(c) $r = 76$ ft, $\theta = \frac{\pi}{16}$

$s = r\theta = 76(\frac{\pi}{16})$

≈ 15 ft

(d) Area of sector with $r = 76$ ft and $\theta = \frac{\pi}{16}$ is

$A = \frac{1}{2}\gamma^2\theta$

$= \frac{1}{2}(76^2)\frac{\pi}{16}$

$\approx 570 \text{ ft}^2$.

49. The angle of inclination of the sun's rays at Alexandria is equal measure to the central angle that subtends the arc from Alexandria to Syene (5000 stades) because these angles are corresponding angles formed by a transversal that intercepts the parallel rays of the sun. To find the radius r of the earth, use $s = r\theta$, with $s = 5000$, and $\theta = \frac{1}{50}(2\pi) = \frac{\pi}{25}$.

$5000 = r(\frac{\pi}{25})$

$r = 39{,}788.736$ stades

$C = 2\pi r = 250{,}000$ stades

$d = 2r$

$= 2(39{,}788.736)$ stades

$= 2(39{,}788.736)(516.7)$ ft

$= \frac{2(39{,}788.736)(516.7)}{5280}$ mi

≈ 7800 mi

51. The area of the sector using the outside radius is

$A_1 = \frac{1}{2}r_1{}^2\theta$.

The area of the sector using the inside radius is

$A_2 = \frac{1}{2}r_2{}^2\theta$.

To find the area of the base of the shell, take the larger area minus the smaller area.

$A = A^1 - A^2$

$= \frac{1}{2}r_1{}^2\theta - \frac{1}{2}r_2{}^2\theta$

$= \frac{1}{2}\theta(r_1{}^2 - r_2{}^2)$

To find the volume, multiply the area of the base by the height.

$V = A \cdot h$

$= \frac{1}{2}\theta(r_1{}^2 - r_2{}^2)h$.

Note that these calculations assume that θ is measured in radians.

Section 3.3

Your calculator must be set in radian mode for calculator exercises in this section.

1. tan .4538

Enter: .4538 [TAN]

Display: .48775041

tan .4538 = .48775041

3. cot 1.0821

Enter: 1.0821 [TAN] [1/x]

Display: .5317473

cot 1.0821 = .53171474

5. sin .8203

Enter: 8203 [SIN]

sin .8203 = .73135046

7. cos .6429

Enter: 6429 [COS]

cos .6429 = .80036052

9. sin 1.5097

Enter: 1.5097 [SIN]

sin 1.5097 = .99813420

11. csc 1.3875

Enter: 1.3875 [SIN] [1/x]

csc 1.3875 = 1.0170372

13. sin 7.5835

Enter: 7.5835 [SIN]

sin 7.5835 = .96364232

15. cot 7.4526

Enter: 7.4526 [TAN] [1/x]

cot 7.4526 = .42442278

17. tan 4.0230

Enter: 4.0230 [TAN]

tan 4.0230 = 1.2131367

19. cos 4.2528

Enter: 4.2528 [COS]

cos 4.2528 = -.44357977

21. sin (-2.2864)

Enter: 2.2864 [+/-] [SIN]

sin (-2.2864) = -.75469733

23. cos (-3.0602)

Enter: 3.0602 [+/-] [COS]

cos (-3.0602) = -.99668945

25. cot 6.0301 = -3.8665127

27. sin 5.9690 = -.30904176

29. cos 13.8143 = .31728667

31. cot 12.9795 = 2.2812466

33. csc (-9.4946) = 14.333769

35. tan (-23.7744) = 4.6358445

37. $\frac{5\pi}{3}$ is in quadrant IV. The reference number is

$$2\pi - \frac{5\pi}{3} = \frac{\pi}{3}.$$

39. $\frac{5\pi}{6}$ is in quadrant II. The reference number is

$$\pi - \frac{5\pi}{6} = \frac{\pi}{6}.$$

41. $\frac{13\pi}{3}$ is coterminal with $\frac{\pi}{3}$, which is in quadrant I.

The reference number is $\frac{\pi}{3}$.

43. $-\frac{3\pi}{4}$ is coterminal with $\frac{5\pi}{4}$, which is in quadrant III.

The reference number is

$$\frac{5\pi}{4} - \pi = \frac{\pi}{4}.$$

45. $-\frac{5\pi}{6}$ is coterminal with $\frac{7\pi}{6}$, which is in quadrant III.

The reference number is

$$\frac{7\pi}{6} - \pi = \frac{\pi}{6}.$$

47. $\frac{13\pi}{12}$ is in quadrant III.

The reference number is

$$\frac{13\pi}{12} - \pi = \frac{\pi}{12}.$$

49. .28376614 is in quadrant I.

The reference number is

.28376614.

51. 3.5551212 is in quadrant III.

The reference number is

3.5551212 - 3.1415927 = .41352850.

53. Find $\sin \frac{7\pi}{6}$.

The reference number for $\frac{7\pi}{6}$ is

$$\frac{7\pi}{6} - \pi = \frac{\pi}{6}.$$

$\frac{7\pi}{6}$ is in quadrant III, where the sine is negative.

$$\sin \frac{7\pi}{6} = -\sin \frac{\pi}{6} = -\frac{1}{2}.$$

55. Find $\tan \frac{3\pi}{4}$.

The reference number for $\frac{3\pi}{4}$ is

$$\pi - \frac{3\pi}{4} = \frac{\pi}{4}.$$

$\frac{3\pi}{4}$ is in quadrant II, where the tangent is negative.

$$\tan \frac{3\pi}{4} = -\tan \frac{\pi}{4} = -1$$

57. Find $\cos \frac{7\pi}{6}$.

The reference number for $\frac{7\pi}{6}$ is

$$\frac{7\pi}{6} - \pi = \frac{\pi}{6}.$$

$\frac{7\pi}{6}$ is in quadrant III, where the cosine is negative.

$$\cos \frac{7\pi}{6} = -\cos \frac{\pi}{6} = -\frac{\sqrt{3}}{2}$$

59. Find $\sec \frac{2\pi}{3}$.

The reference number for $\frac{2\pi}{3}$ is

$$\pi - \frac{2\pi}{3} = \frac{\pi}{3}.$$

$\frac{2\pi}{3}$ is in quadrant II, where the secant is negative.

$$\sec \frac{2\pi}{3} = -\sec \frac{\pi}{3} = -2$$

61. Find $\cot \frac{5\pi}{6}$.

The reference number for $\frac{5\pi}{6}$ is

$$\pi - \frac{5\pi}{6} = \frac{\pi}{6}.$$

$\frac{5\pi}{6}$ is in quadrant II, where the cotangent is negative.

$$\cot \frac{5\pi}{6} = -\cot \frac{\pi}{6} = -\sqrt{3}$$

63. Find $\sin \left(-\frac{5\pi}{6}\right)$.

$-\frac{5\pi}{6}$ is coterminal with $\frac{7\pi}{6}$.

The reference number for $\frac{7\pi}{6}$ is

$$\frac{7\pi}{6} - \pi = \frac{\pi}{6}.$$

$-\frac{5\pi}{6}$ is in quadrant III, where the sine is negative.

$$\sin \left(-\frac{5\pi}{6}\right) = \sin \frac{7\pi}{6} = -\sin \frac{\pi}{6} = -\frac{1}{2}$$

65. Find $\sec \frac{23\pi}{6}$.

$\frac{23\pi}{6}$ is coterminal with $\frac{11\pi}{6}$. The reference number of $\frac{11\pi}{6}$ is

$$2\pi - \frac{11\pi}{6} = \frac{\pi}{6}.$$

$\frac{23\pi}{6}$ is in quadrant IV, where the secant is positive.

$$\sec \frac{23\pi}{6} = \sec \frac{11\pi}{6} = \sec \frac{\pi}{6} = \frac{2\sqrt{3}}{3}$$

67. Find $\cos \frac{13\pi}{4}$.

$\frac{13\pi}{4}$ is coterminal with $\frac{5\pi}{4}$.

The reference number for $\frac{5\pi}{4}$ is

$$\frac{5\pi}{4} - \pi = \frac{\pi}{4}.$$

$\frac{13\pi}{4}$ is in quadrant III, where the cosine is negative.

$$\cos \frac{13\pi}{4} = \cos \frac{5\pi}{4} = -\cos \frac{\pi}{4} = -\frac{\sqrt{2}}{2}$$

In Exercises 69–79, make sure the calculator is set in radians.

69. $\tan s = .21264138$

Enter: .21264138

[INV] [TAN]

Display: .20952066

$s = .20952066$

71. $\sin s = .99184065$

Enter: .99184065

[INV] [SIN]

Display: 1.4429646

$s = 1.4429646$

73. cot s = .62084613

Enter: .62084613

[1/x] [INV] [TAN]

Display: 1.0151896

s = 1.0151896

75. cos s = .57834328

Enter: .57834328

[INV] [COS]

Display: .95409991

s = .95409991

77. cot s = .09637041

Enter: .09637041

[1/x] [INV] [TAN]

Display: 1.4747226

s = 1.4747226

79. tan s = 1.6213129

Enter: 1.6213129

[INV] [TAN]

Display: 1.0181269

s = 1.0181269

81. $\left[0, \frac{\pi}{2}\right]$; $\sin s = \frac{\sqrt{3}}{2}$

Because $\sin s = \sqrt{3}/2$, the reference number for s must be $\pi/3$ since $\sin \pi/3 = \sqrt{3}/2$. Since $\pi/3$ is in the interval $[0, \pi/2]$, $s = \pi/3$.

83. $\left[\frac{\pi}{2}, \pi\right]$; $\sin s = \frac{1}{2}$

Because $\sin s = 1/2$, the reference number for s must be $\pi/6$ since $\sin \pi/6 \doteq 1/2$. For s to be in the interval $[\pi/2, \pi]$, we must subtract the reference number from π. Therefore,

$$s = \pi - \frac{\pi}{6} = \frac{5\pi}{6}.$$

85. $\left[\pi, \frac{3\pi}{2}\right]$; $\tan s = \sqrt{3}$

Because $\tan s = \sqrt{3}$, the reference number for s must be $\pi/3$ since $\tan \pi/3 = \sqrt{3}$. For s to be in the interval $[\pi, 3\pi/2]$, we must add the reference number to π. Therefore,

$$s = \pi + \frac{\pi}{3} = \frac{4\pi}{3}.$$

87. $\left[\frac{3\pi}{2}, 2\pi\right]$; $\tan s = -1$

Because $\tan s = -1$, the reference number for s must be $\frac{\pi}{4}$ since $\tan \frac{\pi}{4} = 1$. For s to be in the interval $[\frac{3\pi}{2}, 2\pi]$, we must subtract the reference number from 2π. Therefore,

$$s = 2\pi - \frac{\pi}{4} = \frac{7\pi}{4}.$$

89. $\left[\frac{\pi}{2}, \pi\right]$; $\cos s = -\frac{\sqrt{2}}{2}$

Because $\cos s = -\sqrt{2}/2$, the reference number for s must be $\pi/4$ since $\cos \pi/4 = \sqrt{2}/2$. For s to be in the interval $[\pi/2, \pi]$, we must subtract the reference number from π. Therefore,

$$s = \pi - \frac{\pi}{4} = \frac{3\pi}{4}.$$

For Exercises 91–97, have calculator in radian mode.

91. s = the length of an arc on the unit circle = 2.5

$x = \cos s$	$y = \sin s$
$x = \cos 2.5$	$y = \sin 2.5$
$x = -.80114362$	$y = .59847214$

$(-.80114362, .59847214)$

93. $s = -7.4$

$x = \cos s$	$y = \sin s$
$x = \cos(-7.4)$	$y = \sin(-7.4)$
$x = .43854733$	$y = -.89870810$

$(.43854733, -.89870810)$

95. $s = 45$

$x = \cos s$	$y = \sin s$
$x = \cos 45$	$y = \sin 45$
$x = .52532199$	$y = .85090353$

Since both x and y are positive, the point (x, y) is in quadrant I, and the arc of length 45 terminates in quadrant I.

97. $s = -32.8$

$x = \cos s$	$y = \sin s$
$x = \cos(-32.8)$	$y = \sin(-32.8)$
$x = .18563972$	$y = -.98261788$

Since the x-coordinate is positive and the y-coordinate is negative, the point (x, y) is in quadrant IV, and the arc of length −32.8 terminates in quadrant IV.

99. $t = 60 - 30 \cos\left[x\left(\frac{\pi}{6}\right)\right]$

(a) January: $x = 0$

$t = 60 - 30 \cos\left[0\left(\frac{\pi}{6}\right)\right]$

$t = 60 - 30 \cos 0$

$t = 60 - 30(1) = 60 - 30$

$t = 30°$

(b) April: $x = 3$

$t = 60 - 30 \cos\left[3\left(\frac{\pi}{6}\right)\right]$

$t = 60 - 30 \cos \frac{\pi}{2}$

$t = 60 - 30(0) = 60 - 0$

$t = 60°$

(c) May: $x = 4$

$t = 60 - 30 \cos\left[4\left(\frac{\pi}{6}\right)\right]$

$t = 60 - 30 \cos \frac{2\pi}{3}$

$t = 60 - 30\left(-\frac{1}{2}\right) = 60 + 15$

$t = 75°$

(d) June: x = 5

$t = 60 - 30 \cos \left[5\left(\frac{\pi}{6}\right)\right]$

$t = 60 - 30 \cos \frac{5\pi}{6}$

$t = 60 - 30\left(-\frac{\sqrt{3}}{2}\right) = 60 + 15\sqrt{3}$

$t \approx 86°$

(e) August: x = 7

$t = 60 - 30 \cos \left[7\left(\frac{\pi}{6}\right)\right]$

$t = 60 - 30 \cos \frac{7\pi}{6}$

$t = 60 - 30\left(-\frac{\sqrt{3}}{2}\right) = 60 + 15\sqrt{3}$

$t \approx 86°$

(f) October: x = 9

$t = 60 - 30 \cos \left[9\left(\frac{\pi}{6}\right)\right]$

$t = 60 - 30 \cos \frac{3\pi}{2}$

$t = 60 - 30(0) = 60 - 0$

$t = 60°$

Section 3.4

1. $\omega = \frac{\pi}{4}$ radians per min, t = 5 min

$\omega = \frac{\theta}{t}$

$\frac{\pi}{4} = \frac{\theta}{5}$

$\theta = 5\left(\frac{\pi}{4}\right)$

$= \frac{5\pi}{4}$ radians

3. $\theta = \frac{2\pi}{5}$ radians, t = 10 sec

$\omega = \frac{\theta}{t}$

$\omega = \frac{\frac{2\pi}{5}}{10}$

$= \frac{2\pi}{50}$

$= \frac{\pi}{25}$ radians per sec

5. $\theta = \frac{3\pi}{8}$ radians, $\omega = \frac{\pi}{24}$ radians per min

$\omega = \frac{\theta}{t}$

$t = \frac{\theta}{\omega}$

$t = \frac{\frac{3\pi}{8}}{\frac{\pi}{24}}$

$= \frac{3\pi}{8} \cdot \frac{24}{\pi}$

$= 9$ min

7. $\omega = .90674$ radians per in, t = 11.876 min

$\omega = \frac{\theta}{t}$

$.90674 = \frac{\theta}{11.876}$

$\theta = (11.876)(.90674)$

$= 10.768$ radians

9. r = 8 cm, $\omega = \frac{9\pi}{5}$ radians per sec

$v = r\omega$

$v = 8\left(\frac{9\pi}{5}\right)$

$= \frac{72\pi}{5}$ cm per sec

11. $v = 18$ ft per sec, $r = 3$ ft

$$v = r\omega$$
$$18 = 3\omega$$
$$\omega = \frac{18}{3}$$
$$= 6 \text{ radians per sec}$$

13. $r = 24.93215$ cm, $\omega = .372914$ radians per sec

$$v = r\omega$$
$$v = (24.93215)(.372914)$$
$$= 9.29755 \text{ cm per sec}$$

15. $r = 6$ cm, $\omega = \frac{\pi}{3}$ radians per sec,

$t = 9$ sec

$$s = r\omega t$$
$$s = 6\left(\frac{\pi}{3}\right)9$$
$$= 18\pi \text{ cm}$$

17. $s = 6\pi$ cm, $r = 2$ cm, $\omega = \frac{\pi}{4}$ radians per sec

$$s = r\omega t$$
$$6\pi = 2\left(\frac{\pi}{4}\right)t$$
$$t = \frac{6\pi}{\frac{\pi}{2}} = 12 \text{ sec}$$

19. $s = \frac{3\pi}{4}$ km, $r = 2$ km, $t = 4$ sec

$$s = r\omega t$$
$$\frac{3\pi}{4} = 2(\omega)4$$
$$\omega = \frac{\frac{3\pi}{4}}{8}$$
$$= \frac{3\pi}{32} \text{ radians per sec}$$

21. Both d and s are distances. d is a linear distance, whereas s is a circular distance, that is, the distance along an arc. In both cases, t is time where r and v are both rates, r being a rate along a straight line and v being a rate along an arc.

23. The hour hand of a clock moves through an angle of 2π radians in 12 hours, so

$$\omega = \frac{\theta}{t} = \frac{2\pi}{12}$$
$$= \frac{\pi}{6} \text{ radians per hr.}$$

25. The second hand of a clock moves through an angle of 2π radians in 60 sec, so

$$\omega = \frac{\theta}{t} = \frac{2\pi}{60}$$
$$= \frac{\pi}{30} \text{ radians per sec.}$$

27. The minute hand of a clock moves through an angle of 2π radians in 60 min, and, at the tip of the minute hand, $r = 7$ cm, so

$$v = \frac{r\theta}{t} = \frac{7(2\pi)}{60}$$
$$= \frac{7\pi}{30} \text{ cm per min.}$$

29. The flywheel making 42 rotations per min turns through an angle $42(2\pi)$ or 84π radians in 1 min with $r = 2$ m.

$$v = \frac{r\theta}{t} = \frac{2(84\pi)}{1} = 168\pi \text{ m per min}$$

31. At 500 rotations per min, the propeller turns through an angle of

$$\theta = 500(2\pi) = 1000\pi \text{ radians}$$

in 1 min with r = 3/2 = 1.5 m.

$$v = \frac{r\theta}{t} = \frac{1.5(1000\pi)}{1}$$

$$= 1500\pi \text{ m per min}$$

33. Let s = the length of the track on the arc.

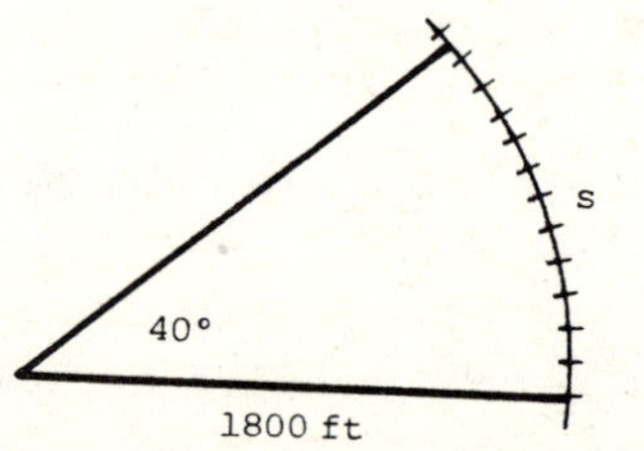

$$40° = 40°\left(\frac{\pi}{180°}\right) = \frac{2\pi}{9}$$

The length of track, s, is the arc length when $\theta = \frac{2\pi}{9}$ and r = 1800 ft.

$$s = r\theta$$

$$s = 1800\left(\frac{2\pi}{9}\right) = 400\pi \text{ ft}$$

Express the velocity v = 30 mph in ft per sec.

$$v = 30 \text{ mph}$$

$$= \frac{30}{3600} \text{ mi per sec}$$

$$= \frac{(30)5280}{3600} \text{ ft per sec}$$

$$= 44 \text{ ft per sec}$$

$$v = \frac{s}{t}$$

$$t = \frac{s}{v}$$

$$t = \frac{400\pi}{44} = \frac{100\pi}{11}$$

$$\approx 29 \text{ sec}$$

35. (a) The earth completes one revolution per day, so it turns through $\theta = 2\pi$ radians in time t = 1 day = 24 hr.

$$\omega = \frac{\theta}{t} = \frac{2\pi}{1}$$

$$= 2\pi \text{ radians per day}$$

$$\omega = \frac{2\pi}{24}$$

$$= \frac{\pi}{12} \text{ radians per hr}$$

(b) At the poles, r = 0, so

$$v = r\omega = 0.$$

(c) At the equator, r = 6400 km.

$$v = r\omega$$

$$= 6400(2\pi)$$

$$= 12{,}800\pi \text{ km per day}$$

$$= \frac{12{,}800\pi}{24} \text{ km per hr}$$

$$\approx 533\pi \text{ km per hr}$$

(d) Salem rotates about the axis in a circle of radius r at angular velocity $\omega = 2\pi$ radians per day.

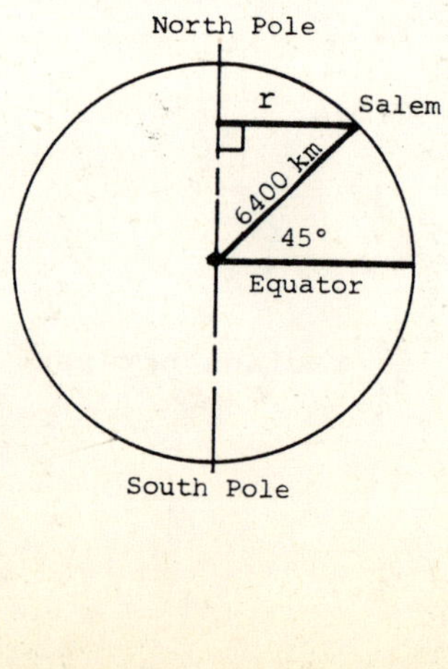

$\sin 45° = \frac{r}{6400}$

$r = 6400 \sin 45°$

$= 6400(\frac{\sqrt{2}}{2})$

$r = 3200\sqrt{2}$ km

$v = r\omega$

$= 3200\sqrt{2}(2\pi)$

$\approx 9050\pi$ km per day

$= \frac{9050\pi}{24}$ km per hr

$\approx 377\pi$ km per hr

37. Since $s = 56$ cm of belt go around in $t = 18$ sec, the linear velocity is

$v = \frac{s}{t}$

$= \frac{56}{18}$

$= \frac{28}{9}$ cm per sec.

$v = r\omega$, $r = 12.96$ cm

$\frac{28}{9} = (12.96)\omega$

$\omega = \frac{\frac{28}{9}}{12.96}$

$= .24$ radian per sec

39. $v = 1.46$ m per sec and 46 revolutions per min is

$\omega = \frac{46(2\pi)}{60} = \frac{23\pi}{15}$ radians sec.

So,

$v = r\omega$

$1.46 = r(\frac{23\pi}{15})$

$r = \frac{1.46}{\frac{23\pi}{15}} = .303$ m.

Chapter 3 Review Exercises

1. $45° = 45°(\frac{\pi}{180°}) = \frac{\pi}{4}$

3. $80° = 80°(\frac{\pi}{180°}) = \frac{4\pi}{9}$

5. $330° = 330°(\frac{\pi}{180°}) = \frac{11\pi}{6}$

7. $1020° = 1020°(\frac{\pi}{180°}) = \frac{17\pi}{3}$

9. $\frac{5\pi}{4} = \frac{5\pi}{4}(\frac{180°}{\pi}) = 225°$

11. $\frac{8\pi}{3} = \frac{8\pi}{3}(\frac{180°}{\pi}) = 480°$

13. $-\frac{11\pi}{18} = -\frac{11\pi}{18}(\frac{180°}{\pi}) = -110°$

15. $\frac{14\pi}{15} = \frac{14\pi}{15}(\frac{180°}{\pi}) = 168°$

17. $\tan \frac{\pi}{3} = \tan 60° = \sqrt{3}$

19. $\sin(-\frac{5\pi}{6}) = \sin(-\frac{5\pi}{6} \cdot \frac{180°}{\pi})$

$= \sin(-150°)$

$= \sin(-150° + 360°)$

$= \sin 210°$

$= -\frac{1}{2}$

21. $\tan(-\frac{7\pi}{3}) = \tan(-\frac{7\pi}{3} \cdot \frac{180°}{\pi})$

$= \tan(-420°)$

$= \tan[-420° + 2(360°)]$

$= \tan 300°$

$= -\sqrt{3}$

3. $\csc\left(-\frac{11\pi}{6}\right) = \csc\left(-\frac{11\pi}{6} \cdot \frac{180°}{\pi}\right)$

$= \csc(-330°)$

$= \csc(-330° + 360°)$

$= \csc 30°$

$= 2$

25. $r = 15.2 \text{ cm}, \ \theta = \frac{3\pi}{4}$

$s = r\theta = 15.2\left(\frac{3\pi}{4}\right)$

$= 11.4\pi$

$= 35.8 \text{ cm}$

27. $r = 8.973 \text{ cm},$

$\theta = 49.06°$

$= 49.06°\left(\frac{\pi}{180°}\right)$

$s = r\theta$

$= 8.973(49.06°)\left(\frac{\pi}{180°}\right)$

$= 7.683 \text{ cm}$

29. $r = 38.0 \text{ m}, \ \theta = 21° \ 40' = 21\frac{2}{3}°$

$= \frac{65°}{3}\left(\frac{\pi}{180°}\right)$

$A = \frac{1}{2}r^2\theta$

$= \frac{1}{2}(38.0)^2\left(\frac{65°}{3}\right)\left(\frac{\pi}{180°}\right)$

$= 273 \text{ m}^2$

31. 28° N, 12° S

12° S = − 12° N

$\theta = 28° - (-12°) = 40°$

$= 40°\left(\frac{\pi}{180°}\right) = \frac{2\pi}{9}$

$s = r\theta$

$= 6400\left(\frac{2\pi}{9}\right)$

$\approx 4500 \text{ km}$

33. sin 1.0472

Enter: 1.0472 [SIN]

sin 1.0472 = .86602663

35. cos (−.2443)

Enter: .2443 [+/−] [COS]

cos (−.2443) = .97030688

37. tan 7.3159 = 1.6755332

39. sec .4864

Enter: .4864

[COS] [1/x]

sec .4864 = 1.1311944

41. $\frac{7\pi}{6}$ is in quadrant III.

The reference number is

$$\frac{7\pi}{6} - \pi = \frac{\pi}{6}.$$

43. $-\frac{4\pi}{4}$ is coterminal with $\frac{2\pi}{3}$, which is in quadrant II.

The reference number is

$$\pi - \frac{2\pi}{3} = \frac{\pi}{3}.$$

45. 1.9169294 is in quadrant II.

The reference number is

3.1415927 − 1.9169294 = 1.2246633.

47. Find $\tan \frac{\pi}{3}$.

The reference number for $\frac{\pi}{3}$ is $\frac{\pi}{3}$.

$\frac{\pi}{3}$ is in quadrant I, where the tangent is positive.

$$\tan \frac{\pi}{3} = \sqrt{3}$$

49. $\sin\left(-\frac{5\pi}{6}\right)$

$-\frac{5\pi}{6}$ is coterminal with $\frac{7\pi}{6}$.

The reference number for $\frac{7\pi}{6}$ is

$$\frac{7\pi}{6} - \pi = \frac{\pi}{6}$$

$-\frac{5\pi}{6}$ is in quadrant III, where the sine is negative.

$$\sin\left(-\frac{5\pi}{6}\right) = \sin\left(\frac{7\pi}{6}\right) = -\sin\frac{\pi}{6} = -\frac{1}{2}$$

51. $\tan\left(-\frac{7\pi}{3}\right)$

$-\frac{7\pi}{3}$ is coterminal with $\frac{5\pi}{3}$.

The reference number for $\frac{5\pi}{3}$ is

$$2\pi - \frac{5\pi}{3} = \frac{\pi}{3}.$$

$-\frac{7\pi}{3}$ is in quadrant IV, where the tangent is negative.

$$\tan\left(-\frac{7\pi}{3}\right) = \tan\left(\frac{5\pi}{3}\right)$$

$$= -\tan\frac{\pi}{3} = -\sqrt{3}$$

53. $\csc\left(-\frac{11\pi}{6}\right)$

$-\frac{11\pi}{6}$ is coterminal with $\frac{\pi}{6}$.

The reference number for $\frac{\pi}{6}$ is $\frac{\pi}{6}$.

$-\frac{11\pi}{6}$ is in quadrant I, where the cosecant is positive.

$$\csc\left(-\frac{11\pi}{6}\right) = \csc\frac{\pi}{6} = 2$$

In Exercises 55–59, be sure the calculator is set in radians.

55. $\cos s = .92500448$

Enter: .92500448

[INV] [COS]

Display: .38974894

$s = .38974894$

57. $\sin s = .49244294$

Enter: .49244294

[INV] [SIN]

Display: .51489440

$s = .51489440$

59. $\cot s = .50221761$

Enter: .50221761

[1/x] [INV] [COS]

Display: 1.1053762

$s = 1.1053762$

61. $\left[0, \frac{\pi}{2}\right]$, $\cos s = \frac{\sqrt{2}}{2}$

Because $\cos s = \sqrt{2}/2$, the reference number for s must be $\pi/4$ since $\cos \pi/4 = \sqrt{2}/2$. For s to be in the interval $[0, \pi/2]$, s must be the reference number. Therefore, $s = \pi/4$.

63. $\left[\pi, \frac{3\pi}{2}\right]$, $\sec s = -\frac{2\sqrt{3}}{3}$

Because $\sec s = -2\sqrt{3}/3$, the reference number for s must be $\pi/6$ since $\sec \pi/6 = 2\sqrt{3}/3$. For s to be in the interval $[\pi, 3\pi/2]$, we must add the reference number to π. Therefore,

$$s = \pi + \frac{\pi}{6} = \frac{7\pi}{6}.$$

65. $\theta = \frac{5\pi}{12}$, $\omega = \frac{8\pi}{9}$ radians per sec

$$\omega = \frac{\theta}{t}$$

$$\frac{8\pi}{9} = \frac{\frac{5\pi}{12}}{t}$$

$$t = \frac{5\pi}{12} \cdot \frac{9}{8\pi}$$

$$= \frac{45}{96} = \frac{15}{32} \text{ sec}$$

67. $t = 8$ sec, $\theta = \frac{2\pi}{5}$

$$\omega = \frac{\theta}{t}$$

$$= \frac{\frac{2\pi}{5}}{8}$$

$$= \frac{2\pi}{40} = \frac{\pi}{20} \text{ radians per sec}$$

69. $r = 11.46$ cm, $\omega = 4.283$ radians per sec, $t = 5.813$ sec

$$v = r\omega$$

$$= 11.46(4.283)$$

$$= 49.08 \text{ cm per sec}$$

$$v = \frac{s}{t}$$

$$49.08 = \frac{s}{5.813}$$

$$s = (49.08)(5.813)$$

$$= 285.3 \text{ cm}$$

CHAPTER 3 TEST

Do not use a calculator for Problems 1–14.

[3.1] Convert each of the following degree measures to radians. Leave answers as multiples of π.

1. $45°$ 2. $360°$ 3. $240°$ 4. $495°$

[3.1] Convert each of the following radian measures to degrees.

5. $\frac{3\pi}{2}$ 6. $-\frac{\pi}{3}$ 7. $\frac{7\pi}{3}$ 8. $\frac{7\pi}{18}$

[3.1] Evaluate each of the following. Give exact values.

9. $\sin \frac{2\pi}{3}$ 10. $\cos\left(-\frac{3\pi}{4}\right)$ 11. $\cot \frac{\pi}{3}$

12. $\csc \frac{3\pi}{2}$ 13. $\tan\left(-\frac{5\pi}{6}\right)$ 14. $\sec \frac{5\pi}{3}$

[3.2] Solve the following problems.

15. The radius of a circle is 1.35 cm. Find the length of the arc of the circle intercepted by a central angle of $3\pi/4$ radians.

16. Find the area of a sector of a circle intercepted by a central angle of $210°$ in a circle of radius 12.5 ft.

17. Find the distance between Kampala, Uganda, and Belem, Brazil, if the two cities are on the equator and Kampala is at $30°$ E longitude and Belem is at $50°$ W longitude. Assume that the radius of the earth is 6400 km.

[3.3] Use a calculator to find each of the following. Be sure that your calculator is set in radian mode.

18. $\sin .9104$ 19. $\cos 1.415$ 20. $\tan 3.005$

[3.3] **Find the value of s in the interval [0, π/2] that makes each of the following true.**

21. cot s = 1.3821596

22. cos s = .92846507

23. sec s = 3.8421657

[3.4] 24. Find the linear velocity of a point on the tip of the minute hand of a clock, if the hand is 5 cm long.

[3.4] 25. Explain how you change radians per second to radians per hour.

CHAPTER 3 TEST ANSWERS

1. $\pi/4$ **2.** 2π **3.** $4\pi/3$ **4.** $11\pi/4$ **5.** 270° **6.** -60°

7. 420° **8.** 70° **9.** $\sqrt{3}/2$ **10.** $-\sqrt{2}/2$ **11.** $\sqrt{3}/3$ **12.** -1

13. $\sqrt{3}/3$ **14.** 2 **15.** 3.18 cm **16.** 286 ft^2 **17.** 8900 km

18. .78974917 **19.** .15516683 **20.** -.13744854 **21.** .62632786

22. .38053767 **23.** 1.3074946 **24.** $\pi/6$ cm per min

25. Multiply radians per second by 3600 to get radians per hour.

CHAPTER 4 GRAPHS OF THE CIRCULAR FUNCTIONS

Section 4.1

For Exercises 1–23, see the answer graphs in the back of the textbook. Some of the data for plotting the graphs are given here.

1. $y = 2 \cos x$

Amplitude: $|2| = 2$

x	0	$\frac{\pi}{2}$	π	$\frac{3\pi}{2}$	2π
$2 \cos x$	2	0	−2	0	2

This table gives five values for graphing one period of the function. Repeat this cycle for the interval $[-2\pi, 0]$.

3. $y = \frac{2}{3} \sin x$

Amplitude: $\left|\frac{2}{3}\right| = \frac{2}{3}$

x	0	$\frac{\pi}{2}$	π	$\frac{3\pi}{2}$	2π
$\frac{2}{3} \sin x$	0	.7	0	−.7	0

5. $y = -\cos x$

Amplitude: $|-1| = 1$

x	0	$\frac{\pi}{2}$	π	$\frac{3\pi}{2}$	2π
$-\cos x$	−1	0	1	0	−1

7. $y = -2 \sin x$

Amplitude: $|-2| = 2$

x	0	$\frac{\pi}{2}$	π	$\frac{3\pi}{2}$	2π
$-2 \sin x$	0	−2	0	2	9

9. $y = \sin \frac{1}{2}x$

Period: $\dfrac{2\pi}{\frac{1}{2}} = 4\pi$

Amplitude: $|1| = 1$

Divide the interval $[0, 4\pi]$ into four equal parts to get x-values that will yield minimum and maximum points and x-intercepts. Then make a table.

x	0	π	2π	3π	4π
$\frac{1}{2}x$	0	$\frac{\pi}{2}$	π	$\frac{3\pi}{2}$	2π
$\sin \frac{1}{2}x$	0	1	0	−1	0

11. $y = \cos \frac{1}{3}x$

Period: $\dfrac{2\pi}{\left(\frac{1}{3}\right)} = 6\pi$

Amplitude: 1

x	0	$\frac{3\pi}{4}$	3π	$\frac{9\pi}{2}$	6π
$\frac{1}{3}x$	0	$\frac{\pi}{2}$	π	$\frac{3\pi}{2}$	2π
$\cos \frac{1}{3}x$	1	0	−1	0	1

13. $y = \sin 3x$

Period: $\frac{2\pi}{3}$

Amplitude: $|1| = 1$

x	0	$\frac{\pi}{6}$	$\frac{\pi}{3}$	$\frac{\pi}{2}$	$\frac{2\pi}{3}$
3x	0	$\frac{\pi}{2}$	π	$\frac{3\pi}{2}$	2π
sin 3x	0	1	0	−1	0

15. $y = \cos 2x$

Period: $\frac{2\pi}{2} = \pi$

Amplitude: $|1| = 1$

x	0	$\frac{\pi}{4}$	$\frac{\pi}{2}$	$\frac{3\pi}{4}$	π
2x	0	$\frac{\pi}{2}$	π	$\frac{3\pi}{2}$	2π
cos 2x	1	0	−1	0	1

17. $y = -\sin 4x$

Period: $\frac{2\pi}{4} = \frac{\pi}{2}$

Amplitude: $|-1| = 1$

x	0	$\frac{\pi}{8}$	$\frac{\pi}{4}$	$\frac{3\pi}{4}$	$\frac{\pi}{2}$
4x	0	$\frac{\pi}{2}$	π	$\frac{3\pi}{2}$	2π
−sin 4x	0	−1	0	1	0

19. $y = 2 \sin \frac{1}{4}x$

Period: $\frac{2\pi}{\frac{1}{4}} = 8\pi$

Amplitude: 2

x	0	2π	4π	6π	8π
$\frac{1}{4}x$	0	$\frac{\pi}{2}$	π	$\frac{3\pi}{2}$	2π
$2 \sin \frac{1}{4}x$	0	2	0	−2	0

21. $y = -2 \cos 3x$

Period: $\frac{2\pi}{3}$

Amplitude: 2

x	0	$\frac{\pi}{6}$	$\frac{\pi}{3}$	$\frac{\pi}{2}$	$\frac{2\pi}{3}$
3x	0	$\frac{\pi}{2}$	π	$\frac{3\pi}{2}$	2π
−2 cos 3x	−2	0	2	0	−2

23. $y = \cos \pi x$

Period: $\frac{2\pi}{\pi} = 2$

Amplitude: 1

x	0	$\frac{1}{2}$	1	$\frac{3}{2}$	2
cos πx	1	0	−1	0	1

25. $y = \sin x$

The graph is a sinusoidal curve with an amplitude of 1 and a period of 2π. Since $\sin 0 = 0$, the point (0, 0) is on the graph. This matches with graph G.

27. $y = -\sin x$

The graph is a sinusoidal curve with an amplitude of 1 and a period of 2π. Because $a = -1$, the graph is a reflection of $y = \sin x$ in the x-axis. This matches with graph E.

29. $y = \sin 2x$

The graph is a sinusoidal curve with a period of π and an amplitude of 1. Since $\sin 2(0) = 0$, the point (0, 0) is on the graph. This matches with graph B.

31. $y = 2 \sin x$

The graph is a sinusoidal curve with a period of 2π and an amplitude of 2. Since $2 \sin 0 = 0$, the point (0, 0) is on the graph. This matches with graph F.

33. The graph of $y = \sin 2x$ has an amplitude of 1 and a period of π. Over the interval $[0, 2\pi]$ there will be three complete cycles.

The graph of $y = 2 \sin x$ has an amplitude of 2 and a period of 2π. Over the interval $[0, 2\pi]$ there will be one complete cycle. Thus, $\sin 2x \neq 2 \sin x$, and, in general, $\sin bx \neq b \sin x$.

For Exercises 35 and 37, see the answer graphs in the back of the textbook.

35. $y = (\sin x)^2$

Since $(\sin x)^2 = \sin x \cdot \sin x \geq 0$ for all values of π, the period is π.

x	0	$\frac{\pi}{2}$	π	$\frac{3\pi}{2}$	2π
sin x	0	1	0	-1	0
$(\sin x)^2$	0	1	0	1	0

37. $y = (\sin 2x)^2$

The period for $\sin 2x$ is π. Since $(\sin 2x)^2 \geq 0$, for all values of x, the period is

$$\frac{1}{2}(\pi) = \frac{\pi}{2}.$$

x	0	$\frac{\pi}{4}$	$\frac{\pi}{2}$	$\frac{3\pi}{4}$	π
sin x	0	1	0	-1	0
$(\sin 2x)^2$	0	1	0	1	0

39. (a) The amplitude is

$$\frac{120 - 80}{2} = \frac{40}{2}$$
$$= 20.$$

(b) Since the period is .8 sec, there are

$$\frac{1}{.8} = 1.25 \text{ beats per sec}$$

and the pulse rate is

$$(60)(1.25) = 75 \text{ beats per min.}$$

41. (a) The latest time that the animals begin their evening activity is 8:00; the earliest time is 4:00.

$$4{:}00 \leq y \leq 8{:}00$$

Since there is a difference of 4 hours in these times, the amplitude is

$$\frac{1}{2}(4) = 2 \text{ hr.}$$

(b) The length of this period is 1 year.

43. $-1 \leq y \leq 1$

Amplitude: 1

Period: 8 squares = (8)(30°)

$$= 240° \text{ or } \frac{4\pi}{3}$$

45. $E = 5 \cos 120\pi t$

(a) Amplitude: $|5| = 5$

Period: $\frac{2\pi}{120\pi} = \frac{1}{60}$ sec

(b) Number of cycles per second

$= \frac{1}{\text{period}}$

$= \frac{1}{\frac{1}{60}} = 60$

(c) $t = 0$

$E = 5 \cos 120\pi(0) = 5$

$t = .03$

$E = 5 \cos 120\pi(.03) = 1.545$

$t = .06$

$E = 5 \cos 120\pi(.06) = -4.045$

$t = .09$

$E = 5 \cos 120\pi(.09) = -4.045$

$t = .12$

$E = 5 \cos 120\pi(.12) = 1.545$

(d) See the answer graph at the back of the textbook.

Section 4.2

1. $y = \sin\left(x - \frac{\pi}{4}\right)$

The graph is a sinusoidal curve $y = \sin x$ shifted $\pi/4$ units to the right. This matches graph D.

3. $y = \cos\left(x - \frac{\pi}{4}\right)$

The graph is a sinusoidal curve $y = \cos x$ shifted $\pi/4$ units to the right. This matches graph H.

5. $y = 1 + \sin x$

The graph is a sinusoidal curve $y = \sin x$ translated vertically 1 unit up. This matches graph B.

7. $y = 1 + \cos x$

The graph is a sinusoidal curve $y = \cos x$ translated vertically 1 unit up. This matches graph F.

9. The function $y = -4 + \sin 4\left(x + \pi/2\right)$ has amplitude $\underline{1}$, period $\underline{\pi/2}$, phase shift $\underline{\pi/2}$ units to the $\underline{\text{left}}$, and has vertical translation $\underline{4}$ units $\underline{\text{down}}$.

11. $y = 2 \sin(x - \pi)$

The amplitude is $|2|$, which is 2.
The period is $2\pi/1$ which is 2π.
There is no vertical translation .
The phase shift is π units to the right.

13. $y = 4 \cos\left(\frac{x}{2} + \frac{\pi}{2}\right)$

$y = 4 \cos \frac{1}{2}(x + \pi)$

The amplitude is $|4|$, which is 4.

The period is $\frac{2\pi}{\frac{1}{2}}$ which is 4π.

There is no vertical translation.
The phase shift is π units to the left.

15. $y = 3 \cos 2(x - \frac{\pi}{4})$

The amplitude is $|3|$, which is 3.

The period is $2\pi/2$ which is π.

There is no vertical translation.

The phase shift is $\pi/4$ units to the right.

17. $y = 2 - \sin (3x - \frac{\pi}{5})$

$y = -1 \sin 3(x - \frac{\pi}{15}) + 2$

The amplitude is $|-1|$, which is 1.

The period is $2\pi/3$.

The vertical translation is 2 units up. The phase shift is $\pi/15$ units to the right.

For Exercises 19–43, see the answer graphs in the back of the textbook.

19. $y = \cos (x - \frac{\pi}{2})$

The amplitude is 1.

The period is 2π.

There is no vertical translation.

The phase shift is $\pi/2$ units to the right.

21. $y = \sin (x + \frac{\pi}{4})$

The amplitude is 1.

The period is 2π.

There is no vertical translation.

The phase shift is $\pi/4$ units to the left.

23. $y = 2 \cos (x - \frac{\pi}{3})$

The amplitude is 2.

The period is 2π.

There is no vertical translation.

The phase shift is $\pi/3$ units to the right.

25. $y = \frac{3}{2} \sin 2(x + \frac{\pi}{4})$

The amplitude is 3/2.

The period is $2\pi/2$, which is π.

There is no vertical translation.

The phase shift is $\pi/4$ units to the left.

27. $y = -4 \sin (2x - \pi)$

$y = -4 \sin 2(x - \frac{\pi}{2})$

The amplitude is $|-4|$, which is 4.

The period is $2\pi/2$, which is π.

There is no vertical translation.

The phase shift is $\pi/2$ units to the right.

29. $y = \frac{1}{2} \cos (\frac{1}{2}x - \frac{\pi}{4})$

$y = \frac{1}{2} \cos \frac{1}{2}(x - \frac{\pi}{2})$

The amplitude is 1/2.

The period is $\frac{2\pi}{\frac{1}{2}}$ which is 4π.

This is no vertical translation.

The phase shift is $\pi/2$ units to the right.

31. $y = -3 + 2 \sin x$

The amplitude is 2.

The period is 2π.

The vertical translation is 3 units down.

There is no phase shift.

33. $y = 1 - \frac{2}{3} \sin \frac{3}{4}x$

The amplitude is $|-2/3|$, which is 2/3.

The period is $\frac{2\pi}{\frac{3}{4}}$, which is $8\pi/3$.

The vertical translation is 1 unit up.

There is no phase shift.

35. $y = 2 - \cos x$

The amplitude is $|-1|$, which is 1.

The period is 2π.

The vertical translation is 2 units up.

There is no phase shift.

37. $y = 1 - 2 \cos \frac{1}{2}x$

The amplitude is $|-2|$, which is 2.

The period is $\frac{2\pi}{\frac{1}{2}}$, which is 4π.

The vertical translation is 1 unit up.

There is no phase shift.

39. $y = -2 + \frac{1}{2} \sin 3x$

The amplitude is 1/2.

The period is $2\pi/3$.

The vertical translation is 2 units down.

There is no phase shift.

41. $y = -3 + 2 \sin (x + \frac{\pi}{2})$

The amplitude is 2.

The period is 2π.

The vertical translation is 3 units down.

The phase shift is $\pi/2$ units to the left.

43. $y = \frac{1}{2} + \sin 2(x + \frac{\pi}{4})$

The amplitude is 1.

The period is $2\pi/2$, which is π.

The vertical translation is 1/2 unit up.

The phase shift is $\pi/4$ units to the left.

45. $y = -4 - 3 \sin 2(x - \frac{\pi}{6})$

-4 causes the graph of $y = \sin x$ to move down 4 units.

-3 cause the graph of $y = \sin x$ to be reflected in the x-axis and changes the amplitude to 3.

2 changes the period of $y = \sin x$, from 2π to π.

$\pi/6$ shifts the graph of $y = \sin x$ $\pi/6$ units to the right.

Section 4.3

1. $y = -\csc x$

The graph is the reflection of the graph of $y = \csc x$ about the x-axis. This matches with graph B.

3. $y = -\tan x$

The graph is the reflection of the graph of $y = \tan x$ about the x-axis. This matches with graph E.

5. $y = \tan\left(x - \frac{\pi}{4}\right)$

The graph is the graph of $y = \tan x$ shifted $\pi/4$ units to the right. This matches with graph D.

7. Parabolas do not have asymptotes. Portions of the graphs of $y = \sec x$ or $y = \csc x$ are not actually parabolas.

For Exercises 9–37, see the answer graphs in the back of the textbook.

9. $y = \csc\left(x - \frac{\pi}{4}\right)$

We graph this function by first graphing the corresponding reciprocal function

$$y = \sin\left(x - \frac{\pi}{4}\right).$$

The period is 2π.

11. $y = \sec\left(x + \frac{\pi}{4}\right)$

We graph this function by first graphing the corresponding reciprocal function

$$y = \cos\left(x + \frac{\pi}{4}\right).$$

The period is 2π.

13. $y = \sec\left(\frac{1}{2}x + \frac{\pi}{3}\right)$

$y = \sec \frac{1}{2}\left(x + \frac{2\pi}{3}\right)$

We graph this function by first graphing the corresponding reciprocal function

$$y = \cos \frac{1}{2}\left(x + \frac{2\pi}{3}\right).$$

The period is 4π.

15. $y = 2 + 3 \sec (2x - \pi)$

$y = 2 + 3 \sec 2\left(x - \frac{\pi}{2}\right)$

We graph this function by first graphing the corresponding reciprocal function

$$y = 2 + 3 \cos 2\left(x - \frac{\pi}{2}\right).$$

The period is π.

17. $y = 1 - \frac{1}{2} \csc\left(x - \frac{3\pi}{4}\right)$

We graph this function by first graphing the corresponding reciprocal function

$y = 1 - \frac{1}{2}\sin(x - \frac{3\pi}{4})$.

The period is 2π.

19. $y = 2\tan x$

Two adjacent vertical asymptotes are $x = -\pi/2$ and $x = \pi/2$. The graph is "stretched" because $a = 2$ and $|2| > 1$.

21. $y = \frac{1}{2}\cot x$

Two adjacent vertical asymptotes are $x = 0$ and $x = \pi$. The graph is "compressed" because $a = 1/2$ and $|1/2| < 1$.

23. $y = \cot 3x$

Two adjacent vertical asymptotes are

$3x = 0$ and $3x = \pi$

or $x = 0$ and $x = \frac{\pi}{3}$.

25. $y = \tan(2x - \pi)$

$y = \tan 2(x - \frac{\pi}{2})$

The period is $\pi/2$.

Two adjacent vertical asymptotes are

$2(x - \frac{\pi}{2}) = -\frac{\pi}{2}$ and $2(x - \frac{\pi}{2}) = \frac{\pi}{2}$

or $x - \frac{\pi}{2} = -\frac{\pi}{4}$ and $x - \frac{\pi}{2} = \frac{\pi}{4}$

or $x = \frac{\pi}{4}$ and $x = \frac{3\pi}{4}$.

27. $y = \cot(3x + \frac{\pi}{4})$

$y = \cot 3(x + \frac{\pi}{12})$

The period is $\pi/3$.

Two adjacent vertical asymptotes are

$3(x + \frac{\pi}{12}) = 0$ and $3(x + \frac{\pi}{12}) = \pi$

or $x + \frac{\pi}{12} = 0$ and $x + \frac{\pi}{12} = \frac{\pi}{3}$

or $x = -\frac{\pi}{12}$ and $x = \frac{\pi}{4}$.

29. $y = 1 + \tan x$

This is the graph of $y = \tan x$ translated vertically 1 unit up.

31. $y = 1 - \cot x$

This is the graph of $y = \cot x$ reflected about the x-axis and then translated vertically 1 unit up.

33. $y = -1 + 2\tan x$

This is the graph of $y = 2\tan x$ translated vertically 1 unit down.

35. $y = -1 + \frac{1}{2}\cot(2x - 3\pi)$

$y = -1 + \frac{1}{2}\cot 2(x - \frac{3\pi}{2})$

The period is $\pi/2$.

Two adjacent vertical asymptotes are

$2(x - \frac{3\pi}{2}) = 0$ and $2(x - \frac{3\pi}{2}) = \pi$

or $x - \frac{3\pi}{2} = 0$ and $x - \frac{3\pi}{2} = \pi$

or $x = \frac{3\pi}{2}$ and $x = \frac{5\pi}{2}$.

This is the graph of $y = \frac{1}{2} \cot 2x$ translated vertically 1 unit down.

37. $y = \frac{2}{3} \tan \left(\frac{3}{4}x - \pi\right) - 2$

$y = -2 + \frac{2}{3} \tan \frac{3}{4} \left(x - \frac{4\pi}{3}\right)$

The period is $\frac{\pi}{\frac{3}{4}} = \frac{4\pi}{3}$.

Two adjacent vertical asymptotes are

$\frac{3}{4}\left(x - \frac{4\pi}{3}\right) = -\frac{\pi}{2}$ and $\frac{3}{4}\left(x - \frac{4\pi}{3}\right) = \frac{\pi}{2}$

or $x - \frac{4\pi}{3} = -\frac{2\pi}{3}$ and $x - \frac{4\pi}{3} = \frac{2\pi}{3}$

or $x = \frac{2\pi}{3}$ and $x = 2\pi$.

This is the graph of $y = \frac{2}{3} \tan \frac{3\pi}{4}$ translated vertically 2 units down.

39. $d = \tan 4 \tan 2\pi t$

(a) $d = \tan 2\pi\ (0) = 4 \tan 0 = 0$ m

(b) $d = 4 \tan 2\pi(.4)$
$= 4 \tan (.8\pi)$
$= 4 \tan (2.5133)$
$= 4(-.7265) = -2.9$ m

(c) $d = 4 \tan 2\pi(.8) = 4 \tan 1.6\pi$
$= 4 \tan 5.0265$
$= 4(-3.0782) = -12.3$ m

(d) $d = 4 \tan 2\pi(1.2)$
$= 4 \tan (2.4\pi)$
$= 4 \tan (7.5398)$
$= 4(3.077) = 12.3$ m

(e) $t = .25$ leads to $\tan \pi/2$, which is undefined.

(f) The domain is $\{t \mid t \neq .25\,n$, where n is an integer.$\}$

Chapter 4 Review Exercises

1. $y = 2 \sin x$

Amplitudes: 2

Period: 2π

Vertical translation: none

Phase shift: none

3. $y = -\frac{1}{2} \cos 3x$

Amplitude: $\frac{1}{2}$

Period: $\frac{2\pi}{3}$

Vertical translation: none

Phase shift: none

5. $y = 1 + 2 \sin \frac{1}{4}x$

Amplitude: 2

Period: $\frac{2\pi}{\frac{1}{4}} = 8\pi$

Vertical translation: up 1 unit

Phase shift: none

7. $y = 3 \cos \left(x + \frac{\pi}{2}\right)$

Amplitude: 3

Period: 2π

Vertical translation: none

Phase shift: $\frac{\pi}{2}$ units to the left

9. $y = \frac{1}{2} \csc (2x - \frac{\pi}{4})$

Amplitude: not applicable

Period: $\frac{2\pi}{\pi} = \pi$

Vertical translation: none

Phase shift: $\frac{\pi}{8}$ units to the right

11. $y = \frac{1}{3} \tan \left(3x - \frac{\pi}{3}\right)$

$y = \frac{1}{3} \tan 3\left(x - \frac{\pi}{9}\right)$

Amplitude: not applicable

Period: $\frac{\pi}{3}$

Vertical translation: none

Phase shift: $\frac{\pi}{9}$ units to the right

13. The tangent function has a period of π and x-intercepts at integral multiples of π.

15. The cosine function has a period of 2π and has the value 0 when $x = \pi/2$.

17. The cotangent function has a period of π and decreases on the interval $0 < x < \pi$.

For Exercises 19–39, see the answer graphs in the back of the textbook.

19. $y = 3 \sin x$

Period: 2π

Amplitude: 3

21. $y = -\tan x$

Period: π

Asymptotes: $x = -\frac{\pi}{2}$, $x = \frac{\pi}{2}$

23. $y = 2 + \cot x$

Period: π

Vertical translation: up 2 units

Asymptotes: $x = 0$, $x = \pi$

25. $y = \sin 2x$

Period: $\frac{2\pi}{2} = \pi$

Amplitude: 1

27. $y = 3 \cos 2x$

Amplitude: 3

Period: $\frac{2\pi}{2} = \pi$

29. $y = \cos \left(x - \frac{\pi}{4}\right)$

Amplitude: 1

Period: 2π

Phase shift: $\frac{\pi}{4}$ units to the right

31. $y = \sec \left(2x + \frac{\pi}{3}\right)$

$= \sec \left[2\left(x + \frac{\pi}{6}\right)\right]$

Period: $\frac{2\pi}{2} = \pi$

Phase shift: $\frac{\pi}{6}$ units to the left

33. $y = 1 + 2 \cos 3x$

Amplitude: 2

Period: $\frac{2\pi}{3}$

Vertical translation: 1 unit up

35. $y = 2 \sin \pi x$

Amplitude: 2

Period: 2

37. $y = 1 - 2 \sec \left(x - \frac{\pi}{4}\right)$

No amplitude

Period: $\frac{2\pi}{1} = 2$

Vertical translation: 1 unit up

Phase shift: $\frac{\pi}{4}$ units to the right

39. The shorter leg of the right triangle has length $h_2 - h_1$.

(a) $\cot \theta = \frac{d}{h_2 - h_1}$. So,

$$d = (h_2 - h_1) \cot \theta.$$

(b) When $h_2 = 55$ and $h_1 = 5$,

$$d = (55 - 5) \cot \theta = 50 \cot \theta.$$

The period is π, but the graph wanted is d for $0 < \theta < \pi/2$. The asymptote is the line $\theta = 0$.

When $\theta = \frac{\pi}{4}$,

$$d = 50(1) = 50.$$

See the answer graph in the back of the textbook.

41. (a) From the graph one period is about 20 years.

(b) The population of hares fluctuates between about 5000 to about 150,000.

CHAPTER 4 TEST

[4.1] 1. Give the amplitude for $y = -\frac{2}{3} \cos x$.

Give the period for each of the following functions.

[4.2] 2. $y = 3 \sin 4x$

[4.3] 3. $y = \cot (2x - \pi)$

Give the vertical translation (if any) for each of the following functions.

[4.2] 4. $y = 4 + 2 \cos \left(\frac{1}{2}x - \pi\right)$

[4.3] 5. $y = \sec (3x - \pi)$

Give the phase shift (if any) for each of the following functions.

[4.2] 6. $y = -1 - \sin \left(x + \frac{\pi}{2}\right)$

[4.3] 7. $y = 3 + \csc (4x - \pi)$

[4.3] 8. Which circular function has a period of π, and the function is increasing on the interval $-\pi/2 < x < \pi/2$?

Graph each of the following functions over a one-period interval.

[4.1] 9. $y = \frac{7}{4} \sin x$

[4.2] 10. $y = -2 - 2 \cos x$

[4.3] 11. $y = -\sec x$

[4.3] 12. $y = 2 \cot x$

[4.2] 13. $y = -\sin (x + \pi)$

[4.3] 14. $1 + \tan \left(\frac{1}{2}x - \frac{\pi}{2}\right)$

[4.3] 15. $y = \csc (x - \pi)$

[4.2] 16. $y = 3 \cos \left(x - \frac{\pi}{2}\right) - 1$

CHAPTER 4 TEST ANSWERS

1. 2/3

2. $\pi/2$

3. $\pi/2$

4. 4 units up

5. No vertical translation

6. $\pi/2$ to the left

7. $\pi/4$ to the right

8. tangent

9.

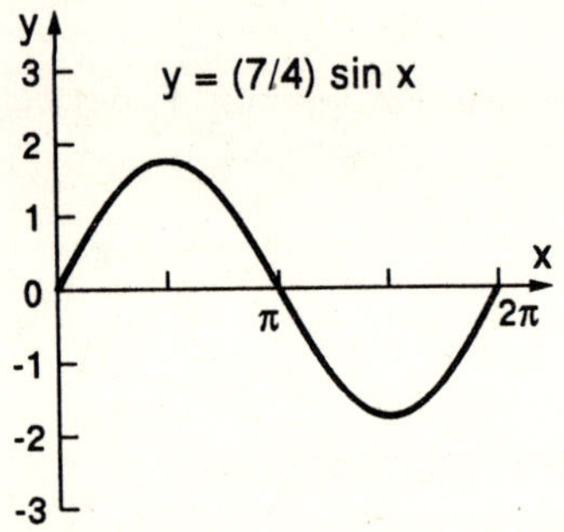

10.

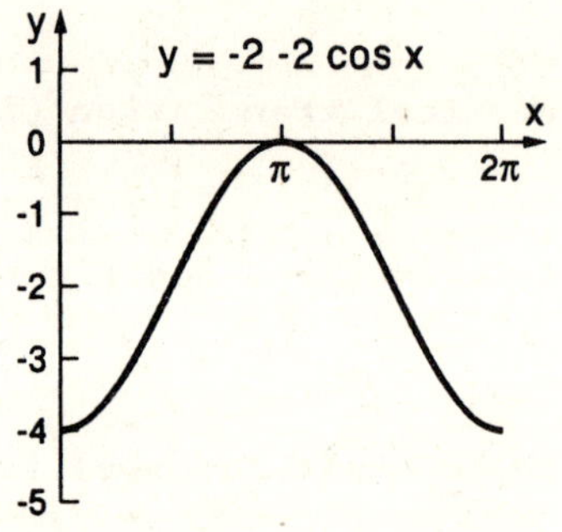

11.

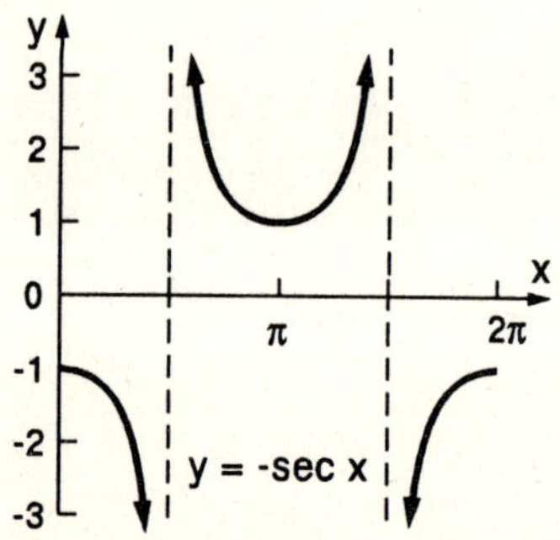

12.

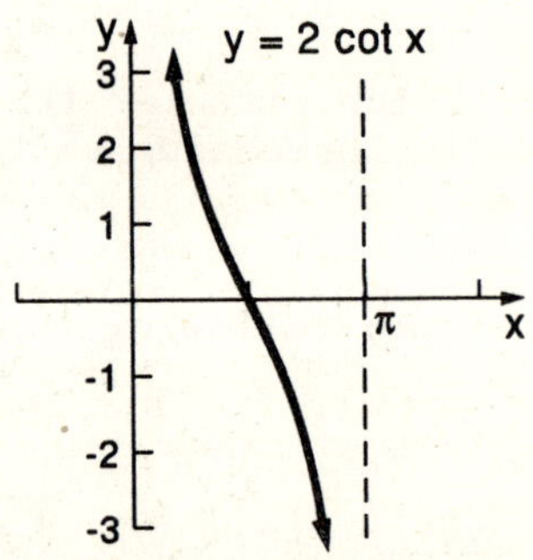

13.

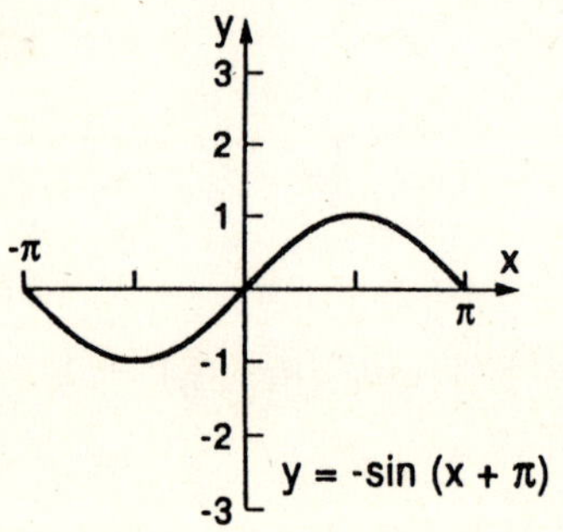

14.

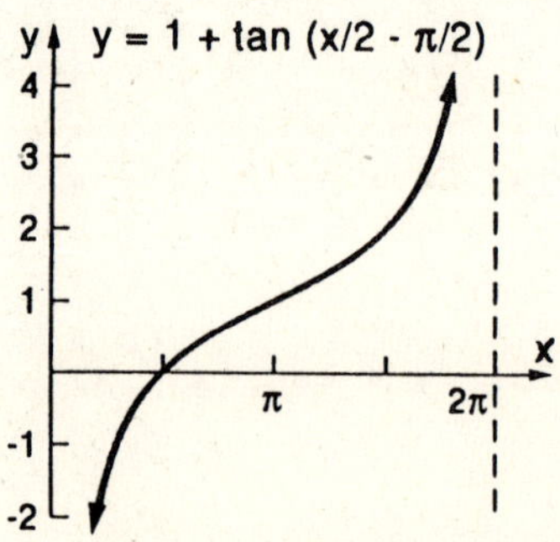

15.

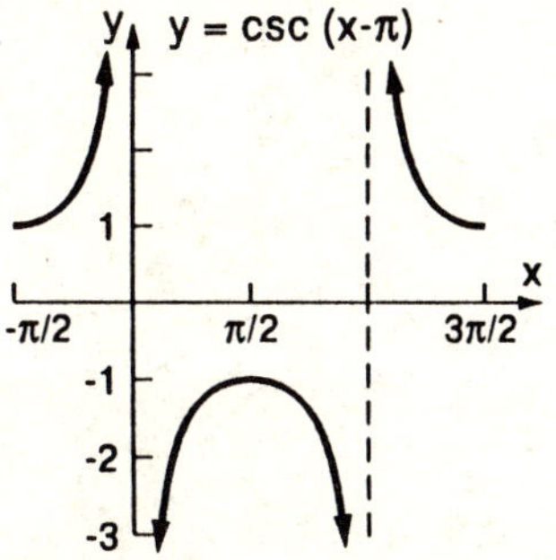

16.

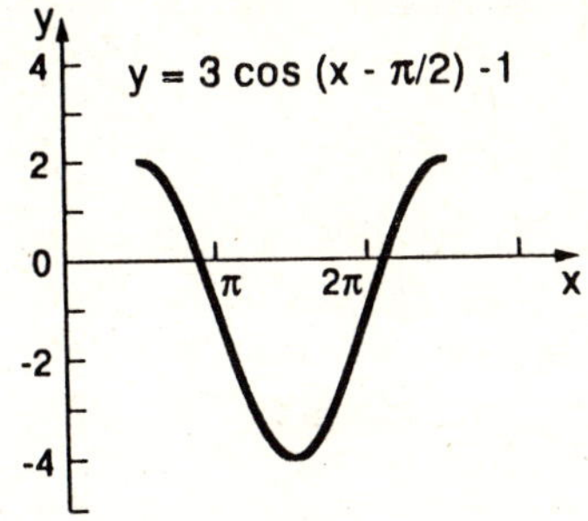

CHAPTER 5 TRIGONOMETRIC IDENTITIES

Section 5.1

1. $\cos s = \frac{3}{4}$, s in quadrant I

$$\sin^2 s + \cos^2 s = 1$$
$$\sin^2 s + \left(\frac{3}{4}\right)^2 = 1$$
$$\sin^2 s = 1 - \frac{9}{16} = \frac{7}{16}$$
$$\sin s = \pm\frac{\sqrt{7}}{4}$$

Since s is in quadrant I,

$$\sin s = \frac{\sqrt{7}}{4}.$$

3. $\cos s = \frac{\sqrt{5}}{5}$, $\tan s < 0$ implies s is in quadrant IV.

$$\sin^2 s + \cos^2 s = 1$$
$$\sin^2 s + \left(\frac{\sqrt{5}}{5}\right)^2 = 1$$
$$\sin^2 s = 1 - \frac{5}{25} = \frac{20}{25}$$
$$= \frac{4}{5}$$
$$\sin s = \pm\frac{2}{\sqrt{5}}$$

Since s is in quadrant IV,

$$\sin s = -\frac{2}{\sqrt{5}}$$
$$= -\frac{2\sqrt{5}}{5}.$$

5. $\sec s = \frac{11}{4}$, $\tan s < 0$ implies s is in quadrant IV.

Since $\sec s = \frac{1}{\cos s}$,

$$\frac{11}{4} = \frac{1}{\cos s}$$
$$\frac{4}{11} = \cos s.$$

$$\sin^2 s + \cos^2 s = 1$$
$$\sin^2 s + \left(\frac{4}{11}\right)^2 = 1$$
$$\sin^2 s + \frac{16}{121} = 1$$
$$\sin s = \frac{105}{121}$$
$$\sin s = \pm\frac{\sqrt{105}}{11}$$

Since s is in quadrant IV,

$$\sin s = -\frac{\sqrt{105}}{11}.$$

7. Since the $\csc s = \frac{1}{\sin s}$, we know that if the csc s is positive, sin s is also positive. Also, if csc s is negative, sin s is also negative. Therefore, a quadrant does not have to be stipulated, since the sine and cosecant have the same sign no matter which quadrant s is in.

9. Find $\tan \theta$ if $\cos \theta = -\frac{2}{5}$, and $\sin \theta < 0$.

$$\sin^2 \theta + \cos^2 \theta = 1$$
$$\sin^2 \theta + \left(-\frac{2}{5}\right)^2 = 1$$
$$\sin^2 \theta = 1 - \frac{4}{25} = \frac{21}{25}$$
$$\sin \theta = \pm\sqrt{\frac{21}{25}} = \pm\frac{\sqrt{21}}{5}$$

Since $\sin \theta < 0$,

$$\sin \theta = -\frac{\sqrt{21}}{5}.$$
$$\tan \theta = \frac{\sin \theta}{\cos \theta}$$
$$= \frac{-\frac{\sqrt{21}}{5}}{-\frac{2}{5}}$$
$$= \frac{\sqrt{21}}{2}$$

11. $\sin\theta = \frac{2}{3}$, θ in quadrant II

$$\sin^2\theta + \cos^2\theta = 1$$
$$\cos^2\theta = 1 - \sin^2\theta$$
$$= 1 - \left(\frac{2}{3}\right)^2 = \frac{5}{9}$$
$$\cos\theta = \pm\frac{\sqrt{5}}{3}$$

Since θ is in quadrant II, $\cos\theta = -\sqrt{5}/3$.

$$\tan\theta = \frac{\sin\theta}{\cos\theta} = \frac{\frac{2}{3}}{-\frac{\sqrt{5}}{3}}$$
$$= -\frac{2}{\sqrt{5}} = -\frac{2\sqrt{5}}{5}$$

$$\cot\theta = \frac{1}{\tan\theta}$$
$$= \frac{1}{-\frac{2}{\sqrt{5}}} = -\frac{\sqrt{5}}{2}$$

$$\sec\theta = \frac{1}{\cos\theta} = \frac{1}{-\frac{\sqrt{5}}{3}}$$
$$= -\frac{3}{\sqrt{5}} = -\frac{3\sqrt{5}}{5}$$

$$\csc\theta = \frac{1}{\sin\theta} = \frac{1}{\frac{2}{3}} = \frac{3}{2}$$

13. $\tan\theta = -\frac{1}{4}$, θ in quadrant IV

$$\sec^2\theta = 1 + \tan^2\theta$$
$$= 1 + \left(-\frac{1}{4}\right)^2$$
$$= \frac{17}{16}$$
$$\sec\theta = \pm\frac{\sqrt{17}}{4}$$

Since θ is in quadrant IV, $\sec\theta = \sqrt{17}/4$.

$$\cos\theta = \frac{1}{\sec\theta}$$
$$= \frac{1}{\frac{\sqrt{17}}{4}} = \frac{4\sqrt{17}}{17}$$

$$\sin\theta = \tan\theta \cdot \cos\theta$$
$$= -\frac{1}{4} \cdot \frac{4}{\sqrt{17}} = -\frac{1}{\sqrt{17}} = -\frac{\sqrt{17}}{17}$$

$$\csc\theta = \frac{1}{\sin\theta} = \frac{1}{-\frac{1}{\sqrt{17}}} = -\sqrt{17}$$

$$\cot\theta = \frac{1}{\tan\theta} = \frac{1}{-\frac{1}{4}} = -4$$

15. $\sec\theta = -3$, θ in quadrant II

$$\cos\theta = \frac{1}{\sec\theta}$$
$$= \frac{1}{-3} = -\frac{1}{3}$$

$$\sin^2\theta + \cos^2\theta = 1$$
$$\sin^2\theta = 1 - \cos^2\theta$$
$$= 1 - \left(-\frac{1}{3}\right)^2 = \frac{8}{9}$$
$$\sin\theta = \pm\frac{\sqrt{8}}{3}$$

Since θ is in quadrant II,

$$\sin\theta = \frac{\sqrt{8}}{3} = \frac{2\sqrt{2}}{3}.$$

$$\tan\theta = \frac{\sin\theta}{\cos\theta} = \frac{\frac{2\sqrt{2}}{3}}{-\frac{1}{3}} = -2\sqrt{2}$$

$$\cot\theta = \frac{1}{\tan\theta} = \frac{1}{-2\sqrt{2}} = -\frac{\sqrt{2}}{4}$$

$$\csc\theta = \frac{1}{\sin\theta} = \frac{1}{\frac{2\sqrt{2}}{3}}$$
$$= \frac{3}{2\sqrt{2}} = \frac{3\sqrt{2}}{4}$$

17. $\cot\theta = \frac{4}{3}$, $\sin\theta > 0$

Since $\cot\theta > 0$ and $\sin\theta > 0$, θ is in quadrant I and all the functions are positive.

$$\tan\theta = \frac{1}{\cot\theta} = \frac{1}{\frac{4}{3}} = \frac{3}{4}$$

$$\sec^2\theta = 1 + \tan^2\theta$$

$$= 1 + \left(\frac{3}{4}\right)^2 = \frac{25}{16}$$

$$\sec\theta = \frac{5}{4}$$

$$\cos\theta = \frac{1}{\sec\theta} = \frac{1}{\frac{5}{4}} = \frac{4}{5}$$

$$\sin^2\theta + \cos^2\theta = 1$$

$$\sin^2\theta = 1 - \cos^2\theta$$

$$= 1 - \left(\frac{4}{5}\right)^2 = \frac{9}{25}$$

$$\sin\theta = \frac{3}{5}$$

$$\csc\theta = \frac{1}{\sin\theta} = \frac{5}{3}$$

19. $\sec\theta = \frac{4}{3}$, $\sin\theta < 0$

Since $\sec\theta > 0$ and $\sin\theta < 0$, θ is in quadrant IV.

$$\cos\theta = \frac{1}{\sec\theta}$$

$$= \frac{1}{\frac{4}{3}} = \frac{3}{4}$$

$$\sin^2\theta + \cos^2\theta = 1$$

$$\sin^2\theta = 1 - \cos^2\theta$$

$$= 1 - \left(\frac{3}{4}\right)^2 = \frac{7}{16}$$

Since $\sin\theta < 0$,

$$\sin\theta = -\frac{\sqrt{7}}{4}.$$

$$\tan\theta = \frac{\sin\theta}{\cos\theta} = \frac{-\frac{\sqrt{7}}{4}}{\frac{3}{4}} = -\frac{\sqrt{7}}{3}$$

$$\cot\theta = \frac{1}{\tan\theta} = \frac{1}{-\frac{\sqrt{7}}{3}} = -\frac{3\sqrt{7}}{7}$$

$$\csc\theta = \frac{1}{\sin\theta} = \frac{1}{-\frac{\sqrt{7}}{4}} = -\frac{4\sqrt{7}}{7}$$

21. $\frac{\cos x}{\sin x} = \cot x$ (b)

23. $\cos(-x) = \cos x$ (e)

25. $1 = \sin^2 x + \cos^2 x$ (a)

27. $\sec^2 x - 1 = \tan^2 x = \frac{\sin^2 x}{\cos^2 x}$ (a)

29. $1 + \sin^2 x$

$= \csc^2 x - \cot^2 x + \sin^2 x$ (d)

31. It is incorrect to write $1 + \cot^2 = \csc^2$ because there is no angle given as argument to the functions cotangent and cosecant. You need to take the cotangent of something for computational purposes. A correct way to write this statement is $1 + \cot^2\theta = \csc^2\theta$.

33. Express $\sin\theta$ in terms of $\cot\theta$ and in terms of $\sec\theta$.

$$\sin\theta = \frac{1}{\csc\theta} = \frac{1}{\pm\sqrt{1 + \cot^2\theta}}$$

$$= \frac{\pm\sqrt{1 + \cot^2\theta}}{1 + \cot^2\theta}$$

$$\sin\theta = \cos\theta\tan\theta$$

$$= \frac{1}{\sec\theta} \cdot \pm\sqrt{\sec^2\theta - 1}$$

$$= \frac{\pm\sqrt{\sec^2\theta - 1}}{\sec\theta}$$

35. Express $\tan\theta$ in terms of $\sin\theta$, $\cos\theta$, $\sec\theta$, and $\csc\theta$.

$$\tan\theta = \frac{\sin\theta}{\cos\theta} = \frac{\sin\theta}{\pm\sqrt{1-\sin^2\theta}} = \frac{\pm\sin\theta}{\sqrt{1-\sin^2\theta}} = \frac{\pm\sin\theta\sqrt{1-\sin^2\theta}}{1-\sin^2\theta}$$

$$\tan\theta = \frac{\sin\theta}{\cos\theta} = \frac{\pm\sqrt{1-\cos^2\theta}}{\cos\theta}$$

$$\tan\theta = \pm\sqrt{\sec^2\theta - 1}$$

$$\tan\theta = \frac{1}{\cot\theta} = \frac{1}{\pm\sqrt{\csc^2\theta - 1}} = \frac{\pm\sqrt{\csc^2\theta - 1}}{\csc^2\theta - 1}$$

37. Express $\sec\theta$ in terms of $\sin\theta$, $\tan\theta$, $\cot\theta$, and $\csc\theta$.

$$\sec\theta = \frac{1}{\cos\theta} = \frac{1}{\pm\sqrt{1-\sin^2\theta}} = \frac{\pm\sqrt{1-\sin^2\theta}}{1-\sin^2\theta}$$

$$\sec\theta = \pm\sqrt{\tan^2\theta + 1}$$

Since $\tan\theta = \frac{1}{\cot\theta}$,

$$\sec\theta = \pm\sqrt{\tan^2\theta + 1} = \pm\sqrt{\left(\frac{1}{\cot^2\theta}\right) + 1} = \frac{\pm\sqrt{1+\cot^2\theta}}{\cot\theta}$$

$$\sec\theta = \frac{1}{\cos\theta} = \frac{1}{\pm\sqrt{1-\sin^2\theta}} = \frac{1}{\pm\sqrt{1-\frac{1}{\csc^2\theta}}} = \frac{\sqrt{\csc^2\theta}}{\pm\sqrt{\csc^2\theta - 1}} = \frac{\pm\csc\theta\sqrt{\csc^2\theta - 1}}{\csc^2\theta - 1}$$

39. Find $\sin\theta$ if $\cos\theta = x/(x+1)$.

$$\cos\theta = \frac{x}{x+1}$$

Since $\sin^2\theta + \cos^2\theta = 1$,

$$\sin^2\theta = 1 - \cos^2\theta$$

$$\sin^2\theta = 1 - \left(\frac{x}{x+1}\right)^2 = \frac{(x+1)^2 - x^2}{(x+1)^2} = \frac{x^2 + 2x + 1 - x^2}{(x+1)^2} = \frac{2x+1}{(x+1)^2}$$

$$\sin\theta = \frac{\pm\sqrt{2x+1}}{x+1}$$

41. $\csc^2\beta - \cot^2\beta$

$$= \frac{1}{\sin^2\beta} - \frac{\cos^2\beta}{\sin^2\beta} = \frac{1-\cos^2\beta}{\sin^2\beta} = \frac{\sin^2\beta}{\sin^2\beta} = 1$$

43. $\tan(-\alpha)\cos(-\alpha)$

$$= \frac{\sin(-\alpha)}{\cos(-\alpha)} \cdot \cos(-\alpha) = \sin(-\alpha) = -\sin\alpha$$

45. $\tan^2\theta - \dfrac{\sec^2\theta}{\csc^2\theta}$

$= \dfrac{\sin^2\theta}{\cos^2\theta} - \dfrac{\frac{1}{\cos^2\theta}}{\frac{1}{\sin^2\theta}}$

$= \dfrac{\sin^2\theta}{\cos^2\theta} - \dfrac{\sin^2\theta}{\cos^2\theta}$

$= 0$

47. $\sec\theta + \tan\theta$

$= \dfrac{1}{\cos\theta} + \dfrac{\sin\theta}{\cos\theta}$

$= \dfrac{1 + \sin\theta}{\cos\theta}$

49. $\sec^2 t - \tan^2 t$

$= \dfrac{1}{\cos^2 t} - \dfrac{\sin^2 t}{\cos^2 t}$

$= \dfrac{1 - \sin^2 t}{\cos^2 t}$

$= \dfrac{\cos^2 t}{\cos^2 t}$

$= 1$

51. $\cot^2\beta - \csc^2\beta$

$= \dfrac{\cos^2\beta}{\sin^2\beta} - \dfrac{1}{\sin^2\beta}$

$= \dfrac{-(1 - \cos^2\beta)}{\sin^2\beta}$

$= \dfrac{-(\sin^2\beta)}{\sin^2\beta}$

$= -1$

53. $\dfrac{1 + \tan^2\theta}{\cot^2\theta}$

$= \dfrac{1 + \frac{\sin^2\theta}{\cos^2\theta}}{\frac{\cos^2\theta}{\sin^2\theta}}$

$= \dfrac{\cos^2\theta + \sin^2\theta}{\cos^2\theta} \cdot \dfrac{\sin^2\theta}{\cos^2\theta}$

$= \dfrac{1(\sin^2\theta)}{\cos^4\theta}$

$= \dfrac{\sin^2\theta}{\cos^4\theta}$

55. $\cot^2\beta\sin^2\beta + \tan^2\beta\cos^2\beta$

$= \dfrac{\cos^2\beta}{\sin^2\beta}\sin^2\beta + \dfrac{\sin^2\beta}{\cos^2\beta}\cos^2\beta$

$= \cos^2\beta + \sin^2\beta$

$= 1$

57. $\dfrac{\cot^2\alpha + \csc^2\alpha}{\cos^2\alpha}$

$= \dfrac{\frac{\cos^2\alpha}{\sin^2\alpha} + \frac{1}{\sin^2\alpha}}{\cos^2\alpha}$

$= \dfrac{\cos^2\alpha + 1}{\sin^2\alpha\cos^2\alpha}$

59. $1 - \cot^4 s$

$= 1 - \dfrac{\cos^4 s}{\sin^4 s}$

$= \dfrac{\sin^4 s - \cos^4 s}{\sin^4 s}$

$= \dfrac{(\sin^2 s - \cos^2 s)(\sin^2 s + \cos^2 s)}{\sin^4 s}$

$= \dfrac{\sin^2 s - \cos^2 s}{\sin^4 s}$

For Exercises 61–65, since θ is in the interval $(0, \pi/2)$, all the trigonometric functions are positive.

61. $\sqrt{16 + 9x^2}$, let $x = \frac{4}{3}\tan\theta$.

Then

$\sqrt{16 + 9\left(\frac{4}{3}\tan\theta\right)^2}$

$= \sqrt{16 + 16\tan^2\theta}$

$= 4\sqrt{1 + \tan^2\theta}$

$= 4\sec\theta$

$\sqrt{16 + 9x^2} = 4\sec\theta$

$\sec\theta = \dfrac{\sqrt{16 + 9x^2}}{4}$

Label a right triangle in quadrant I using the ratio for $\sec\theta$ and the Pythagorean theorem.

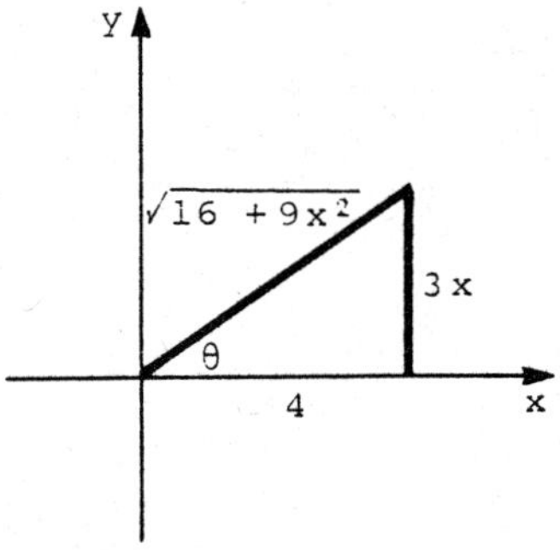

From the triangle,

$$\sin\theta = \frac{3x}{\sqrt{16 + 9x^2}}$$

$$\cos\theta = \frac{4}{\sqrt{16 + 9x^2}}$$

$$= \frac{4\sqrt{16 + 9x^2}}{16 + 9x^2}.$$

63. $\sqrt{(1 - x^2)^3}$, let $x = \cos\theta$.

$$\sqrt{(1 - x^2)^3} = \sqrt{(1 - \cos^2\theta)^3}$$

$$= \sqrt{(\sin^2\theta)^3}$$

$$= \sin^3\theta$$

$$\sin\theta = \sqrt{1 - x^2}$$

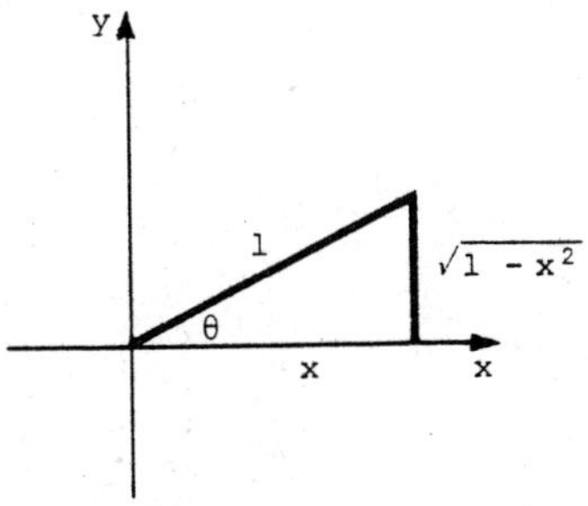

From the triangle,

$$\sin\theta = \frac{\sqrt{1 - x^2}}{1}$$

$$\tan\theta = \frac{\sqrt{1 - x^2}}{x}.$$

65. $x^2\sqrt{1 + 16x^2}$, let $x = \frac{1}{4}\tan\theta$.

$$x^2\sqrt{1 + 16x^2}$$

$$= \left(\frac{1}{4}\tan\theta\right)^2\sqrt{1 + 16\left(\frac{1}{4}\tan\theta\right)^2}$$

$$= \frac{1}{16}\tan^2\theta\sqrt{1 + \tan^2\theta}$$

$$= \frac{1}{16}\tan^2\theta\sec\theta$$

$$\sec\theta = \frac{x^2\sqrt{1 + 16x^2}}{\frac{1}{16}\tan^2\theta}$$

$$= \frac{x^2\sqrt{1 + 16x^2}}{x^2}$$

$$= \sqrt{1 + 16x^2}$$

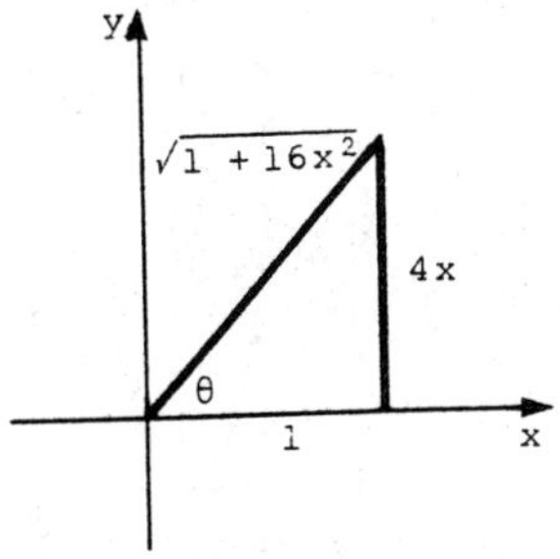

From the triangle,

$$\sin\theta = \frac{4x}{\sqrt{1 + 16x^2}} = \frac{4x\sqrt{1 + 16x^2}}{1 + 16x^2}$$

$$\cos\theta = \frac{1}{\sqrt{1 + 16x^2}} = \frac{\sqrt{1 + 16x^2}}{1 + 16x^2}.$$

67. Let $\cos x = \frac{1}{5}$. Then x is in quadrant I or quadrant IV. Then,

$$\sin x = \pm\sqrt{1 - \cos^2 x}$$

$$= \pm\sqrt{1 - \left(\frac{1}{5}\right)^2}$$

$$= \pm\sqrt{\frac{24}{25}} = \pm\frac{2\sqrt{6}}{5}$$

$$\tan x = \frac{\sin x}{\cos x} = \frac{\pm\frac{2\sqrt{6}}{5}}{\frac{1}{5}} = \pm 2\sqrt{6}$$

$$\sec x = \frac{1}{\cos x} = 5$$

If x is in quadrant I,

$$\frac{\sec x - \tan x}{\sin x} = \frac{5 - 2\sqrt{6}}{\frac{2\sqrt{6}}{5}}$$

$$= \frac{25 - 10\sqrt{6}}{2\sqrt{6}}$$

$$= \frac{25\sqrt{6} - 60}{12}.$$

If x is in quadrant IV,

$$\frac{\sec x - \tan x}{\sin x} = \frac{5 - (-2\sqrt{6})}{-\frac{2\sqrt{6}}{5}}$$

$$= \frac{25 + 10\sqrt{6}}{-2\sqrt{6}}$$

$$= \frac{-25\sqrt{6} - 60}{12}.$$

69. (a) $y = \sin x$

$\sin(-x) = -\sin(x)$

This is an odd function.

(b) $y = \cos x$

$\cos(-x) = \cos x$

This is an even function.

(c) $y = \tan x$

$$\tan(-x) = \frac{\sin(-x)}{\cos(-x)} = \frac{-\sin(x)}{\cos x}$$

$$= -\tan x$$

This is an odd function.

(d) $y = \cot x$

$$\cot(-x) = \frac{\cos(-x)}{\sin(-x)} = \frac{\cos x}{-\sin x}$$

$$= -\cot x$$

This is an odd function.

(e) $y = \csc x$

$$\csc(-x) = \frac{1}{\sin(-x)}$$

$$= \frac{1}{-\sin x}$$

$$= -\csc x$$

This is an odd function.

(f) $y = \sec x$

$$\sec(-x) = \frac{1}{\cos(-x)}$$

$$= \frac{1}{\cos x}$$

$$= \sec x$$

This is an even function.

For Exercises 71 and 73, see the answer graphs at the back of the textbook.

71. $y = \cos(-4x) = \cos 4x$

73. $y = 3 \sin(-4x) = -3 \sin 4x$

75. $y = \sec(-x) = \frac{1}{\cos(-x)}$

$$= \frac{1}{\cos x} = \sec x$$

Since $\sec(-x) = \sec x$, the graph of $y = \sec(-x)$ is exactly the same as the graph of $y = \sec x$.

77. $y = \tan(-x) = \frac{\sin(-x)}{\cos(-x)}$

$$= \frac{-\sin x}{\cos x} = -\tan x$$

Since $\tan(-x) = -\tan x$, the graph of $y = \tan(-x)$ is a reflection across the x-axis as compared to the graph of $y = \tan x$.

79. Is $(\sin s + \cos s)^2 = 1$ an identity?

$$(\sin s + \cos s)^2$$
$$= \sin^2 s + \cos^2 s + 2 \sin s \cos s$$

For example, let $s = \frac{\pi}{4}$.

$$\left(\sin \frac{\pi}{4} + \cos \frac{\pi}{4}\right)^2$$
$$= \sin^2 \frac{\pi}{4} + \cos^2 \frac{\pi}{4} + 2 \sin \frac{\pi}{4} \cos \frac{\pi}{4}$$
$$= \left(\frac{\sqrt{2}}{2}\right)^2 + \left(\frac{\sqrt{2}}{2}\right)^2 + 2\frac{\sqrt{2}}{2} \cdot \frac{\sqrt{2}}{2}$$
$$= 1 + 2$$
$$= 3$$

Since $\left(\sin \frac{\pi}{4} + \cos \frac{\pi}{4}\right)^2 \neq 1$, $(\sin s + \cos s)^2 = 1$ is not an identity.

81. Is $2 \sin s = \sin 2s$ an identity?

Let $s = \frac{\pi}{6}$.

$$2 \sin \frac{\pi}{6} = 2 \cdot \frac{1}{2} = 1$$
$$\sin\left(2 \cdot \frac{\pi}{6}\right) = \sin \frac{\pi}{3} = \frac{\sqrt{3}}{2}$$

Since $1 \neq \frac{\sqrt{3}}{2}$, $2 \sin s \neq \sin 2s$, and $2 \sin s = \sin 2s$ is not an identity.

83. Is $\sin^3 x + \cos^3 x = 1$ an identity?

Let $x = \frac{\pi}{4}$.

$$\sin^3 \frac{\pi}{4} + \cos^3 \frac{\pi}{4} = \left(\frac{\sqrt{2}}{2}\right)^3 + \left(\frac{\sqrt{2}}{2}\right)^3$$
$$= \frac{\sqrt{2}}{4} + \frac{\sqrt{2}}{4} = \frac{\sqrt{2}}{2}$$
$$\frac{\sqrt{2}}{2} \neq 1$$

$\sin^3 x + \cos^3 x \neq 1$. Thus, it is not an identity.

85. Verify $\log \sin s = -\log \csc s$.

Since $\sin s = \frac{1}{\csc s}$,

$$\log \sin s = \log \frac{1}{\csc s}$$
$$= \log 1 - \log \csc s$$
$$= 0 - \log \csc s$$
$$= -\log \csc s$$

87. Verify $\log \sec s = \frac{1}{2} \log (\tan^2 s + 1)$.

Work with the right side.

Since $\tan^2 s + 1 = \sec^2 s$,

$$\frac{1}{2} \log (\tan^2 s + 1)$$
$$= \frac{1}{2} \log \sec^2 s$$
$$= \log (\sec^2 s)^{1/2}$$
$$= \log \sec s$$

Section 5.2

1. $\tan \theta + \frac{1}{\tan \theta}$

$$= \frac{\sin \theta}{\cos \theta} + \frac{\cos \theta}{\sin \theta} = \frac{\sin^2 \theta + \cos^2 \theta}{\sin \theta \cos \theta}$$
$$= \frac{1}{\sin \theta \cos \theta}$$

or $\csc \theta \sec \theta$

3. $\cot s (\tan s + \sin s)$

$$= \frac{\cos s}{\sin s}\left(\frac{\sin s}{\cos s} + \sin s\right)$$
$$= 1 + \cos s$$

5. $\frac{1}{\csc^2 \theta} + \frac{1}{\sec^2 \theta}$

$$= \sin^2 \theta + \cos^2 \theta$$
$$= 1$$

7. $\frac{\cos x}{\sec x} + \frac{\sin x}{\csc x}$

$= \cos x (\cos x) + \sin x (\sin x)$

$= \cos^2 x + \sin^2 x$

$= 1$

9. $(1 + \sin t)^2 + \cos^2 t$

$= 1 + \sin^2 t + 2 \sin t + \cos^2 t$

$= 2 + 2 \sin t$

11. $\frac{1}{1 + \cos x} - \frac{1}{1 - \cos x}$

$= \frac{1 - \cos x - 1 - \cos x}{1 - \cos^2 x}$

$= \frac{-2 \cos x}{\sin^2 x}$ or $\frac{-2 \cos x}{1 - \cos^2 x}$

$= \frac{-2 \cos x}{(1 + \cos x)(1 - \cos x)}$

13. $\sin^2 \gamma - 1$

$= (\sin \gamma - 1)(\sin \gamma + 1)$

15. $(\sin x + 1)^2 - (\sin x - 1)^2$

Let $a = \sin x + 1$ and $b = \sin x - 1$.

$(\sin x + 1)^2 - (\sin x - 1)^2$

$= a^2 - b^2 = (a - b)(a + b)$

$= [(\sin x + 1) - (\sin x - 1)] \cdot$

$[(\sin x + 1) + (\sin x - 1)]$

$= 2(2 \sin x)$

$= 4 \sin x$

17. $2 \sin^2 x + 3 \sin x + 1$

Let $a = \sin x$.

$2 \sin^2 x + 3 \sin x + 1$

$= 2a^2 + 3a + 1$

$= (2a + 1)(a + 1)$

$= (2 \sin x + 1)(\sin x + 1)$

19. $4 \sec^2 x + 3 \sec x - 1$

Let $\sec x = a$.

$4 \sec^2 x + 3 \sec x - 1$

$= 4a^2 + 3a - 1$

$= (4a - 1)(a + 1)$

$= (4 \sec x - 1)(\sec x + 1)$

21. $\cos^4 x + 2 \cos^2 x + 1$

Let $\cos^2 x = a$.

$\cos^4 x + 2 \cos^2 x + 1$

$= a^2 + 2a + 1 = (a + 1)^2$

$= (\cos^2 x + 1)^2$

23. $\sin^3 x - \cos^3 x$

Let $a = \sin x$, $b = \cos x$.

$\sin^3 x - \cos^3 x$

$= a^3 - b^3$

$= (a - b)(a^2 + ab + b^2)$

$= (\sin x + \cos x) \cdot$

$(\sin^2 x + \sin x \cos x + \cos^2 x)$

$= (\sin x - \cos x)(1 + \sin x \cos x)$

25. $\tan \theta \cos \theta = \frac{\sin \theta}{\cos \theta} \cos \theta = \sin \theta$

27. $\sec r \cos r = \frac{1}{\cos r} \cos r = 1$

29. $\frac{\sin \beta \tan \beta}{\cos \beta} = \tan \beta \tan \beta = \tan^2 \beta$

31. $\sec^2 x - 1 = \frac{1}{\cos^2 x} - 1$

$= \frac{1 - \cos^2 x}{\cos^2 x}$

$= \frac{\sin^2 x}{\cos^2 x} = \tan^2 x$

33. $\frac{\sin^2 x}{\cos^2 x} + \sin x \csc x$

$= \tan^2 x + \sin x \frac{1}{\sin x}$

$= \tan^2 x + 1$

$= \sec^2 x$

35. Verify $\frac{\cot\theta}{\csc\theta} = \cos\theta$.

$$\frac{\cot\theta}{\csc\theta} = \frac{\frac{\cos\theta}{\sin\theta}}{\frac{1}{\sin\theta}} = \cos\theta$$

37. Verify $\frac{1-\sin^2\beta}{\cos\beta} = \cos\beta$.

$$\frac{1-\sin^2\beta}{\cos\beta} = \frac{\cos^2\beta}{\cos\beta} = \cos\beta$$

39. Verify $\cos^2\theta\,(\tan^2\theta + 1) = 1$.

$$\cos^2\theta\,(\tan^2\theta + 1)$$
$$= \cos^2\theta\left(\frac{\sin^2\theta}{\cos^2\theta} + 1\right)$$
$$= \cos^2\theta\left(\frac{\sin^2\theta + \cos^2\theta}{\cos^2\theta}\right)$$
$$= 1$$

41. Verify $\sin^2\alpha + \tan^2\alpha + \cos^2\alpha = \sec^2\alpha$.

$$\sin^2\alpha + \tan^2\alpha + \cos^2\alpha$$
$$= 1 + \tan^2\alpha = \sec^2\alpha$$

43. Verify $\frac{\sin^2\gamma}{\cos\gamma} = \sec\gamma - \cos\gamma$.

$$\frac{\sin^2\gamma}{\cos\gamma} = \frac{1-\cos^2\gamma}{\cos\gamma}$$
$$= \frac{1}{\cos\gamma} - \frac{\cos^2\gamma}{\cos\gamma}$$
$$= \sec\gamma - \cos\gamma$$

45. Verify $\frac{\cos\theta}{\sin\theta\cot\theta} = 1$.

$$\frac{\cos\theta}{\sin\theta\cot\theta} = \frac{\cos\theta}{\sin\theta\left(\frac{\cos\theta}{\sin\theta}\right)}$$
$$= \frac{\sin\theta\cos\theta}{\sin\theta\cos\theta} = 1$$

47. Verify $\tan^2\gamma\sin^2\gamma = \tan^2\gamma + \cos^2\gamma - 1$.

Work with the left side.

$$\tan^2\gamma\sin^2\gamma$$
$$\tan^2\gamma\,(1-\cos^2\gamma)$$
$$= \tan^2\gamma - \tan^2\gamma\cos^2\gamma$$
$$= \tan^2\gamma - \sin^2\gamma$$

Now work with the right side.

$$\tan^2\gamma + \cos^2\gamma - 1$$
$$= \tan^2\gamma - (1-\cos^2\gamma)$$
$$= \tan^2\gamma - \sin^2\gamma$$

49. Verify $\frac{(\sec\theta-\tan\theta)^2+1}{\sec\theta\csc\theta - \tan\theta\csc\theta} = 2\tan\theta$.

$$\frac{(\sec\theta-\tan\theta)^2+1}{\sec\theta\csc\theta - \tan\theta\csc\theta}$$
$$= \frac{\sec^2\theta - 2\sec\theta\tan\theta + \tan^2\theta + 1}{\csc\theta\,(\sec\theta - \tan\theta)}$$
$$= \frac{2\sec^2\theta - 2\sec\theta\tan\theta}{\csc\theta\,(\sec\theta-\tan\theta)}$$
$$= \frac{2\sec\theta\,(\sec\theta-\tan\theta)}{\csc\theta\,(\sec\theta-\tan\theta)}$$
$$= \frac{2\sec\theta}{\csc\theta}$$
$$= 2\,\frac{\sin\theta}{\cos\theta}$$
$$= 2\tan\theta$$

51. Verify $\frac{1}{\sec\alpha - \tan\alpha} = \sec\alpha + \tan\alpha$.

$$\frac{1}{\sec\alpha-\tan\alpha}$$
$$= \frac{1}{\sec\alpha-\tan\alpha}\cdot\frac{\sec\alpha+\tan\alpha}{\sec\alpha+\tan\alpha}$$
$$= \frac{\sec\alpha+\tan\alpha}{\sec^2\alpha-\tan^2\alpha}$$
$$= \frac{\sec\alpha+\tan\alpha}{\frac{1}{\cos^2\alpha}-\frac{\sin^2\alpha}{\cos^2\alpha}}$$
$$= \frac{\sec\alpha+\tan\alpha}{\frac{\cos^2\alpha}{\cos^2\alpha}}$$
$$= \sec\alpha + \tan\alpha$$

53. Verify $\frac{1 - \cos x}{1 + \cos x} = (\cot x - \csc x)^2$.

Work with the left side.

$$\frac{1 - \cos x}{1 + \cos x}$$
$$= \frac{(1 - \cos x)(1 - \cos x)}{(1 + \cos x)(1 - \cos x)}$$
$$= \frac{1 - 2\cos x + \cos^2 x}{1 - \cos^2 x}$$
$$= \frac{1 - 2\cos x + \cos^2 x}{\sin^2 x}$$

Work with the right side.

$$(\cot x - \csc x)^2$$
$$= \left(\frac{\cos x}{\sin x} - \frac{1}{\sin x}\right)^2$$
$$= \left(\frac{\cos x - 1}{\sin x}\right)^2$$
$$= \frac{\cos^2 x - 2\cos x + 1}{\sin^2 x}$$

55. Verify $\frac{1}{\tan \alpha - \sec \alpha} + \frac{1}{\tan \alpha + \sec \alpha}$ $= -2\tan \alpha$.

$$\frac{1}{\tan \alpha - \sec \alpha} + \frac{1}{\tan \alpha + \sec \alpha}$$
$$= \frac{\tan \alpha + \sec \alpha + \tan \alpha - \sec \alpha}{\tan^2 \alpha - \sec^2 \alpha}$$
$$= \frac{2\tan \alpha}{\frac{\sin^2 \alpha}{\cos^2 \alpha} - \frac{1}{\cos^2 \alpha}}$$
$$= \frac{2\tan \alpha}{\frac{-(1 - \sin^2 \alpha)}{\cos^2 \alpha}}$$
$$= \frac{2\tan \alpha}{-\frac{\cos^2 \alpha}{\cos^2 \alpha}} = -2\tan \alpha$$

57. Verify $\frac{\csc \theta + \cot \theta}{\tan \theta + \sin \theta} = \cot \theta \csc \theta$.

$$\frac{\csc \theta + \cot \theta}{\tan \theta + \sin \theta}$$
$$= \frac{\frac{1}{\sin \theta} + \frac{\cos \theta}{\sin \theta}}{\frac{\sin \theta}{\cos \theta} + \sin \theta}$$
$$= \frac{\frac{1 + \cos \theta}{\sin \theta}}{\frac{\sin \theta(1 + \cos \theta)}{\cos \theta}}$$
$$= \frac{(1 + \cos \theta)\cos \theta}{\sin^2 \theta(1 + \cos \theta)}$$
$$= \frac{\cos \theta}{\sin^2 \theta}$$
$$= \frac{1}{\sin \theta} \cdot \frac{\cos \theta}{\sin \theta}$$
$$= \csc \theta \cot \theta$$

59. Verify $\sec^4 x - \sec^2 x$ $= \tan^4 x + \tan^2 x$.

Simplify left side.

$$\sec^4 x - \sec^2 x$$
$$= \sec^2 x(\sec^2 x - 1)$$
$$= \sec^2 x \tan^2 x$$

Simplify right side.

$$\tan^4 x + \tan^2 x$$
$$= \tan^2 x(\tan^2 x + 1)$$
$$= \tan^2 x \sec^2 x$$

61. Verify $\sin \theta + \cos \theta$ $= \frac{\sin \theta}{1 - \frac{\cos \theta}{\sin \theta}} + \frac{\cos \theta}{1 - \frac{\sin \theta}{\cos \theta}}$.

Work with the right side.

$$\frac{\sin \theta}{1 - \frac{\cos \theta}{\sin \theta}} + \frac{\cos \theta}{1 - \frac{\sin \theta}{\cos \theta}}$$
$$= \frac{\sin^2 \theta}{\sin \theta - \cos \theta} + \frac{\cos^2 \theta}{\cos \theta - \sin \theta}$$
$$= \frac{\sin^2 \theta}{\sin \theta - \cos \theta} + \frac{\cos^2 \theta}{-(\sin \theta - \cos \theta)}$$
$$= \frac{\sin^2 \theta - \cos^2 \theta}{\sin \theta - \cos \theta}$$
$$= \frac{(\sin \theta + \cos \theta)(\sin \theta - \cos \theta)}{(\sin \theta - \cos \theta)}$$
$$= \sin \theta + \cos \theta$$

63. Verify $\dfrac{\sec^4 s - \tan^4 s}{\sec^2 s + \tan^2 s} = \sec^2 s - \tan^2 s.$

$$\frac{\sec^4 s - \tan^4 s}{\sec^2 s + \tan^2 s} = \frac{(\sec^2 s - \tan^2 s)(\sec^2 s + \tan^2 s)}{\sec^2 s + \tan^2 s} = \sec^2 s - \tan^2 s$$

65. Verify $\dfrac{\tan^2 t - 1}{\sec^2 t} = \dfrac{\tan t - \cot t}{\tan t + \cot t}.$

Work with the right side.

$$\frac{\tan t - \cot t}{\tan t + \cot t} = \frac{\tan t - \dfrac{1}{\tan t}}{\tan t + \dfrac{1}{\tan t}} = \frac{\dfrac{\tan^2 t - 1}{\tan t}}{\dfrac{\tan^2 t + 1}{\tan t}} = \frac{\tan^2 t - 1}{\tan^2 t + 1} = \frac{\tan^2 t - 1}{\sec^2 t}$$

67. Verify $(\sin s + \cos s)^2 \cdot \csc s = 2 \cos s + \dfrac{1}{\sin s}.$

$$\begin{aligned}(\sin s + \cos s)^2 \cdot \csc s &= \csc s\,(\sin^2 s + \cos^2 s + 2 \cos s \sin s)\\ &= \csc s\,(1 + 2 \cos s \sin s)\\ &= \csc s + 2 \cos s \sin s \csc s\\ &= \frac{1}{\sin s} + 2 \cos s\end{aligned}$$

69. Verify $\dfrac{1 + \cos x}{1 - \cos x} - \dfrac{1 - \cos x}{1 + \cos x} = 4 \cot x \csc x.$

$$\begin{aligned}\frac{1 + \cos x}{1 - \cos x} - \frac{1 - \cos x}{1 + \cos x} &= \frac{1 + \cos^2 x + 2 \cos x - 1 - \cos^2 x + 2 \cos x}{1 - \cos^2 x}\\ &= \frac{4 \cos x}{\sin^2 x} = 4 \cot x \csc x\end{aligned}$$

71. Verify $(\sec \alpha + \csc \alpha)(\cos \alpha - \sin \alpha) = \cot \alpha - \tan \alpha.$

$$\begin{aligned}(\sec \alpha + \csc \alpha)(\cos \alpha - \sin \alpha) &= \left(\frac{1}{\cos \alpha} + \frac{1}{\sin \alpha}\right)(\cos \alpha - \sin \alpha)\\ &= 1 + \cot \alpha - \tan \alpha - 1 = \cot \alpha - \tan \alpha\end{aligned}$$

73. Verify $\dfrac{\cot^2 x + \sec^2 x + 1}{\cot^2 x} = \sec^4 x.$

$$\begin{aligned}\frac{\cot^2 x + \sec^2 x + 1}{\cot^2 x} &= \frac{\dfrac{1}{\tan^2 x} + \sec^2 x + 1}{\dfrac{1}{\tan^2 x}} = 1 + \tan^2 x + \sec^2 x \tan^2 x\\ &= \sec^2 x + \sec^2 x \tan^2 x = \sec^2 x(1 + \tan^2 x)\\ &= \sec^2 x \sec^2 x = \sec^4 x\end{aligned}$$

75. Verify $\dfrac{\cot\theta}{1+\csc\theta}+\dfrac{1+\csc\theta}{\cot\theta}=2\sec\theta$.

$$\frac{\cot\theta}{1+\csc\theta}+\frac{1+\csc\theta}{\cot\theta}=\frac{\cot\theta}{1+\csc\theta}\cdot\frac{\cot\theta}{\cot\theta}+\frac{1+\csc\theta}{\cot\theta}\cdot\frac{1+\csc\theta}{1+\csc\theta}$$

$$=\frac{\cot^2\theta}{\cot\theta\,(1+\csc\theta)}+\frac{1+2\csc\theta+\csc^2\theta}{\cot\theta\,(1+\csc\theta)}$$

$$=\frac{\cot^2\theta+1+2\csc\theta+\csc^2\theta}{\cot\theta\,(1+\csc\theta)}$$

$$=\frac{\csc^2\theta+2\csc\theta+\csc^2\theta}{\cot\theta\,(1+\csc\theta)}$$

$$=\frac{2\csc^2\theta+2\csc\theta}{\cot\theta\,(1+\csc\theta)}=\frac{2\csc\theta(\csc\theta+1)}{\cot\theta\,(1+\csc\theta)}$$

$$=\frac{2\cdot\dfrac{1}{\sin\theta}}{\dfrac{\cos\theta}{\sin\theta}}=\frac{2}{\cos\theta}=2\sec\theta$$

77. Verify $\dfrac{\sec x}{1+\tan x}=\dfrac{\csc x}{1+\cot x}$.

$$\frac{\sec x}{1+\tan x}=\frac{\dfrac{1}{\cos x}}{1+\dfrac{\sin x}{\cos x}}=\frac{1}{\cos x+\sin x}=\frac{1\cdot\dfrac{1}{\sin x}}{(\cos x+\sin x)\cdot\dfrac{1}{\sin x}}$$

$$=\frac{\dfrac{1}{\sin x}}{\dfrac{\cos x}{\sin x}+1}=\frac{\csc x}{\cot x+1}=\frac{\csc x}{1+\cot x}$$

79. Verify $\dfrac{1+\tan x}{1-\tan x}=\dfrac{\sec^2 x+2\tan x}{2-\sec^2 x}$.

$$\frac{1+\tan x}{1-\tan x}=\frac{1+\tan x}{1-\tan x}\cdot\frac{1+\tan x}{1+\tan x}=\frac{1+2\tan x+\tan^2 x}{1-\tan^2 x}$$

$$=\frac{1+\tan^2 x+2\tan x}{1-(\sec^2 x-1)}=\frac{\sec^2 x+2\tan x}{2-\sec^2 x}$$

81. $\cos\theta+\sin\theta=1$ is true when $\theta=90°$. However, an identity is a statement that is true not just for a particular value of θ, but for all values of θ. Notice if $\theta=180°$, $\cos 180°+\sin 180°=-1+0=-1\neq 1$. Since the statement is not true for all values of θ, it is not an identity.

83. $\frac{2 + 5 \cos s}{\sin s} = 2 \csc s + 5 \cot s$

Let s = 1.

$$\frac{2 + 5 \cos 1}{\sin 1} \approx 5.5872533$$

$$2 \csc 1 + 5 \cot 1 \approx 5.5872533$$

Let s = 2.

$$\frac{2 + 5 \cos 2}{\sin 2} \approx -.08878743$$

$$2 \csc 2 + 5 \cot 2 \approx -.08878743$$

The given equation may be an identity. Verify it.

$$\frac{2 + 5 \cos s}{\sin s} = \frac{2}{\sin s} + \frac{5 \cos s}{\sin s}$$

$$= 2 \csc s + 5 \cot s$$

The given statement is an identity.

85. $\frac{\tan s - \cot s}{\tan s + \cot s} = 2 \sin^2 s$

Let s = 1.

$$\frac{\tan 1 - \cot 1}{\tan 1 + \cot 1} \approx .41614683$$

$$2 \sin^2 1 \approx 1.4161468$$

The given statement is not an identity.

87. $\frac{1 - \tan^2 s}{1 + \tan^2 s} = \cos^2 s - \sin s$

Let s = 1.

$$\frac{1 - \tan^2 1}{1 + \tan^2 1} \approx = -.41614684$$

$$\cos^2 1 - \sin 1 \approx -.54954440$$

The given statement is not an identity.

89. $\sin^2 s + \cot^2 s = \frac{1}{2}(1 - \cos 4s)$

$\sin^2 s + \cos^2 s = 1$ for all values of s.

Let s = 1.

$$\frac{1}{2}(1 - \cos 4 \cdot 1) \approx .82682181$$

The given statement is not an identity.

91. Show that $\sin(\csc s) = 1$ is not an identity.

Let s = 2.

$$\sin(\csc 2) \approx .89109401$$

The given statement is not an identity for all real numbers s.

93. Show that $\csc t = \sqrt{1 + \cot^2 t}$ is not an identity.

Let t = 1.

$$\csc 1 \approx 1.1883951$$

$$\sqrt{1 + \cot^2 1} \approx 1.1883951$$

But let t = 4.

$$\csc 4 \approx -1.3213487$$

$$\sqrt{1 + \cot^2 4} \approx 1.3213487$$

Recall that $\sqrt{a}$ denotes the positive square root of a.

The statement is not an identity for all real numbers t.

95. Let $\tan \theta = t$ and show that $\sin \theta \cos \theta = \frac{t}{t^2 + 1}$.

Work with the right side.

$$\frac{t}{t^2 + 1} = \frac{\tan \theta}{\tan^2 \theta + 1}$$

$$= \frac{\tan \theta}{\sec^2 \theta}$$

$$= \tan \theta \cos^2 \theta$$

$$= \frac{\sin \theta}{\cos \theta} \cos^2 \theta$$

$$= \sin \theta \cos \theta$$

97. Verify $\ln e^{|\sin x|} = |\sin x|$.

$$\ln e^{|\sin x|} = |\sin x| \cdot \ln e$$
$$= |\sin x| \cdot 1$$
$$= |\sin x|$$

99. Verify $\ln |\sec x - \tan x|$
$= -\ln |\sec x + \tan x|$.

Work with the right side.

$$-\ln |\sec x + \tan x|$$
$$= \ln |\sec x + \tan x|^{-1}$$
$$= \ln \left|\frac{1}{\sec x + \tan x}\right|$$
$$= \ln \left|\frac{1}{\sec x + \tan x} \cdot \frac{\sec x - \tan x}{\sec x - \tan x}\right|$$
$$= \ln \left|\frac{\sec x - \tan x}{\sec^2 x - \tan^2 x}\right|$$
$$= \ln |\sec x - \tan x|$$

Section 5.3

1. $\cos 75°$

$$= \cos (30° + 45°)$$
$$= \cos 30° \cos 45° - \sin 30° \sin 45°$$
$$= \frac{\sqrt{3}}{2} \cdot \frac{\sqrt{2}}{2} - \frac{1}{2} \cdot \frac{\sqrt{2}}{2}$$
$$= \frac{\sqrt{6} - \sqrt{2}}{4}$$

3. $\cos (-75°)$

$$= \cos [-30° + (-45°)]$$
$$= \cos (-30°) \cos (-45°) - \sin (-30°) \sin (-45°)$$
$$= \cos 30° \cos 45° - (-\sin 30°)(-\sin 45°)$$
$$= \frac{\sqrt{3}}{2} \cdot \frac{\sqrt{2}}{2} - \left(-\frac{1}{2}\right)\left(-\frac{\sqrt{2}}{2}\right)$$
$$= \frac{\sqrt{6}}{4} - \frac{\sqrt{2}}{4} = \frac{\sqrt{6} - \sqrt{2}}{4}$$

5. $\cos (-105°)$

$$= \cos [-45° + (-60°)]$$
$$= \cos (-45°) \cos (-60°) - \sin (-45°) \sin (-60°)$$
$$= \cos 45° \cos 60° - (-\sin 45°)(-\sin 60°)$$
$$= \frac{\sqrt{2}}{2} \cdot \frac{1}{2} - \left(-\frac{\sqrt{2}}{2}\right)\left(-\frac{\sqrt{3}}{2}\right)$$
$$= \frac{\sqrt{2}}{4} - \frac{\sqrt{6}}{4}$$
$$= \frac{\sqrt{2} - \sqrt{6}}{4}$$

7. $\cos \left(-\frac{\pi}{12}\right)$

$$= \cos \left(\frac{\pi}{6} - \frac{\pi}{4}\right)$$
$$= \cos \frac{\pi}{6} \cos \frac{\pi}{4} + \sin \frac{\pi}{6} \sin \frac{\pi}{4}$$
$$= \frac{\sqrt{3}}{2} \cdot \frac{\sqrt{2}}{2} + \frac{1}{2} \cdot \frac{\sqrt{2}}{2}$$
$$= \frac{\sqrt{6}}{4} + \frac{\sqrt{2}}{4}$$
$$= \frac{\sqrt{6} + \sqrt{2}}{4}$$

9. $\cos 40° \cos 50° - \sin 40° \sin 50°$

$$= \cos (40° + 50°)$$
$$= \cos (90°)$$
$$= 0$$

11. $\cos \frac{2\pi}{5} \cos \frac{\pi}{10} - \sin \frac{2\pi}{5} \sin \frac{\pi}{10}$

$$= \cos \left(\frac{2\pi}{5} + \frac{\pi}{10}\right)$$
$$= \cos \left(\frac{5\pi}{10}\right)$$
$$= \cos \left(\frac{\pi}{2}\right)$$
$$= 0$$

13. $\tan 87° = \cot (90° - 87°) = \cot 3°$

15. $\cos \frac{\pi}{12} = \sin \left(\frac{\pi}{2} - \frac{\pi}{12}\right) = \sin \frac{5\pi}{12}$

17. $\csc(-14° \; 24')$

$= \sec[90° - (-14° \; 24')]$

$= \sec 104° \; 24'$

19. $\sin \frac{5\pi}{8} = \cos\left(\frac{\pi}{2} - \frac{5\pi}{8}\right)$

$= \cos\left(-\frac{\pi}{8}\right)$

21. $\sec 146° \; 42' = \csc(90° - 146° \; 42')$

$= \csc(-56° \; 42')$

23. $\cot 176.9814°$

$= \tan(90° - 176.9814°)$

$= \tan(-86.9814°)$

25. $\cot \frac{\pi}{3} =$ _____ $\frac{\pi}{6}$

Since $\frac{\pi}{6} = \frac{\pi}{2} - \frac{\pi}{3}$,

$\cot \frac{\pi}{3} = \tan \frac{\pi}{6}$.

Answer: tan

27. _____ $33° = \sin 57°$

$\sin 57° = \cos(90° - 57°)$

$= \cos 33°$

Answer: cos

29. $\cos 70° = \frac{1}{\text{\_\_\_\_\_ } 20°}$

$\cos 70° = \sin(90° - 70°)$

$= \sin 20°$

$= \frac{1}{\csc 20°}$

Answer: csc

31. $\tan \theta = \cot(45° + 2\theta)$

Since $\tan \theta = \cot(90° - \theta)$,

$90° - \theta = 45° + 2\theta$

$3\theta = 45°$

$\theta = 15°$.

33. $\sec \theta = \csc\left(\frac{\theta}{2} + 20°\right)$

Since $\sec \theta = \csc(90° - \theta)$,

$90° - \theta = \frac{\theta}{2} + 20°$.

$\frac{\theta}{2} + 20° = 90° - \theta$

$\theta + \frac{\theta}{2} + 20° = 90°$

$2\theta + \theta + 40° = 180°$

$3\theta = 140°$

$\theta = \frac{140°}{3}$

35. $\sin(3\theta - 15°) = \cos(\theta + 25°)$

Since $\sin \theta = \cos(90° - \theta)$,

$\sin(3\theta - 15°) = \cos[90° - (3\theta - 15°)]$.

$90° - (3\theta - 15°) = \theta + 25°$

$105° - 3\theta = \theta + 25°$

$80° = 4\theta$

$20° = \theta$

37. $\cos(0° - \theta)$

$= \cos 0° \cos \theta + \sin \theta \sin 0°$

$= 1 \cos \theta + 0 \sin \theta$

$= \cos \theta$

39. $\cos(180° - \theta)$

$= \cos 180° \cos \theta + \sin 180° \sin \theta$

$= (-1) \cos \theta + 0 \sin \theta$

$= -\cos \theta$

41. $\cos(0° + \theta)$

$= \cos 0° \cos \theta - \sin 0° \sin \theta$

$= 1 \cos \theta - 0 \sin \theta$

$= \cos \theta$

43. $\cos(180° + \theta)$

$= \cos 180° \cos \theta - \sin 180° \sin \theta$

$= (-1) \cos \theta - 0 \sin \theta$

$= -\cos \theta$

45. $\cos s = -\frac{1}{5}$ and $\sin t = \frac{3}{5}$, s and t are in quadrant II.

Solve $\sin^2 s + \cos^2 s = 1$ for $\sin s$. Since s is in quadrant II, $\sin s > 0$.

$$\sin s = \sqrt{1 - \left(-\frac{1}{5}\right)^2} = \frac{\sqrt{24}}{5}$$

Solve $\sin^2 t + \cos^2 t = 1$ for $\cos t$. Since t is in quadrant II, $\cos t < 0$.

$$\cos t = -\sqrt{1 - \left(\frac{3}{5}\right)^2} = -\frac{4}{5}$$

$\cos (s + t)$

$= \cos s \cos t - \sin s \sin t$

$= \left(-\frac{1}{5}\right)\left(-\frac{4}{5}\right) - \left(\frac{\sqrt{24}}{5}\right)\left(\frac{3}{5}\right)$

$= \frac{4}{25} - \frac{3\sqrt{24}}{25} = \frac{4 - 6\sqrt{6}}{25}$

$\cos (s - t)$

$= \cos s \cos t + \sin s \sin t$

$= \left(-\frac{1}{5}\right)\left(-\frac{4}{5}\right) + \left(\frac{\sqrt{24}}{5}\right)\left(\frac{3}{5}\right)$

$= \frac{4}{25} + \frac{3\sqrt{24}}{25} = \frac{4 + 6\sqrt{6}}{25}$

47. $\sin s = \frac{3}{5}$ and $\sin t = -\frac{12}{13}$, s is in quadrant I and t is in quadrant III. $\cos s > 0$ and $\cos t < 0$.

$$\cos s = \sqrt{1 - \left(\frac{3}{5}\right)^2} = \frac{4}{5}$$

$$\cos t = -\sqrt{1 - \left(-\frac{12}{13}\right)^2} = -\frac{5}{13}$$

$\cos (s + t)$

$= \cos s \cos t - \sin s \sin t$

$= \left(\frac{4}{5}\right)\left(-\frac{5}{13}\right) - \left(\frac{3}{5}\right)\left(-\frac{12}{13}\right) = \frac{16}{65}$

$\cos (s - t)$

$= \cos s \cos t + \sin s \sin t$

$= \left(\frac{4}{5}\right)\left(-\frac{5}{13}\right) + \left(\frac{3}{5}\right)\left(-\frac{12}{13}\right) = -\frac{56}{65}$

49. $\cos s = -\frac{15}{17}$ and $\sin t = \frac{4}{5}$, s is in quadrant II, t is in quadrant I. $\sin s > 0$ and $\cos t > 0$.

$$\sin s = \sqrt{1 - \left(-\frac{15}{17}\right)^2} = \frac{8}{17}$$

$$\cos t = \sqrt{1 - \left(\frac{4}{5}\right)^2} = \frac{3}{5}$$

$\cos (s + t)$

$= \cos s \cos t - \sin s \sin t$

$= \left(-\frac{15}{17}\right)\left(\frac{3}{5}\right) - \left(\frac{8}{17}\right)\left(\frac{4}{5}\right) = -\frac{77}{85}$

$\cos (s - t)$

$= \cos s \cos t + \sin s \sin t$

$= \left(-\frac{15}{17}\right)\left(\frac{3}{5}\right) + \left(\frac{8}{17}\right)\left(\frac{4}{5}\right) = -\frac{13}{85}$

51. $\sin s = \frac{\sqrt{5}}{7}$ and $\sin t = \frac{\sqrt{6}}{8}$, s and t are in quadrant I. $\cos s > 0$ and $\cos t > 0$.

$$\cos s = \sqrt{1 - \left(\frac{\sqrt{5}}{7}\right)^2} = \frac{\sqrt{44}}{7}$$

$$\cos t = \sqrt{1 - \left(\frac{\sqrt{6}}{8}\right)^2} = \frac{\sqrt{58}}{8}$$

$\cos (s + t)$

$= \cos s \cos t - \sin s \sin t$

$= \left(\frac{\sqrt{44}}{7}\right)\left(\frac{\sqrt{58}}{8}\right) - \left(\frac{\sqrt{5}}{7}\right)\left(\frac{\sqrt{6}}{8}\right)$

$= \frac{2\sqrt{638} - \sqrt{30}}{56}$

$\cos (s - t)$

$= \cos s \cos t + \sin s \sin t$

$= \left(\frac{\sqrt{44}}{7}\right)\left(\frac{\sqrt{58}}{8}\right) + \left(\frac{\sqrt{5}}{7}\right)\left(\frac{\sqrt{6}}{8}\right)$

$= \frac{2\sqrt{638} + \sqrt{30}}{56}$

53. $\cos 42° = \cos (30° + 12°)$

Since $42° = 30° + 12°$, the statement is true.

55. $\cos 74° = \cos 60° \cos 14° + \sin 60° \sin 14°$

$\cos (A + B) = \cos A \cos B - \sin A \sin B$

$\cos 74° = \cos (60° + 14°) = \cos 60° \cos 14° - \sin 60° \sin 14°$

The statement is false.

57. $\cos \frac{\pi}{3} = \cos \frac{\pi}{12} \cos \frac{\pi}{4} - \sin \frac{\pi}{12} \sin \frac{\pi}{4}$

$\cos (A + B) = \cos A \cos B - \sin A \sin B$

$\cos \frac{\pi}{3} = \cos \left(\frac{\pi}{12} + \frac{\pi}{4}\right) = \cos \frac{\pi}{12} \cos \frac{\pi}{4} - \sin \frac{\pi}{12} \sin \frac{\pi}{4}$

The statement is true.

59. $\cos 70° \cos 20° - \sin 70° \sin 20° = 0$

$\cos (A + B) = \cos A \cos B - \sin A \sin B$

$\cos 70° \cos 20° - \sin 70° \sin 20° = \cos (70° + 20°) = \cos 90° = 0$

The statement is true.

61. Verify $\cos \left(\frac{\pi}{2} + x\right) = -\sin x$.

$$\cos \left(\frac{\pi}{2} + x\right) = \cos \frac{\pi}{2} \cos x - \sin \frac{\pi}{2} \sin x = 0 \cos x - 1 \sin x = -\sin x$$

63. Verify $\cos 2x = \cos^2 x - \sin^2 x$.

$$\cos 2x = \cos (x + x) = \cos x \cos x - \sin x \sin x = \cos^2 x - \sin^2 x$$

65. Verify $\dfrac{\cos (\alpha - \theta) - \cos (\alpha + \theta)}{\cos (\alpha - \theta) + \cos (\alpha + \theta)} = \tan \theta \tan \alpha$.

$$\frac{\cos (\alpha - \theta) - \cos (\alpha + \theta)}{\cos (\alpha - \theta) + \cos (\alpha + \theta)}$$

$$= \frac{\cos \alpha \cos \theta + \sin \alpha \sin \theta - (\cos \alpha \cos \theta - \sin \alpha \sin \theta)}{\cos \alpha \cos \theta + \sin \alpha \sin \theta + (\cos \alpha \cos \theta - \sin \alpha \sin \theta)}$$

$$= \frac{\cos \alpha \cos \theta + \sin \alpha \sin \theta - \cos \alpha \cos \theta + \sin \alpha \sin \theta}{\cos \alpha \cos \theta + \sin \alpha \sin \theta + \cos \alpha \cos \theta - \sin \alpha \sin \theta}$$

$$= \frac{2 \sin \alpha \sin \theta}{2 \cos \alpha \cos \theta} = \frac{\sin \alpha}{\cos \alpha} \cdot \frac{\sin \theta}{\cos \theta} = \tan \alpha \tan \theta$$

67. Verify $\cos(\pi + s - t) = -\sin s \sin t - \cos s \cos t$.

$$\cos(\pi + s - t)$$
$$= \cos[\pi + (s - t)] = \cos \pi \cos(s - t) - \sin \pi \sin(s - t)$$
$$= (-1)\cos(s - t) - 0 \sin(s - t) = (-1)(\cos s \cos t + \sin s \sin t)$$
$$= -\cos s \cos t - \sin s \sin t$$

69. Verify $\cos(\alpha + \beta)\cos(\alpha - \beta) = 1 - \sin^2 \alpha - \sin^2 \beta$.

Work with the left side.

$$\cos(\alpha + \beta)\cos(\alpha - \beta)$$
$$= (\cos \alpha \cos \beta - \sin \alpha \sin \beta)(\cos \alpha \cos \beta + \sin \alpha \sin \beta)$$
$$= \cos^2 \alpha \cos^2 \beta - \sin^2 \alpha \sin^2 \beta + \cos \alpha \cos \beta \sin \alpha \sin \beta - \sin \alpha \sin \beta \cos \alpha \cos \beta$$
$$= (1 - \sin^2 \beta)\cos^2 \alpha - (1 - \cos^2 \alpha)\sin^2 \beta$$
$$= \cos^2 \alpha - \sin^2 \beta \cos^2 \alpha - \sin^2 \beta + \cos^2 \alpha \sin^2 \beta = \cos^2 \alpha - \sin^2 \beta$$

Work with the right side.

$$1 - \sin^2 \alpha - \sin^2 \beta$$
$$= \sin^2 \alpha + \cos^2 \alpha - \sin^2 \alpha - \sin^2 \beta = \cos^2 \alpha - \sin^2 \beta$$

71. The cosine of the sum of two angles is equal to the product of the cosines of the angles minus the product of the sines of the angles.

73. $\cos 255° = \cos(180° + 75°) = \cos 180° \cos 75° - \sin 180° \sin 75° = -\cos 75°$

$$= -[\cos(30° + 45°)] = -(\cos 30° \cos 45° - \sin 30° \sin 45°)$$
$$= -\left(\frac{\sqrt{3}}{2} \cdot \frac{\sqrt{2}}{2} - \frac{1}{2} \cdot \frac{\sqrt{2}}{2}\right) = \frac{-\sqrt{6} + \sqrt{2}}{4}$$

75. $\cos 285° = \cos(360° - 75°) = \cos 360° \cos 75° + \sin 360° \sin 75° = \cos 75°$

$$= \cos(30° + 45°) = \cos 30° \cos 45° - \sin 30° \sin 45°$$
$$= \frac{\sqrt{3}}{2} \cdot \frac{\sqrt{2}}{2} - \frac{1}{2} \cdot \frac{\sqrt{2}}{2} = \frac{\sqrt{6} - \sqrt{2}}{4}$$

77. $\cos \frac{11\pi}{12} = \cos\left(\pi + \frac{\pi}{12}\right) = \cos \pi \cos \frac{\pi}{12} - \sin \pi \sin \frac{\pi}{12} = -\cos \frac{\pi}{12}$

$$= -\left[\cos\left(\frac{\pi}{4} - \frac{\pi}{6}\right)\right] = -\left(\cos \frac{\pi}{4} \cos \frac{\pi}{6} + \sin \frac{\pi}{4} \sin \frac{\pi}{6}\right)$$
$$= -\left(\frac{\sqrt{2}}{2} \cdot \frac{\sqrt{3}}{2} + \frac{\sqrt{2}}{2} \cdot \frac{1}{2}\right) = \frac{-\sqrt{6} - 2}{4}$$

79. Verify cos A = sin (90° - A).

Start with the identity cos (90° - θ) = sin θ and replace θ with 90° - A.

$$\cos[90° - (90° - A)] = \sin(90° - A)$$
$$\cos(90° - 90° + A) = \sin(90° - A)$$
$$\cos A = \sin(90° - A)$$

81. Verify $\dfrac{f(x+h) - f(x)}{h}$

$$= \cos x \left(\frac{\cos h - 1}{h}\right) - \sin x \left(\frac{\sin h}{h}\right)$$

Let f(x) = cos x.

$$f(x + h)$$
$$= \cos(x + h)$$
$$= \cos x \cos h - \sin x \sin h$$

$$\frac{f(x+h) - f(x)}{h}$$
$$= \frac{\cos x \cos h - \sin x \sin h - \cos x}{h}$$
$$= \frac{\cos x \cos h - \cos x - \sin x \sin h}{h}$$
$$= \frac{\cos x \cos h - \cos x}{h} - \frac{\sin x \sin h}{h}$$
$$= \cos x \left(\frac{\cos h - 1}{h}\right) - \sin x \left(\frac{\sin h}{h}\right)$$

Section 5.4

1. sin 15°

$$= \sin(45° - 30°)$$
$$= \sin 45° \cos 30° - \cos 45° \sin 30°$$
$$= \frac{\sqrt{2}}{2} \cdot \frac{\sqrt{3}}{2} - \frac{\sqrt{2}}{2} \cdot \frac{1}{2}$$
$$= \frac{\sqrt{6}}{4} - \frac{\sqrt{2}}{4}$$
$$= \frac{\sqrt{6} - \sqrt{2}}{4}$$

3. $\tan 15° = \tan(60° - 45°)$

$$= \frac{\tan 60° - \tan 45°}{1 + \tan 60° \tan 45°}$$
$$= \frac{\sqrt{3} - 1}{1 + \sqrt{3}(1)} = \frac{\sqrt{3} - 1}{1 + \sqrt{3}} \cdot \frac{1 - \sqrt{3}}{1 - \sqrt{3}}$$
$$= \frac{\sqrt{3} - 3 - 1 + \sqrt{3}}{1 - 3} = \frac{-4 + 2\sqrt{3}}{-2}$$
$$= \frac{-2(2 - \sqrt{3})}{-2} = 2 - \sqrt{3}$$

5. sin (-105°)

$$= \sin(45° - 150°)$$
$$= \sin 45° \cos 150° - \cos 45° \sin 150°$$
$$= \left(\frac{\sqrt{2}}{2}\right)\left(-\frac{\sqrt{3}}{2}\right) - \left(\frac{\sqrt{2}}{2}\right)\left(\frac{1}{2}\right)$$
$$= \frac{-\sqrt{6} - \sqrt{2}}{4}$$

7. $\sin \frac{5\pi}{12} = \sin\left(\frac{\pi}{4} + \frac{\pi}{6}\right)$

$$= \sin \frac{\pi}{4} \cos \frac{\pi}{6} + \cos \frac{\pi}{4} \sin \frac{\pi}{6}$$
$$= \frac{\sqrt{2}}{2} \cdot \frac{\sqrt{3}}{2} + \frac{\sqrt{2}}{2} \cdot \frac{1}{2}$$
$$= \frac{\sqrt{6}}{4} + \frac{\sqrt{2}}{4}$$
$$= \frac{\sqrt{6} + \sqrt{2}}{4}$$

9. $\tan \frac{\pi}{12} = \tan\left(\frac{\pi}{4} - \frac{\pi}{6}\right)$

$$= \frac{\tan \frac{\pi}{4} - \tan \frac{\pi}{6}}{1 + \tan \frac{\pi}{4} \tan \frac{\pi}{6}}$$
$$= \frac{1 - \frac{\sqrt{3}}{3}}{1 + (1)\frac{\sqrt{3}}{3}} = \frac{1 - \frac{\sqrt{3}}{3}}{1 + \frac{\sqrt{3}}{3}}$$
$$= \frac{3 - \sqrt{3}}{3 + \sqrt{3}}$$
$$= \frac{3 - \sqrt{3}}{3 + \sqrt{3}} \cdot \frac{3 - \sqrt{3}}{3 - \sqrt{3}}$$
$$= \frac{12 - 6\sqrt{3}}{6}$$
$$= 2 - \sqrt{3}$$

11. $\sin\left(-\frac{7\pi}{12}\right)$

$= \sin\left(-\frac{\pi}{4} - \frac{\pi}{3}\right)$

$= \sin\left(-\frac{\pi}{4}\right)\cos\frac{\pi}{3} - \cos\left(-\frac{\pi}{4}\right)\sin\frac{\pi}{3}$

$= -\sin\frac{\pi}{4}\cos\frac{\pi}{3} - \cos\frac{\pi}{4}\sin\frac{\pi}{3}$

$= -\frac{\sqrt{2}}{2}\cdot\frac{1}{2} - \frac{\sqrt{2}}{2}\cdot\frac{\sqrt{3}}{2} = -\frac{\sqrt{2}}{4} - \frac{\sqrt{6}}{4}$

$= \frac{-\sqrt{2} - \sqrt{6}}{4}$

13. $\sin 76° \cos 31° - \cos 76° \sin 31°$

$= \sin(76° - 31°)$

$= \sin 45°$

$= \frac{\sqrt{2}}{2}$

15. $\frac{\tan 80° + \tan 55°}{1 - \tan 80° \tan 55°}$

$= \tan(80° + 55°)$

$= \tan 135°$

$= -1$

17. $\frac{\tan 100° + \tan 80°}{1 - \tan 100° \tan 80°}$

$= \tan(100° + 80°)$

$= \tan 180°$

$= 0$

19. $\sin\frac{\pi}{5}\cos\frac{3\pi}{10} + \cos\frac{\pi}{5}\sin\frac{3\pi}{10}$

$= \sin\left(\frac{\pi}{5} + \frac{3\pi}{10}\right)$

$= \sin\frac{\pi}{2}$

$= 1$

21. $\cos(30° + \theta)$

$= \cos 30° \cos\theta - \sin 30° \sin\theta$

$= \frac{\sqrt{3}}{2}\cos\theta - \frac{1}{2}\sin\theta$

$= \frac{1}{2}(\sqrt{3}\cos\theta - \sin\theta)$

$= \frac{\sqrt{3}\cos\theta - \sin\theta}{2}$

23. $\cos(60° + \theta)$

$= \cos 60° \cos\theta - \sin 60° \sin\theta$

$= \frac{1}{2}\cos\theta - \frac{\sqrt{3}}{2}\sin\theta$

$= \frac{1}{2}(\cos\theta - \sqrt{3}\sin\theta)$

$= \frac{\cos\theta - \sqrt{3}\sin\theta}{2}$

25. $\cos\left(\frac{3\pi}{4} - x\right)$

$= \cos\frac{3\pi}{4}\cos x + \sin\frac{3\pi}{4}\sin x$

$= -\cos\frac{\pi}{4}\cos x + \sin\frac{\pi}{4}\sin x$

$= -\frac{\sqrt{2}}{2}\cos x + \frac{\sqrt{2}}{2}\sin x$

$= \frac{\sqrt{2}}{2}(-\cos x + \sin x)$

$= \frac{\sqrt{2}(\sin x - \cos x)}{2}$

27. $\sin(45° + \theta)$

$= \sin 45° \cos\theta + \sin\theta \cos 45°$

$= \frac{\sqrt{2}}{2}\sin\theta + \frac{\sqrt{2}}{2}\cos\theta$

$= \frac{\sqrt{2}}{2}(\sin\theta + \cos\theta)$

$= \frac{\sqrt{2}(\sin\theta + \cos\theta)}{2}$

29. $\tan(\theta + 30°)$

$= \frac{\tan\theta + \tan 30°}{1 - \tan\theta \tan 30°}$

$= \frac{\tan\theta + \frac{1}{\sqrt{3}}}{1 - \frac{1}{\sqrt{3}}\tan\theta}$

$= \frac{\sqrt{3}\tan\theta + 1}{\sqrt{3} - \tan\theta}$

31. $\tan\left(\frac{\pi}{4} + x\right)$

$$= \frac{\tan\frac{\pi}{4} + \tan x}{1 - \tan\frac{\pi}{4}\tan x}$$

$$= \frac{1 + \tan x}{1 - \tan x}$$

33. $\sin(180° - \theta)$

$= \sin 180° \cos\theta - \cos 180° \sin\theta$

$= 0(\cos\theta) - (-1)(\sin\theta)$

$= \sin\theta$

35. $\tan(180° + \theta)$

$$= \frac{\tan 180° + \tan\theta}{1 - \tan 180° \tan\theta}$$

$$= \frac{0 + \tan\theta}{1 - 0}$$

$= \tan\theta$

37. $\sin(\pi + \theta)$

$= \sin\pi\cos\theta + \cos\pi\sin\theta$

$= 0 \cdot \cos\theta + (-1)\sin\theta$

$= -\sin\theta$

39. By the identity $\tan(A - B) = \frac{\tan A - \tan B}{1 + \tan A \tan B}$,

$$\tan(270° - \theta) = \frac{\tan 270° - \tan\theta}{1 + \tan 270° \tan\theta}.$$

But tan 270° is undefined. Therefore, we cannot develop a formula for $\tan(270° - \theta)$.

41. $\cos s = \frac{3}{5}$, $\sin t = \frac{5}{13}$, s and t are in quadrant I.

First find sin s, tan s, cos t, and tan t. Since s and t are in quadrant I, all are positive.

$$\sin s = \sqrt{1 - \left(\frac{3}{5}\right)^2} = \frac{4}{5}$$

$$\tan s = \frac{\cos s}{\sin s} = \frac{\frac{4}{5}}{\frac{3}{5}} = \frac{4}{3}$$

$$\cos t = \sqrt{1 - \left(\frac{5}{13}\right)^2} = \frac{12}{13}$$

$$\tan t = \frac{\frac{5}{13}}{\frac{12}{13}} = \frac{5}{12}$$

$\sin(s + t)$

$$= \frac{4}{5}\cdot\frac{12}{13} + \frac{5}{13}\cdot\frac{3}{5} = \frac{63}{65}$$

$\sin(s - t)$

$$= \frac{4}{5}\cdot\frac{12}{13} - \frac{5}{13}\cdot\frac{3}{5} = \frac{33}{65}$$

$\tan(s + t)$

$$= \frac{\tan s + \tan t}{1 - \tan s \tan t}$$

$$= \frac{\frac{4}{3} + \frac{5}{12}}{1 - \left(\frac{4}{3}\right)\left(\frac{5}{12}\right)} = \frac{63}{16}$$

$\tan(s - t)$

$$= \frac{\tan s - \tan t}{1 + \tan s \tan t}$$

$$= \frac{\frac{4}{3} - \frac{5}{12}}{1 + \left(\frac{4}{5}\right)\left(\frac{5}{12}\right)} = \frac{33}{56}$$

To find the quadrant of s + t, notice that $\sin(s + t) > 0$, which implies s + t is in quadrant I or II. $\tan(s + t) > 0$, which implies s + t is in quadrant I or III. Therefore, s + t is in quadrant I. Then, $\sin(s - t) > 0$ and $\tan(s - t) > 0$. From the preceding, s − t must also be in quadrant I.

43. $\sin s = \frac{2}{3}$, $\sin t = -\frac{1}{3}$, s is in quadrant II, and t is in quadrant IV.

Since s is in quadrant II, $\cos s < 0$.

$$\cos s = -\sqrt{1 - \left(\frac{2}{3}\right)^2} = -\frac{\sqrt{5}}{3}$$

Since t is in quadrant IV, $\cos t > 0$.

$$\cos t = \sqrt{1 - \left(-\frac{1}{3}\right)^2} = \frac{\sqrt{8}}{3} = \frac{2\sqrt{2}}{3}$$

$$\tan s = \frac{\frac{2}{3}}{-\frac{\sqrt{5}}{3}} = -\frac{2}{\sqrt{5}} = -\frac{2\sqrt{5}}{5}$$

$$\tan t = \frac{-\frac{1}{3}}{\frac{2\sqrt{2}}{3}} = -\frac{1}{2\sqrt{2}} = -\frac{\sqrt{2}}{4}$$

$\sin (s + t)$

$$= \frac{2}{3}\cdot\left(\frac{2\sqrt{2}}{3}\right) + \left(-\frac{\sqrt{5}}{3}\right)\left(-\frac{1}{3}\right)$$

$$= \frac{4\sqrt{2}}{9} + \frac{\sqrt{5}}{9} = \frac{4\sqrt{2} + \sqrt{5}}{9}$$

$\sin (s - t)$

$$= \frac{2}{3}\left(\frac{2\sqrt{2}}{3}\right) - \left(-\frac{\sqrt{5}}{3}\right)\left(-\frac{1}{3}\right)$$

$$= \frac{4\sqrt{2}}{9} - \frac{\sqrt{5}}{9} = \frac{4\sqrt{2} - \sqrt{5}}{9}$$

Different forms of $\tan (s + t)$ and $\tan (s - t)$ will be obtained depending on whether $\tan s$ and $\tan t$ are written with rationalized denominators.

$\tan (s + t)$

$$= \frac{-\frac{2\sqrt{5}}{5} + \left(-\frac{\sqrt{2}}{4}\right)}{1 - \left(-\frac{2\sqrt{5}}{5}\right)\left(-\frac{\sqrt{2}}{4}\right)}$$

$$= \frac{-8\sqrt{5} - 5\sqrt{2}}{20 - 2\sqrt{10}}$$

or

$\tan (s + t)$

$$= \frac{-\frac{2}{\sqrt{5}} + \left(-\frac{1}{2\sqrt{2}}\right)}{1 - \left(-\frac{2}{\sqrt{5}}\right)\left(-\frac{1}{2\sqrt{2}}\right)}$$

$$= \frac{-4\sqrt{2} - \sqrt{5}}{2\sqrt{10} - 2} = \frac{4\sqrt{2} + \sqrt{5}}{2 - 2\sqrt{10}}$$

$\tan (s - t)$

$$= \frac{-\frac{2\sqrt{5}}{5} - \left(-\frac{\sqrt{2}}{4}\right)}{1 + \left(-\frac{2\sqrt{5}}{5}\right)\left(-\frac{\sqrt{2}}{4}\right)}$$

$$= \frac{-8\sqrt{5} + 5\sqrt{2}}{20 + 2\sqrt{10}}$$

or

$\tan (s - t)$

$$= \frac{-\frac{2}{\sqrt{5}} - \left(-\frac{1}{2\sqrt{2}}\right)}{1 + \left(-\frac{2}{\sqrt{5}}\right)\left(-\frac{2}{2\sqrt{2}}\right)}$$

$$= \frac{-4\sqrt{2} + \sqrt{5}}{2\sqrt{10} + 2}$$

To find the quadrant of $s + t$, notice that $\sin (s + t) > 0$, which implies $s + t$ is in quadrant I or II. $\tan (s + t) < 0$, which implies $s + t$ is in quadrant II or IV. Therefore, $s + t$ is in quadrant II. Also notice that $\sin (s - t) > 0$, which implies $s - t$ is in quadrant I or II, and $\tan (s - t) < 0$, which implies $s - t$ is in quadrant II or IV. Therefore, $s - t$ is in quadrant II.

45. $\cos s = -\frac{8}{17}$, $\cos t = -\frac{3}{5}$, s and t are in quadrant III.

Since s is in quadrant III, $\sin s < 0$.

$$\sin s = -\sqrt{1 - \left(-\frac{8}{17}\right)^2} = -\frac{15}{17}$$

Since t is in quadrant III, $\sin t < 0$.

$$\sin t = -\sqrt{1 - \left(-\frac{3}{5}\right)^2} = -\frac{4}{5}$$

$$\tan s = \frac{-\frac{15}{17}}{-\frac{8}{17}} = \frac{15}{8}$$

$$\tan t = \frac{-\frac{4}{5}}{-\frac{3}{5}} = \frac{4}{3}$$

$\sin (s + t)$

$$= \left(-\frac{15}{17}\right)\left(-\frac{3}{5}\right) + \left(-\frac{4}{5}\right)\left(-\frac{8}{17}\right) = \frac{77}{85}$$

$\sin (s - t)$

$$= \left(-\frac{15}{17}\right)\left(-\frac{3}{5}\right) - \left(-\frac{4}{5}\right)\left(-\frac{8}{17}\right) = \frac{13}{85}$$

$\tan (s + t)$

$$= \frac{\frac{15}{8} + \frac{4}{3}}{1 - \left(\frac{15}{8}\right)\left(\frac{4}{3}\right)} = \frac{\frac{77}{24}}{-\frac{36}{24}} = -\frac{77}{36}$$

$\tan (s - t)$

$$= \frac{\frac{15}{8} - \frac{4}{3}}{1 + \left(\frac{15}{8}\right)\left(\frac{4}{3}\right)} = \frac{13}{84}$$

To find the quadrant of s + t, notice that $\sin (s + t) > 0$, which implies s + t is in quadrant I or II, and $\tan (s + t) < 0$, which implies s + t is in quadrant II or IV. Therefore, s + t is in quadrant II. Also notice that $\sin (s - t) > 0$, which implies s − t is in quadrant I or II, and $\tan (s - t) > 0$, which implies s − t is in quadrant I or III. Therefore, s − t is in quadrant I.

47. $\sin s = -\frac{4}{5}$, $\cos t = \frac{12}{13}$, s is in quadrant III, t is in quadrant IV.

Since s is in quadrant III, $\cos s < 0$.

$$\cos s = -\sqrt{1 - \left(-\frac{4}{5}\right)^2} = -\frac{3}{5}$$

Since t is in quadrant IV, $\sin t < 0$.

$$\sin t = -\sqrt{1 - \left(\frac{12}{13}\right)^2} = -\frac{5}{13}$$

$$\tan s = \frac{-\frac{4}{5}}{-\frac{3}{5}} = \frac{4}{3}$$

$$\tan t = \frac{-\frac{5}{13}}{\frac{12}{13}} = -\frac{5}{12}$$

$\sin (s + t)$

$$= \left(-\frac{4}{5}\right)\left(\frac{12}{13}\right) + \left(-\frac{5}{13}\right)\left(-\frac{3}{5}\right) = -\frac{33}{65}$$

$\sin (s - t)$

$$= \left(-\frac{4}{5}\right)\left(\frac{12}{13}\right) - \left(-\frac{5}{13}\right)\left(-\frac{3}{5}\right) = -\frac{63}{65}$$

$\tan (s + t)$

$$= \frac{\frac{4}{3} - \frac{5}{12}}{1 - \left(\frac{4}{3}\right)\left(-\frac{5}{12}\right)} = \frac{48 - 15}{36 + 20} = \frac{33}{56}$$

$\tan (s - t)$

$$= \frac{\frac{4}{3} - \left(-\frac{5}{12}\right)}{1 + \left(\frac{4}{3}\right)\left(-\frac{5}{12}\right)} = \frac{48 + 15}{36 - 20} = \frac{63}{16}$$

To find the quadrants of s + t and s − t, notice that $\sin (s + t) < 0$ and $\sin (s - t) < 0$, which implies that s + t and s − t are in quadrant III or IV. Also, notice that $\tan (s + t) > 0$ and $\tan (s - t) > 0$, which implies that both s + t and s − t are in quadrant I or III. Therefore, both s + t and s − t are in quadrant III.

49. $\sin s = -\frac{8}{17}$ and $\cos t = -\frac{8}{17}$, s and t are in quadrant III.

Since s is in quadrant III, $\cos s < 0$.

$$\cos s = -\sqrt{1 - \left(-\frac{8}{17}\right)^2} = -\frac{15}{17}$$

Since t is in quadrant III, $\sin t < 0$.

$$\sin t = -\sqrt{1 - \left(-\frac{8}{17}\right)^2} = -\frac{15}{17}$$

$$\tan s = \frac{-\frac{8}{17}}{-\frac{15}{17}} = \frac{8}{15}$$

$$\tan t = \frac{-\frac{15}{17}}{-\frac{8}{17}} = \frac{15}{8}$$

$\sin (s + t)$

$$= \left(-\frac{8}{17}\right)\left(-\frac{8}{17}\right) + \left(-\frac{15}{17}\right)\left(-\frac{15}{17}\right)$$

$$= \frac{64 + 225}{289} = 1$$

$\sin (s - t)$

$$= \left(-\frac{8}{17}\right)\left(-\frac{8}{17}\right) - \left(-\frac{15}{17}\right)\left(-\frac{15}{17}\right)$$

$$= \frac{64 - 225}{289} = -\frac{161}{289}$$

$\tan (s + t)$

$$= \frac{\frac{8}{15} + \frac{15}{8}}{1 - \frac{8}{15} \cdot \frac{15}{8}} = \frac{64 + 225}{120 - 120}$$

$\tan (s + t)$ is undefined.

$\tan (s - t)$

$$= \frac{\frac{8}{15} - \frac{15}{8}}{1 + 1} = \frac{64 - 225}{2(120)} = -\frac{161}{240}$$

To find the quadrant of s + t, notice that $\sin (s + t) = 1$ and $\tan (s + t)$ is undefined. Therefore, $s + t = 90°$, which means s + t is a quadrantal angle. Also notice that $\sin (s - t) < 0$, which implies s − t is in quadrant III or IV, and $\tan (s - t) < 0$, which implies s − t is in quadrant II or IV. Therefore, s − t is in quadrant IV.

51. $\cos s = -\frac{\sqrt{7}}{4}$, $\sin t = \frac{\sqrt{3}}{5}$, s and t are in quadrant II.

Since s is in quadrant II, $\sin s > 0$.

$$\sin s = \sqrt{1 - \left(-\frac{\sqrt{7}}{4}\right)^2} = \frac{3}{4}$$

Since t is in quadrant II, $\cos t < 0$.

$$\cos t = -\sqrt{1 - \left(\frac{\sqrt{3}}{5}\right)^2} = -\frac{\sqrt{22}}{5}$$

$$\tan s = \frac{\frac{3}{4}}{-\frac{\sqrt{7}}{4}} = -\frac{3}{\sqrt{7}} = -\frac{3\sqrt{7}}{7}$$

$$\tan t = \frac{\frac{\sqrt{3}}{5}}{-\frac{\sqrt{22}}{5}} = -\frac{\sqrt{3}}{\sqrt{22}} = -\frac{\sqrt{66}}{22}$$

$\sin (s + t)$

$$= \left(\frac{3}{4}\right)\left(-\frac{\sqrt{22}}{5}\right) + \left(\frac{\sqrt{3}}{5}\right)\left(-\frac{\sqrt{7}}{4}\right)$$

$$= \frac{-3\sqrt{22} - \sqrt{21}}{20} = \frac{-(3\sqrt{22} + \sqrt{21})}{20}$$

$\sin (s - t)$

$$= \left(\frac{3}{4}\right)\left(-\frac{\sqrt{22}}{5}\right) - \left(\frac{\sqrt{3}}{5}\right)\left(-\frac{\sqrt{7}}{4}\right)$$

$$= \frac{-3\sqrt{22} + \sqrt{21}}{20}$$

tan (s + t)

$$= \frac{\left(-\frac{3\sqrt{7}}{7}\right) + \left(-\frac{\sqrt{66}}{22}\right)}{1 - \left(-\frac{3\sqrt{7}}{7}\right)\left(-\frac{\sqrt{66}}{22}\right)}$$

$$= \frac{-(66\sqrt{7} + 7\sqrt{66})}{154 - 3\sqrt{462}}$$

tan (s − t)

$$= \frac{\left(-\frac{3\sqrt{7}}{7}\right) - \left(-\frac{\sqrt{66}}{22}\right)}{1 + \left(-\frac{3\sqrt{7}}{7}\right)\left(-\frac{\sqrt{66}}{22}\right)}$$

$$= \frac{-66\sqrt{7} + 7\sqrt{66}}{154 + 3\sqrt{462}}$$

To find the quadrant of s + t and s − t, notice that sin (s + t) < 0 and sin (s − t) < 0, which mean s + t and s − t are in quadrant III or IV. Also notice that tan (s + t) < 0 and tan (s − t) < 0, which imply that s + t and s − t are in quadrant II or IV. Therefore, both s + t and s − t are in quadrant IV.

53. Verify $\sin\left(\frac{\pi}{2} + x\right) = \cos x$.

$\sin\left(\frac{\pi}{2} + x\right)$

$$= \sin\frac{\pi}{2}\cos x + \sin x \cos\frac{\pi}{2}$$

$$= 1 \cdot \cos x + \sin x \cdot 0$$

$$= \cos x$$

55. Verify $\tan\left(\frac{\pi}{2} + x\right) = -\cot x$.

$\tan\left(\frac{\pi}{2} + x\right)$

$$= \frac{\sin\left(\frac{\pi}{2} + x\right)}{\cos\left(\frac{\pi}{2} + x\right)}$$

$$= \frac{\sin\frac{\pi}{2}\cos x + \cos\frac{\pi}{2}\sin x}{\cos\frac{\pi}{2}\cos x - \sin\frac{\pi}{2}\sin x}$$

$$= \frac{(1)(\cos x) + (0)(\sin x)}{(0)(\cos x) - (1)(\sin x)}$$

$$= \frac{\cos x}{-\sin x}$$

$$= -\cot x$$

57. Verify sin 2x = 2 sin x cos x.

sin 2x

$$= \sin(x + x)$$

$$= \sin x \cos x + \sin x \cos x$$

$$= 2\sin x \cos x$$

59. Verify $\tan(x - y) - \tan(y - x) = \frac{2(\tan x - \tan y)}{1 + \tan x \tan y}$.

tan (x − y) − tan (y − x)

$$= \frac{\tan x - \tan y}{1 + \tan x \tan y} - \frac{\tan y - \tan x}{1 + \tan y \tan x}$$

$$= \frac{\tan x - \tan y - \tan y + \tan x}{1 + \tan x \tan y}$$

$$= \frac{2\tan x - 2\tan y}{1 + \tan x \tan y}$$

$$= \frac{2(\tan x - \tan y)}{1 + \tan x \tan y}$$

61. Verify $\frac{\cos(\alpha - \beta)}{\cos\alpha \sin\beta} = \tan\alpha + \cot\beta$.

$\frac{\cos(\alpha - \beta)}{\cos\alpha \sin\beta}$

$$= \frac{\cos\alpha\cos\beta + \sin\alpha\sin\beta}{\cos\alpha\sin\beta}$$

$$= \frac{\cos\beta}{\sin\beta} + \frac{\sin\alpha}{\cos\alpha}$$

$$= \cot\beta + \tan\alpha$$

63. Verify $\frac{\sin(x - y)}{\sin(x + y)} = \frac{\tan x - \tan y}{\tan x + \tan y}$.

Work with the right side.

$$\frac{\tan x - \tan y}{\tan x + \tan y}$$

$$= \frac{\frac{\sin x}{\cos x} - \frac{\sin y}{\cos y}}{\frac{\sin x}{\cos x} + \frac{\sin y}{\cos y}}$$

$$= \frac{\sin x \cos y - \cos x \sin y}{\sin x \cos y + \cos x \sin y}$$

$$= \frac{\sin(x - y)}{\sin(x + y)}$$

65. Verify $\frac{\sin(s - t)}{\sin t} + \frac{\cos(s - t)}{\cos t} = \frac{\sin s}{\sin t \cos t}$.

$$\frac{\sin(s - t)}{\sin t} + \frac{\cos(s - t)}{\cos t}$$

$$= \frac{\sin s \cos t - \sin t \cos s}{\sin t} + \frac{\cos s \cos t + \sin s \sin t}{\cos t}$$

$$= \frac{\sin s \cos^2 t - \sin t \cos t \cos s}{\sin t \cos t} + \frac{\sin t \cos t \cos s + \sin^2 t \sin s}{\sin t \cos t}$$

$$= \frac{\sin s \cos^2 t + \sin s \sin^2 t}{\sin t \cos t}$$

$$= \frac{\sin s(\cos^2 t + \sin^2 t)}{\sin t \cos t}$$

$$= \frac{\sin s}{\sin t \cos t}$$

67. $\sin 165°$

$$= \sin(180° - 15°)$$

$$= \sin 180° \cos 15° - \cos 180° \sin 15°$$

$$= \sin 15° = \sin(45° - 30°)$$

$$= \sin 45° \cos 30° - \cos 45° \sin 30°$$

$$= \frac{\sqrt{2}}{2} \cdot \frac{\sqrt{3}}{2} - \frac{\sqrt{2}}{2} \cdot \frac{1}{2}$$

$$= \frac{\sqrt{6} - \sqrt{2}}{4}$$

69. $\tan 255°$

$$= \tan(180° + 75°)$$

$$= \frac{\tan 180° + \tan 75°}{1 + \tan 180° \tan 75°}$$

$$= \tan 75° = \tan(45° + 30°)$$

$$= \frac{\tan 45° + \tan 30°}{1 - \tan 45° \tan 30°}$$

$$= \frac{1 + \frac{\sqrt{3}}{3}}{1 - \frac{\sqrt{3}}{3}}$$

$$= \frac{3 + \sqrt{3}}{3 - \sqrt{3}} \cdot \frac{3 + \sqrt{3}}{3 + \sqrt{3}}$$

$$= \frac{12 + 6\sqrt{3}}{6} = 2 + \sqrt{3}$$

71. $\sin 285°$

$$= \sin(360° - 75°)$$

$$= \sin 360° \cos 75° - \cos 360° \sin 75°$$

$$= -\sin 75° = -\sin(30° + 45°)$$

$$= -(\sin 30° \cos 45° + \cos 30° \sin 45°)$$

$$= -\left(\frac{1}{2} \cdot \frac{\sqrt{2}}{2} + \frac{\sqrt{3}}{2} \cdot \frac{\sqrt{2}}{2}\right)$$

$$= \frac{-\sqrt{2} - \sqrt{6}}{4}$$

73. $\sin \frac{11\pi}{12}$

$$= \sin\left(\pi - \frac{\pi}{12}\right)$$

$$= \sin \pi \cos \frac{\pi}{12} - \cos \pi \sin \frac{\pi}{12}$$

$$= \sin \frac{\pi}{12} = \sin\left(\frac{\pi}{4} - \frac{\pi}{6}\right)$$

$$= \sin \frac{\pi}{4} \cos \frac{\pi}{6} - \cos \frac{\pi}{4} \sin \frac{\pi}{6}$$

$$= \frac{\sqrt{2}}{2} \cdot \frac{\sqrt{3}}{2} - \frac{\sqrt{2}}{2} \cdot \frac{1}{2}$$

$$= \frac{\sqrt{6} - \sqrt{2}}{4}$$

75. $\tan\left(-\frac{13\pi}{12}\right)$

$= -\tan\left(\pi + \frac{\pi}{12}\right)$

$= -\dfrac{\tan \pi + \tan \frac{\pi}{12}}{1 + \tan \pi \tan \frac{\pi}{12}}$

$= -\tan \frac{\pi}{12} = -\tan\left(\frac{\pi}{4} - \frac{\pi}{6}\right)$

$= -\dfrac{\tan \frac{\pi}{4} - \tan \frac{\pi}{6}}{1 + \tan \frac{\pi}{4} \tan \frac{\pi}{6}}$

$= -\dfrac{1 - \frac{\sqrt{3}}{3}}{1 + \frac{\sqrt{3}}{3}} = -\dfrac{3 - \sqrt{3}}{3 + \sqrt{3}} \cdot \dfrac{3 - \sqrt{3}}{3 - \sqrt{3}}$

$= -\dfrac{12 - 6\sqrt{3}}{6} = -2 + \sqrt{3}$

77. Verify $\tan (A - B) = \dfrac{\tan A - \tan B}{1 + \tan A \tan B}$.

$\tan (A - B)$

$= \tan [A + (-B)]$

$= \dfrac{\tan A + \tan (-B)}{1 - \tan A \tan (-B)}$

(Remember that $\tan (-B) = -\tan B$.)

$= \dfrac{\tan A - \tan B}{1 - \tan A (-\tan B)}$

$= \dfrac{\tan A - \tan B}{1 + \tan A \tan B}$

79. $\sin (A + B + C)$

$= \sin (A + B) \cos C + \sin C \cos (A + B)$

$= (\sin A \cos B + \sin B \cos A) \cdot \cos C$

$+ \sin C \cdot (\cos A \cos B - \sin A \sin B)$

$= \sin A \cos B \cos C + \sin B \cos A \cos C$

$+ \sin C \cos A \cos B - \sin C \sin A \sin B$

81. Let $f(x) = \sin x$. Show that

$\dfrac{f(x + h) - f(x)}{h}$

$= \sin x \left(\dfrac{\cos h - 1}{h}\right) + \cos x \left(\dfrac{\sin h}{h}\right)$.

$\dfrac{f(x + h) - f(x)}{h}$

$= \dfrac{\sin (x + h) - \sin x}{h}$

$= \dfrac{\sin x \cos h + \sin h \cos x - \sin x}{h}$

$= \sin x \left(\dfrac{\cos h - 1}{h}\right) + \cos x \left(\dfrac{\sin h}{h}\right)$

83. $x + y = 9$, $2x + y = -1$

Change the equations of the given lines to slope-intercept form to find their slopes.

$x + y = 9$

$y = -x + 9$

$m_2 = -1$

$2x + y = -1$

$y = -2x - 1$

$m_1 = -2$

$\tan \theta = \dfrac{m_2 - m_1}{1 + m_2 m_1}$

$= \dfrac{-1 - (-2)}{1 + (-1)(-2)}$

$= \dfrac{1}{1 + 2}$

$= \dfrac{1}{3}$

$\theta = 18.4°$

Section 5.5

1. $\cos 2\theta = \frac{3}{5}$, θ is in quadrant I.

$\cos 2\theta = 2 \cos^2 \theta - 1$

$\frac{3}{5} = 2 \cos^2 \theta - 1$

$2 \cos^2 \theta = \frac{3}{5} + 1 = \frac{8}{5}$

$\cos^2 \theta = \frac{8}{10} = \frac{4}{5}$

Since θ is in quadrant I, $\cos\theta > 0$.

$$\cos\theta = \frac{2}{\sqrt{5}} = \frac{2\sqrt{5}}{5}$$

Since θ is in quadrant I, $\sin\theta > 0$.

$$\sin\theta = \sqrt{1 - \cos^2\theta}$$
$$= \sqrt{1 - \left(\frac{2\sqrt{5}}{5}\right)^2} = \sqrt{\frac{1}{5}}$$
$$= \frac{1}{\sqrt{5}} = \frac{\sqrt{5}}{5}$$

$$\tan\theta = \frac{\sin\theta}{\cos\theta} = \frac{\frac{\sqrt{5}}{5}}{\frac{2\sqrt{5}}{5}} = \frac{1}{2}$$

$$\cot\theta = \frac{1}{\tan\theta} = \frac{1}{\frac{1}{2}} = 2$$

$$\sec\theta = \frac{1}{\cos\theta} = \frac{1}{\frac{2\sqrt{5}}{5}} = \frac{5}{2\sqrt{5}} = \frac{\sqrt{5}}{2}$$

$$\csc\theta = \frac{1}{\sin\theta} = \frac{1}{\frac{\sqrt{5}}{5}} = \sqrt{5}$$

3. $\cos 2x = -\frac{5}{12}, \ \frac{\pi}{2} < x < \pi$

$$\cos 2x = 2\cos^2 x - 1$$
$$2\cos^2 x = \cos 2x + 1$$
$$= -\frac{5}{12} + 1 = \frac{7}{12}$$

$$\cos^2 x = \frac{7}{24}$$

Since $\frac{\pi}{2} < x < \pi$, $\cos x < 0$.

$$\cos x = -\sqrt{\frac{7}{24}} = -\frac{\sqrt{7}}{2\sqrt{6}} = -\frac{\sqrt{42}}{12}$$

$$\sin^2 x = 1 - \cos^2 x = 1 - \frac{7}{24} = \frac{17}{24}$$

Since $\frac{\pi}{2} < x < \pi$, $\sin x > 0$.

$$\sin x = \frac{\sqrt{17}}{2\sqrt{6}} = \frac{\sqrt{102}}{12}$$

$$\tan x = \frac{\frac{\sqrt{17}}{2\sqrt{6}}}{-\frac{\sqrt{7}}{2\sqrt{6}}} = -\frac{\sqrt{17}}{\sqrt{7}} = -\frac{\sqrt{119}}{7}$$

$$\cot x = -\frac{\sqrt{7}}{\sqrt{17}} = -\frac{\sqrt{119}}{17}$$

$$\sec x = -\frac{2\sqrt{6}}{\sqrt{7}} = -\frac{2\sqrt{42}}{7}$$

$$\csc x = \frac{2\sqrt{6}}{\sqrt{17}} = \frac{2\sqrt{102}}{17}$$

5. $\sin\theta = \frac{2}{5}, \ \cos\theta < 0$

$$\cos 2\theta = 1 - 2\sin^2\theta = 1 - 2\left(\frac{2}{5}\right)^2$$
$$= 1 - \frac{8}{25} = \frac{17}{25}$$

$$\cos^2 2\theta + \sin^2 2\theta = 1$$
$$\sin^2 2\theta = 1 - \cos^2 2\theta$$
$$= 1 - \left(\frac{17}{25}\right)^2 = \frac{336}{625}$$

Since $\cos\theta < 0$, $\sin 2\theta < 0$ because $\sin 2\theta = 2\sin\theta\cos\theta < 0$.

$$\sin 2\theta = -\sqrt{\frac{336}{625}} = -\frac{4\sqrt{21}}{25}$$

$$\tan 2\theta = \frac{\sin 2\theta}{\cos 2\theta} = \frac{-\frac{4\sqrt{21}}{25}}{\frac{17}{25}} = -\frac{4\sqrt{21}}{17}$$

$$\cot 2\theta = -\frac{17}{4\sqrt{21}} = -\frac{17\sqrt{21}}{84}$$

$$\sec 2\theta = \frac{25}{17}$$

$$\csc 2\theta = -\frac{25}{4\sqrt{21}} = -\frac{25\sqrt{21}}{84}$$

7. $\tan x = 2$, $\cos x > 0$

$$\tan 2x = \frac{2 \tan x}{1 - \tan^2 x}$$

$$= \frac{2(2)}{1 - (2)^2} = -\frac{4}{3}$$

Since both tan x and cos x are positive, x must be in quadrant I. Thus, 2x must be in quadrant II.

$$\sec^2 2x = 1 + \tan^2 2x = 1 + \frac{16}{9}$$

$$= \frac{25}{9}$$

$$\sec 2x = -\frac{5}{3}$$

$$\cos 2x = \frac{1}{\sec 2x} = -\frac{3}{5}$$

$$\cot 2x = \frac{2}{\tan 2x} = -\frac{3}{4}$$

$$\sin 2x = \tan 2x \cos 2x$$

$$= \left(-\frac{4}{3}\right)\left(-\frac{3}{5}\right) = \frac{4}{5}$$

$$\csc 2x = \frac{1}{\sin 2x} = \frac{5}{4}$$

9. $\sin \alpha = -\frac{\sqrt{5}}{7}$, $\cos \alpha > 0$

$$\cos^2 \alpha = 1 - \sin^2 \alpha$$

$$= 1 - \left(-\frac{\sqrt{5}}{7}\right)^2 = \frac{44}{49}$$

Since $\cos \alpha > 0$,

$$\cos \alpha = \sqrt{\frac{44}{49}} = \frac{\sqrt{44}}{7} = \frac{2\sqrt{11}}{7}.$$

$$\cos 2\alpha = 1 - 2 \sin^2 \alpha$$

$$= 1 - 2\left(-\frac{\sqrt{5}}{7}\right)^2 = \frac{39}{49}$$

$$\sin 2\alpha = 2 \sin \alpha \cos \alpha$$

$$= 2\left(-\frac{\sqrt{5}}{7}\right)\left(\frac{2\sqrt{11}}{7}\right) = -\frac{4\sqrt{55}}{49}$$

$$\tan 2\alpha = \frac{\sin 2\alpha}{\cos 2\alpha} = \frac{-\frac{4\sqrt{55}}{49}}{\frac{39}{49}} = -\frac{4\sqrt{55}}{39}$$

$$\cot 2\alpha = \frac{1}{\tan 2\alpha} = -\frac{39\sqrt{55}}{220}$$

$$\sec 2\alpha = \frac{1}{\cos 2\alpha} = \frac{49}{39}$$

$$\csc 2\alpha = \frac{1}{\sin 2\alpha} = -\frac{49\sqrt{55}}{220}$$

11. $2 \cos^2 15° - 1$

Since $\cos 2A = 2 \cos^2 A - 1$,

$$2 \cos^2 15° - 1 = \cos 2(15°)$$

$$= \cos 30° = \frac{\sqrt{3}}{2}.$$

13. $\dfrac{2 \tan 15°}{1 - \tan^2 15°}$

$$= \frac{\tan 15° + \tan 15°}{1 - \tan^2 15°}$$

$$= \tan (15° + 15°)$$

$$= \tan 30° = \frac{1}{\sqrt{3}} = \frac{\sqrt{3}}{3}$$

15. $2 \sin \frac{\pi}{3} \cos \frac{\pi}{3} = \sin 2\left(\frac{\pi}{3}\right)$

$$= \sin \frac{2\pi}{3} = \frac{\sqrt{3}}{2}$$

17. $1 - 2 \sin^2 22\frac{1}{2}°$

$$= \cos 2\left(22\frac{1}{2}°\right) = \cos 45° = \frac{\sqrt{2}}{2}$$

19. $2 \cos^2 67\frac{1}{2}° - 1$

$$= \cos^2 67\frac{1}{2}° - \sin^2 67\frac{1}{2}°$$

$$= \cos 2\left(67\frac{1}{2}°\right) = \cos 135° = -\frac{\sqrt{2}}{2}$$

21. $\sin \frac{\pi}{8} \cos \frac{\pi}{8}$

Since $\sin 2A = 2 \sin A \cos A$,

$$\frac{1}{2} \sin 2A = \sin A \cos A.$$

$$\sin \frac{\pi}{8} \cos \frac{\pi}{8} = \frac{1}{2} \sin 2\left(\frac{\pi}{8}\right)$$
$$= \frac{1}{2} \sin \frac{\pi}{4}$$
$$= \frac{1}{2} \cdot \frac{\sqrt{2}}{2} = \frac{\sqrt{2}}{4}$$

23. $\dfrac{\tan 51°}{1 - \tan^2 51°}$

Since $\dfrac{2 \tan A}{1 - \tan^2 A} = \tan 2A$,

then

$$\frac{1}{2}\left(\frac{2 \tan A}{1 - \tan^2 A}\right) = \frac{1}{2} \tan 2A$$
$$\frac{\tan A}{1 - \tan^2 A} = \frac{1}{2} \tan 2A.$$
$$\frac{\tan 51°}{1 - \tan^2 51°} = \frac{1}{2} \tan 2(51°)$$
$$= \frac{1}{2} \tan 102°$$

25. $\frac{1}{4} - \frac{1}{2} \sin^2 47.1° = \frac{1}{4}(1 - 2 \sin^2 47.1°)$

Since $\cos 2A = 1 - 2 \sin^2 A$,

$$\frac{1}{4}(1 - 2 \sin^2 47.1°)$$
$$= \frac{1}{4} \cos 2(47.1°)$$
$$= \frac{1}{4} \cos 94.2°.$$

27. $\sin^2 \frac{2\pi}{5} - \cos^2 \frac{2\pi}{5} = -\left(\cos^2 \frac{2\pi}{5} - \sin^2 \frac{2\pi}{5}\right)$

Since $\cos^2 A - \sin^2 A = \cos 2A$,

$$-\left(\cos^2 \frac{2\pi}{5} - \sin^2 \frac{2\pi}{5}\right)$$
$$= -\cos\left(2 \cdot \frac{2\pi}{5}\right) = -\cos \frac{4\pi}{5}.$$

29. $2 \sin 5x \cos 5x$

Since $2 \sin A \cos A = \sin 2A$,

$$2 \sin 5x \cos 5x = \sin 2(5x)$$
$$= \sin 10x.$$

31. $\cos^2 2\alpha - \sin^2 2\alpha$

Since $\cos^2 A - \sin^2 A = \cos 2A$,

$$\cos^2 2\alpha - \sin^2 2\alpha = \cos 2(2\alpha)$$
$$= \cos 4\alpha.$$

33. $\dfrac{2 \tan \frac{x}{9}}{1 - \tan^2 \frac{x}{9}} = \tan \dfrac{2x}{9}$

37. $\sin 2(45°) = \sin 90° = 1$

$$\sin 2(45°) = 2 \sin 45° \cos 45°$$
$$= 2\left(\frac{\sqrt{2}}{2} \cdot \frac{\sqrt{2}}{2}\right) = 1$$

39. $\cos 2(60°) = \cos 120° = -\frac{1}{2}$

$$\cos 2(60°) = \cos^2 60 - \sin^2 60°$$
$$= \left(\frac{1}{2}\right)^2 - \left(\frac{\sqrt{3}}{2}\right)^2 = -\frac{1}{2}$$

41. $\cos 2\left(\frac{5\pi}{3}\right) = \cos \frac{10\pi}{3} = \cos \frac{4\pi}{3}$

$$= -\cos \frac{\pi}{3}$$
$$= -\frac{1}{2}$$

$$\cos 2\left(\frac{5\pi}{3}\right) = \cos^2\left(\frac{5\pi}{3}\right) - \sin^2\left(\frac{5\pi}{3}\right)$$
$$= \cos^2\left(\frac{\pi}{3}\right) - \left(-\sin \frac{\pi}{3}\right)^2$$
$$= \left(\frac{1}{2}\right)^2 - \left(-\frac{\sqrt{3}}{2}\right)^2$$
$$= \frac{1}{4} - \frac{3}{4}$$
$$= -\frac{2}{4}$$
$$= -\frac{1}{2}$$

43. $\tan 2\left(-\frac{\pi}{3}\right) = \tan\left(-\frac{2\pi}{3}\right)$

$$= \sqrt{3}$$

$$\tan 2\left(-\frac{\pi}{3}\right) = -\tan 2\left(\frac{\pi}{3}\right)$$

$$= -\frac{2\tan\frac{\pi}{3}}{1-\tan^2\frac{\pi}{3}} = -\frac{2\sqrt{3}}{1-(\sqrt{3})^2}$$

$$= -\frac{2\sqrt{3}}{1-3} = \sqrt{3}$$

45. $\tan 2\left(-\frac{4\pi}{3}\right) = \tan\left(-\frac{8\pi}{3}\right) = \sqrt{3}$

$$\tan 2\left(-\frac{4\pi}{3}\right) = -\tan 2\left(\frac{4\pi}{3}\right)$$

$$= -\frac{2\tan\frac{4\pi}{3}}{1-\tan^2\frac{4\pi}{3}}$$

$$= -\frac{2\sqrt{3}}{1-(\sqrt{3})^2} = -\frac{2\sqrt{3}}{1-3}$$

$$= \sqrt{3}$$

47. $\sin 2\left(-\frac{11\pi}{2}\right) = \sin(-11\pi) = 0$

$$\sin 2\left(-\frac{11\pi}{2}\right) = 2\sin\left(-\frac{11\pi}{2}\right)\cos\left(-\frac{11\pi}{2}\right)$$

$$= 2(0)(-1) = 0$$

49. Verify $(\sin\gamma + \cos\gamma)^2 = \sin 2\gamma + 1$.

$$(\sin\gamma + \cos\gamma)^2$$

$$= \sin^2\gamma + \cos^2\gamma + 2\sin\gamma\cos\gamma$$

$$= 1 + \sin 2\gamma$$

51. Verify $\sec 2x = \frac{\sec^2 x + \sec^4 x}{2 + \sec^2 x - \sec^4 x}$.

Work with the right side.

$$\frac{\sec^2 x + \sec^4 x}{2 + \sec^2 x - \sec^4 x}$$

$$= \frac{\left(\frac{1}{\cos^2 x} + \frac{1}{\cos^4 x}\right)\cos^4 x}{\left(2 + \frac{1}{\cos^2 x} - \frac{1}{\cos^4 x}\right)\cos^4 x}$$

$$= \frac{\cos^2 x + 1}{2\cos^4 x + \cos^2 x - 1}$$

$$= \frac{\cos^2 x + 1}{(2\cos^2 x - 1)(\cos^2 x + 1)}$$

$$= \frac{1}{2\cos^2 x - 1} = \frac{1}{\cos 2x} = \sec 2x$$

53. Verify $\cot 2\beta = \frac{\cot^2\beta - 1}{2\cot\beta}$.

$$\cot 2\beta = \frac{1}{\tan 2\beta}$$

$$= \frac{1}{\frac{2\tan\beta}{1-\tan^2\beta}}$$

$$= \frac{(1-\tan^2\beta)\cot^2\beta}{2\tan\beta\cot^2\beta}$$

$$= \frac{\cot^2\beta - 1}{2\cot\beta}$$

55. Verify $\sin 2\gamma = \frac{2\tan\gamma}{1+\tan^2\gamma}$.

Work with the right side.

$$\frac{2\tan\gamma}{1+\tan^2\gamma} = \frac{2\tan\gamma}{1+\frac{\sin^2\gamma}{\cos^2\gamma}}$$

$$= \frac{2\frac{\sin\gamma}{\cos\gamma}}{\frac{\cos^2\gamma + \sin^2\gamma}{\cos^2\gamma}}$$

$$= 2\sin\gamma\cos\gamma$$

$$= \sin 2\gamma$$

57. Verify $\cos 2y = \frac{2 - \sec^2 y}{\sec^2 y}$.

Work with the right side.

$$\frac{2 - \sec^2 y}{\sec^2 y}$$

$$= \frac{\left(2 - \frac{1}{\cos^2 y}\right)\cos^2 y}{\left(\frac{1}{\cos^2 y}\right)\cos^2 y}$$

$$= 2\cos^2 y - 1$$

$$= \cos 2y$$

59. Verify $\sin 4\alpha = 4 \sin \alpha \cos \alpha \cos 2\alpha$.

Work with the right side.

$$4 \sin \alpha \cos \alpha \cos 2\alpha$$
$$= 2(2 \sin \alpha \cos \alpha) \cos 2\alpha$$
$$= 2 \sin 2\alpha \cos 2\alpha$$
$$= \sin 2(2\alpha)$$
$$= \sin 4\alpha$$

61. Verify $\tan(\theta - 45°) + \tan(\theta + 45°) = 2 \tan 2\theta$.

$$\tan(\theta - 45°) + \tan(\theta + 45°)$$
$$= \frac{\tan \theta - \tan 45°}{1 + \tan \theta \tan 45°} + \frac{\tan \theta + \tan 45°}{1 - \tan \theta \tan 45°}$$
$$= \frac{\tan \theta - 1}{1 + \tan \theta} + \frac{\tan \theta + 1}{1 - \tan \theta}$$
$$= \frac{\tan \theta - 1}{\tan \theta + 1} - \frac{\tan \theta + 1}{\tan \theta - 1}$$
$$= \frac{\tan^2 \theta - 2 \tan \theta + 1}{(\tan \theta + 1)(\tan \theta - 1)} - \frac{\tan^2 \theta + 2 \tan \theta + 1}{(\tan \theta + 1)(\tan \theta - 1)}$$
$$= \frac{-4 \tan \theta}{\tan^2 \theta - 1} = \frac{4 \tan \theta}{1 - \tan^2 \theta}$$
$$= \frac{2(2 \tan \theta)}{1 - \tan^2 \theta}$$
$$= 2 \tan 2\theta$$

63. Verify $\frac{2 \cos 2\alpha}{\sin 2\alpha} = \cot \alpha - \tan \alpha$.

Work with the right side.

$$\cot \alpha - \tan \alpha$$
$$= \frac{\cos \alpha}{\sin \alpha} - \frac{\sin \alpha}{\cos \alpha}$$
$$= \frac{\cos^2 \alpha - \sin^2 \alpha}{\sin \alpha \cos \alpha}$$
$$= \frac{2(\cos^2 \alpha - \sin^2 \alpha)}{2 \sin \alpha \cos \alpha}$$
$$= \frac{2 \cos 2\alpha}{\sin 2\alpha}$$

65. Verify $\sin 2\alpha \cos 2\alpha = \sin 2\alpha - 4 \sin^3 \alpha \cos \alpha$.

$$\sin 2\alpha \cos 2\alpha$$
$$= (2 \sin \alpha \cos \alpha)(1 - 2 \sin^2 \alpha)$$
$$= 2 \sin \alpha \cos \alpha - 4 \sin^3 \alpha \cos \alpha$$
$$= \sin 2\alpha - 4 \sin^3 \alpha \cos \alpha$$

67. Verify $\cos 2x = \frac{1 - \tan^2 x}{1 + \tan^2 x}$.

Work with the right side.

$$\frac{1 - \tan^2 x}{1 + \tan^2 x} = \frac{1 - \frac{\sin^2 x}{\cos^2 x}}{1 + \frac{\sin^2 x}{\cos^2 x}}$$
$$= \frac{\frac{\cos^2 x - \sin^2 x}{\cos^2 x}}{\frac{\cos^2 x + \sin^2 x}{\cos^2 x}}$$
$$= \cos^2 x - \sin^2 x$$
$$= \cos 2x$$

69. Verify $\tan s + \cot s = 2 \csc 2s$.

$$\tan s + \cot s$$
$$= \frac{\sin s}{\cos s} + \frac{\cos s}{\sin s}$$
$$= \frac{\sin^2 s + \cos^2 s}{\cos s \sin s}$$
$$= \frac{1}{\cos s \sin s}$$
$$= \frac{2}{2 \sin s \cos s}$$
$$= \frac{2}{\sin 2s}$$
$$= 2 \csc 2s$$

71. Verify $1 + \tan x \tan 2x = \sec 2x$.

$$1 + \tan x \tan 2x$$
$$= 1 + \tan x\left(\frac{2 \tan x}{1 - \tan^2 x}\right)$$
$$= 1 + \frac{2 \tan^2 x}{1 - \tan^2 x}$$
$$= \frac{1 - \tan^2 x + 2 \tan^2 x}{1 - \tan^2 x}$$

$$= \frac{1 + \tan^2 x}{1 - \tan^2 x}$$

$$= \frac{\frac{\cos^2 x + \sin^2 x}{\cos^2 x}}{\frac{\cos^2 x - \sin^2 x}{\cos^2 x}}$$

$$= \frac{1}{\cos^2 x - \sin^2 x}$$

$$= \frac{1}{\cos 2x}$$

$$= \sec 2x$$

In Exercises 73–79, other forms may be possible.

73. $\tan^2 2x = \left(\frac{2 \tan x}{1 - \tan^2 x}\right)^2$

$$= \frac{4 \tan^2 x}{1 - 2 \tan^2 x + \tan^4 x}$$

75. $\cos 3x$

$$= \cos (x + 2x)$$
$$= \cos x \cos 2x - \sin x \sin 2x$$
$$= \cos x (2 \cos^2 x - 1) - \sin x (2 \sin x \cos x)$$
$$= 2 \cos^3 x - \cos x - 2 \sin^2 x \cos x$$
$$= 2 \cos^3 x - \cos x - 2(1 - \cos^2 x) \cos x$$
$$= 2 \cos^3 x - \cos x - 2 \cos x + 2 \cos^3 x$$
$$= 4 \cos^3 x - 3 \cos x$$

77. $\tan 3x$

$$= \tan (2x + x)$$

$$= \frac{\tan 2x + \tan x}{1 - \tan 2x \tan x}$$

$$= \frac{\frac{2 \tan x}{1 - \tan^2 x} + \tan x}{1 - \frac{2 \tan x}{1 - \tan^2 x} \tan x}$$

$$= \frac{2 \tan x + \tan x - \tan^3 x}{1 - \tan^2 x - 2 \tan^2 x}$$

$$= \frac{3 \tan x - \tan^3 x}{1 - 3 \tan^2 x}$$

79. $\tan 4x$

$$= \frac{2 \tan 2x}{1 - \tan^2 2x}$$

$$= \frac{2\left(\frac{2 \tan x}{1 - \tan^2 x}\right)}{1 - \left(\frac{2 \tan x}{1 - \tan^2 x}\right)^2}$$

$$= \frac{\frac{4 \tan x}{1 - \tan^2 x}}{1 - \frac{4 \tan^2 x}{1 - 2 \tan^2 x + \tan^4 x}}$$

$$= \frac{4 \tan x(1 - \tan^2 x)}{1 - 2 \tan^2 x + \tan^4 x - 4 \tan^2 x}$$

$$= \frac{4 \tan x - 4 \tan^3 x}{1 - 6 \tan^2 x + \tan^4 x}$$

$$= \frac{4(\tan x - \tan^3 x)}{1 - 6 \tan^2 x + \tan^4 x}$$

81. $\sin^2 2x + \cos^2 2x$

Use the Pythagorean identity $\sin^2 \theta + \cos^2 \theta = 1$, and let $\theta = 2x$.

$$\sin^2 2x + \cos^2 2x = 1.$$

83. $\cot^2 3r + 1$

Use the Pythagorean identity $1 + \cot^2 \theta = \csc^2 \theta$, with $\theta = 3r$.

$$\cot^2 3r + 1 = \csc^2 3r$$

85. $\phi = 47° 12'$, $h = 387$ ft

Use a calculator and substitute these values in

$$g = 978.0524(1 + .005297 \sin^2 \phi - .0000059 \sin^2 2\phi) - .000094h.$$

$$g = 978.0524[1 + .005297 \sin^2 47° 12' - .0000059 \sin^2 (2 \cdot 47° 12')] - .000094(387)$$

$$= 980.799 \text{ cm per sec per sec}$$

Section 5.6

1. $\sin 15° = \sin\left(\frac{30°}{2}\right)$

Since 15° is in quadrant I, $\sin 15° > 0$.

$$\sin 15° = \sqrt{\frac{1 - \cos 30°}{2}}$$

$$= \sqrt{\frac{1 - \frac{\sqrt{3}}{2}}{2}}$$

$$= \sqrt{\frac{2 - \sqrt{3}}{4}}$$

$$= \frac{\sqrt{2 - \sqrt{3}}}{2}$$

3. $\cos \frac{\pi}{8} = \cos\left(\frac{\frac{\pi}{4}}{2}\right)$

Since $\pi/8$ is in quadrant I, $\cos \pi/8 > 0$.

$$\cos \frac{\pi}{8} = \sqrt{\frac{1 + \cos\left(\frac{\pi}{4}\right)}{2}}$$

$$= \sqrt{\frac{1 + \frac{\sqrt{2}}{2}}{2}}$$

$$= \sqrt{\frac{2 + \sqrt{2}}{4}}$$

$$= \frac{\sqrt{2 + \sqrt{2}}}{2}$$

5. $\tan 67.5° = \tan\left(\frac{135°}{2}\right)$

$$= \frac{\sin 135°}{1 + \cos 135°}$$

$$= \frac{\sin 45°}{1 + (-\cos 45°)}$$

$$= \frac{\frac{\sqrt{2}}{2}}{1 - \frac{\sqrt{2}}{2}}$$

$$= \frac{\sqrt{2}}{2 - \sqrt{2}} \cdot \frac{2 + \sqrt{2}}{2 + \sqrt{2}}$$

$$= \frac{2\sqrt{2} + 2}{2}$$

$$= \sqrt{2} + 1$$

7. $\sin 67.5° = \sin\left(\frac{135°}{2}\right)$

Since 67.5° is in quadrant I, $\sin 67.5° > 0$.

$$\sin 67.5° = \sqrt{\frac{1 - \cos 135°}{2}}$$

$$= \sqrt{\frac{1 + \cos 45°}{2}}$$

$$= \sqrt{\frac{1 + \frac{\sqrt{2}}{2}}{2}} = \sqrt{\frac{2 + \sqrt{2}}{4}}$$

$$= \frac{\sqrt{2 + \sqrt{2}}}{2}$$

9. $\cos 195° = \cos\left(\frac{390°}{2}\right)$

Since 195° is in quadrant III, $\cos 195° < 0$.

$$\cos 195° = -\sqrt{\frac{1 + \cos 390°}{2}}$$

$$= -\sqrt{\frac{1 + \cos 30°}{2}}$$

$$= -\sqrt{\frac{1 + \frac{\sqrt{3}}{2}}{2}} = -\sqrt{\frac{2 + \sqrt{3}}{4}}$$

$$= \frac{-\sqrt{2 + \sqrt{3}}}{2}$$

11. $\cos 165° = \cos\left(\frac{330°}{2}\right)$

Since 165° is in quadrant II, $\cos 165° < 0$.

$$\cos 165° = -\sqrt{\frac{1 + \cos 330°}{2}}$$

$$= -\sqrt{\frac{1 + \cos 30°}{2}}$$

$$= -\sqrt{\frac{1 + \frac{\sqrt{3}}{2}}{2}} = -\sqrt{\frac{2 + \sqrt{3}}{4}}$$

$$= \frac{-\sqrt{2 + \sqrt{3}}}{2}$$

13. To find sin 7.5°, first notice that 7.5° is in quadrant I. Therefore $\sin 7.5° > 0$. Next notice that $7.5° = \frac{1}{2} \cdot \frac{1}{2} \cdot 30°$.

$$\sin 15° = \sin\left(\frac{30°}{2}\right)$$

$$= \sqrt{\frac{1 - \cos 30°}{2}}$$

$$= \sqrt{\frac{1 - \frac{\sqrt{3}}{2}}{2}}$$

$$= \frac{\sqrt{2 - \sqrt{3}}}{2}$$

Since

$$\sin 7.5° = \sin\left(\frac{15°}{2}\right) = \sqrt{\frac{1 - \cos 15°}{2}},$$

we can find cos 15° by using the identity $\sin^2 \theta + \cos^2 \theta = 1$ and using

$$\sin 15° = \frac{\sqrt{2 - \sqrt{3}}}{2}.$$

15. Find $\cos \theta/2$, given $\cos \theta = 1/4$, with $0 < \theta < \pi/2$.

$$0 < \theta < \frac{\pi}{2}$$

$$0 < \frac{\theta}{2} < \frac{\pi}{4}$$

Thus, $\cos \theta/2 > 0$.

$$\cos \frac{\theta}{2} = \sqrt{\frac{1 + \cos \theta}{2}}$$

$$= \sqrt{\frac{1 + \frac{1}{4}}{2}} = \sqrt{\frac{5}{8}}$$

$$= \frac{\sqrt{10}}{4}$$

17. Find $\tan \theta/2$, given $\sin \theta = 3/5$, with $90° < \theta < 180°$.

Since $90° < \theta < 180°$, $\cos \theta < 0$.

$$\cos \theta = -\sqrt{1 - \left(\frac{3}{5}\right)^2} = -\frac{4}{5}$$

$$\tan \frac{\theta}{2} = \frac{1 - \cos \theta}{\sin \theta}$$

$$= \frac{1 - \left(-\frac{4}{5}\right)}{\frac{3}{5}} = 3$$

19. Find $\sin \alpha/2$, given $\tan \alpha = 2$, with $0 < \alpha < \pi/2$.

Since α is in quadrant I, $\sec \alpha > 0$.

$$\sec^2 \alpha = \tan^2 \alpha + 1$$

$$= (2)^2 + 1$$

$$= 5$$

$$\sec \alpha = \sqrt{5}$$

$$\cos \alpha = \frac{1}{\sec \alpha} = \frac{1}{\sqrt{5}} = \frac{\sqrt{5}}{5}$$

$$0 < \alpha < \frac{\pi}{2}$$

$$0 < \frac{\alpha}{2} < \frac{\pi}{4}$$

Thus, $\sin \alpha/2 > 0$.

$$\sin \frac{\alpha}{2} = \sqrt{\frac{1 - \cos \alpha}{2}}$$

$$= \sqrt{\frac{1 - \frac{\sqrt{5}}{5}}{2}} = \frac{\sqrt{50 - 10\sqrt{5}}}{10}$$

21. Find $\tan \beta/2$, given $\tan \beta = \sqrt{7}/3$, with $180° < \beta < 270°$.

$$\sec^2 \beta = \tan^2 \beta + 1$$

$$= \left(\frac{\sqrt{7}}{3}\right)^2 + 1 = \frac{16}{9}$$

Since β is in quadrant III, $\sec \beta < 0$ and $\sin \beta < 0$.

$$\sec \beta = -\sqrt{\frac{16}{9}} = -\frac{4}{3}$$

$$\cos \beta = \frac{1}{\sec \beta} = \frac{1}{-\frac{4}{3}} = -\frac{3}{4}$$

$$\sin \beta = -\sqrt{1 - \cos^2 \beta}$$

$$= -\sqrt{1 - \frac{9}{16}} = -\frac{\sqrt{7}}{4}$$

$$\tan \frac{\beta}{2} = \frac{\sin \beta}{1 + \cos \beta}$$

$$= \frac{-\frac{\sqrt{7}}{4}}{1 + \left(-\frac{3}{4}\right)} = -\sqrt{7}$$

23. Find $\sin \theta$, given $\cos 2\theta = 3/5$, θ is in quadrant I.

Since θ is in quadrant I, $\sin \theta > 0$.

$$\sin \theta = \sin \frac{1}{2}(2\theta)$$

$$= \sqrt{\frac{1 - \cos 2\theta}{2}}$$

$$= \sqrt{\frac{1 - \frac{3}{5}}{2}} = \frac{\sqrt{5}}{5}$$

25. Find $\cos \theta$, given $\cos 2x = -5/12$, $\frac{\pi}{2} < x < \pi$.

Since $\frac{\pi}{2} < x < \pi$, $\cos x < 0$.

$$\cos x = \cos \frac{1}{2}(2x)$$

$$= -\sqrt{\frac{1 + \cos 2x}{2}}$$

$$\cos x = -\sqrt{\frac{1 + \left(-\frac{5}{12}\right)}{2}}$$

$$= -\frac{\sqrt{42}}{12}$$

27. $$\sqrt{\frac{1 - \cos 40°}{2}} = \sin \frac{40°}{2}$$

$$= \sin 20°$$

29. $$\sqrt{\frac{1 - \cos 147°}{1 + \cos 147°}} = \tan \frac{147°}{2}$$

$$= \tan 73.5°$$

31. $$\frac{1 - \cos 59.74°}{\sin 59.74°} = \tan \frac{59.74°}{2}$$

$$= \tan 29.87°$$

33. $$\pm\sqrt{\frac{1 + \cos 18x}{2}} = \cos \frac{18x}{2}$$

$$= \cos 9x$$

35. $$\pm\sqrt{\frac{1 - \cos 8\theta}{1 + \cos 8\theta}} = \tan \frac{8\theta}{2}$$

$$= \tan 4\theta$$

37. $$\pm\sqrt{\frac{1 + \cos\left(\frac{x}{4}\right)}{2}} = \cos \frac{\frac{x}{4}}{2}$$

$$= \cos \frac{x}{8}$$

39. Verify $\sec^2 \frac{x}{2} = \frac{2}{1 + \cos x}$.

$$\sec^2 \frac{x}{2} = \frac{1}{\cos^2 \frac{x}{2}}$$

$$= \frac{1}{\left(\pm\sqrt{\frac{1 + \cos x}{2}}\right)^2}$$

$$= \frac{1}{\frac{1 + \cos x}{2}}$$

$$= \frac{2}{1 + \cos x}$$

41. Verify $\sin^2 \frac{x}{2} = \frac{\tan x - \sin x}{2 \tan x}$.

Work with the left side.

$$\sin^2 \frac{x}{2} = \left(\pm\sqrt{\frac{1 - \cos x}{2}}\right)^2$$

$$= \frac{1 - \cos x}{2}$$

Work with the right side.

$$\frac{\tan x - \sin x}{2 \tan x}$$

$$= \frac{\frac{\sin x}{\cos x} - \sin x}{2\frac{\sin x}{\cos x}}$$

$$= \frac{\sin x - \cos x \sin x}{2 \sin x}$$

$$= \frac{1 - \cos x}{2}$$

43. Verify $\frac{2}{1 + \cos x} - \tan^2 \frac{x}{2} = 1$.

$$\frac{2}{1 + \cos x} - \tan^2 \frac{x}{2}$$

$$= \frac{2}{1 + \cos x} - \left(\pm\sqrt{\frac{1 - \cos x}{1 + \cos x}}\right)^2$$

$$= \frac{2}{1 + \cos x} - \frac{1 - \cos x}{1 + \cos x}$$

$$= \frac{2 - 1 + \cos x}{1 + \cos x} = \frac{1 + \cos x}{1 + \cos x}$$

$$= 1$$

45. Verify $\tan \frac{\alpha}{2} = \frac{1 - \cos \alpha}{\sin \alpha}$.

$$\tan \frac{\alpha}{2} = \sqrt{\frac{1 - \cos \alpha}{1 + \cos \alpha}}$$

$$= \frac{\sqrt{1 - \cos \alpha}}{\sqrt{1 + \cos \alpha}} \cdot \frac{\sqrt{1 - \cos \alpha}}{\sqrt{1 - \cos \alpha}}$$

$$= \frac{1 - \cos \alpha}{\sqrt{1 - \cos^2 \alpha}}$$

$$= \frac{1 - \cos \alpha}{\sin \alpha}$$

47. Verify $\dfrac{\tan \frac{x}{2} + \cos \frac{x}{2}}{\cot \frac{x}{2} - \tan \frac{x}{2}} = \sec x$.

$$\frac{\tan \frac{x}{2} + \cos \frac{x}{2}}{\cot \frac{x}{2} - \tan \frac{x}{2}}$$

$$= \frac{\frac{\sin \frac{x}{2}}{\cos \frac{x}{2}} + \frac{\cos \frac{x}{2}}{\sin \frac{x}{2}}}{\frac{\cos \frac{x}{2}}{\sin \frac{x}{2}} - \frac{\sin \frac{x}{2}}{\cos \frac{x}{2}}}$$

$$= \frac{\sin^2 \frac{x}{2} + \cos^2 \frac{x}{2}}{\cos^2 \frac{x}{2} - \sin^2 \frac{x}{2}}$$

$$= \frac{1}{\cos x} = \sec x$$

49. Verify $\cos x = \dfrac{1 - \tan^2 \frac{x}{2}}{1 + \tan^2 \frac{x}{2}}$.

Work with the right side.

$$\frac{1 - \tan^2 \frac{x}{2}}{1 + \tan^2 \frac{x}{2}}$$

$$= \frac{1 - \frac{1 - \cos x}{1 + \cos x}}{1 + \frac{1 - \cos x}{1 + \cos x}}$$

$$= \frac{(1 + \cos x) - (1 - \cos x)}{(1 + \cos x) + (1 - \cos x)}$$

$$= \frac{2 \cos x}{2} = \cos x$$

51. Verify $8 \sin^2 \frac{\gamma}{2} \cos^2 \frac{\gamma}{2} = 1 - \cos 2\gamma$.

Work with the left side.

$$8 \sin^2 \frac{\gamma}{2} \cos^2 \frac{\gamma}{2}$$

$$= 2 \cdot 4 \sin^2 \frac{\gamma}{2} \cos^2 \frac{\gamma}{2}$$

$$= 2\left(2 \sin \frac{\gamma}{2} \cos \frac{\gamma}{2}\right)^2$$

$$= 2\left(\sin 2 \frac{\gamma}{2}\right)^2 = 2 \sin^2 \gamma$$

Work with the right side.

$$1 - \cos 2\gamma$$

$$= 1 - (1 - 2 \sin^2 \gamma)$$

$$= 2 \sin^2 \gamma$$

53. $\tan \frac{A}{2} = \frac{\sin A}{1 + \cos A}$

$$= \frac{\sin A}{1 + \cos A} \cdot \frac{1 - \cos A}{1 - \cos A}$$

$$= \frac{\sin A\,(1 - \cos A)}{1 - \cos^2 A}$$

$$= \frac{\sin A\,(1 - \cos A)}{\sin^2 A}$$

$$= \frac{1 - \cos A}{\sin A}$$

55. $\sin \frac{\alpha}{2} = \frac{1}{m}$, $m = \frac{3}{2}$

$$\sin \frac{\alpha}{2} = \frac{1}{\frac{3}{2}} = \frac{2}{3}$$

$$\sin^2 \frac{\alpha}{2} = \frac{1 - \cos \alpha}{2}$$

so

$$\left(\frac{2}{3}\right)^2 = \frac{1 - \cos \alpha}{2}$$

$$\left(\frac{4}{9}\right)^2 - 1 = -\cos \alpha$$

$$-\frac{1}{9} = -\cos \alpha$$

$$\cos \alpha = \frac{1}{9}$$

$$\alpha \approx 84°$$

57. $\sin \frac{\alpha}{2} = \frac{1}{m}$, $m = 2$

$$\sin \frac{\alpha}{2} = \frac{1}{2}$$

$$\sin^2 \frac{\alpha}{2} = \frac{1 - \cos \alpha}{2}$$

$$\left(\frac{1}{2}\right)^2 = \frac{1 - \cos \alpha}{2}$$

$$2\left(\frac{1}{4}\right) = 1 - \cos \alpha$$

$$\frac{1}{2} - 1 = -\cos \alpha$$

$$-\frac{1}{2} = -\cos \alpha$$

$$\frac{1}{2} = \cos \alpha$$

$$\alpha = 60°$$

59. $\alpha = 30°$

$$\sin \frac{30°}{2} = \frac{1}{m}$$

$$\sin 15° = \frac{1}{m}$$

$$m = \frac{1}{\sin 15°}$$

$$\approx 3.9$$

Section 5.7

1. $\cos 45° \sin 25°$

$$= \frac{1}{2}[\sin (45° + 25°) - \sin (45° - 25°)]$$

$$= \frac{1}{2}(\sin 70° - \sin 20°)$$

3. $3 \cos 5x \cos 3x$

$= \frac{3}{2}[\cos (5x + 3x) + \cos (5x - 3x)]$

$= \frac{3}{2}(\cos 8x + \cos 2x)$

5. $\sin (-\theta) \sin (-3\theta)$

$= \frac{1}{2}\{\cos [-\theta - (-3\theta)]$

$- \cos [-\theta + (-3\theta)]\}$

$= \frac{1}{2}[\cos 2\theta - \cos (-4\theta)]$

$= \frac{1}{2}(\cos 2\theta - \cos 4\theta)$

7. $-8 \cos 4y \cos 5y$

$= -\frac{8}{2}[\cos (4y + 5y) + \cos (4y - 5y)]$

$= -4[\cos 9y + \cos (-y)]$

$= -4(\cos 9y + \cos y)$

9. $\sin 60° - \sin 30°$

$= 2 \cos \left(\frac{60° + 30°}{2}\right) \sin \left(\frac{60° - 30°}{2}\right)$

$= 2 \cos 45° \sin 15°$

11. $\cos 42° + \cos 148°$

$= 2 \cos \left(\frac{42° + 148°}{2}\right) \cdot \cos \left(\frac{42° - 148°}{2}\right)$

$= 2 \cos 95° \cos (-53°)$

$= 2 \cos 95° \cos 53°$

13. $\sin 12\beta - \sin 3\beta$

$= 2 \cos \left(\frac{12\beta + 3\beta}{2}\right) \cdot \sin \left(\frac{12\beta - 3\beta}{2}\right)$

$= 2 \cos \frac{15\beta}{2} \sin \frac{9\beta}{2}$

15. $-3 \sin 2x + 3 \sin 5x$

$= -3(\sin 2x - \sin 5x)$

$= -3(2) \cos \left(\frac{2x + 5x}{2}\right) \cdot \sin \left(\frac{2x - 5x}{2}\right)$

$= -6 \cos \frac{7x}{2} \sin \left(-\frac{3x}{2}\right)$

$= 6 \cos \frac{7x}{2} \sin \frac{3x}{2}$

17. Verify $\tan x = \frac{\sin 3x - \sin x}{\cos 3x + \cos x}$.

Work with the right side.

$\frac{\sin 3x - \sin x}{\cos 3x + \cos x}$

$= \frac{2 \cos \left(\frac{3x + x}{2}\right) \sin \left(\frac{3x - x}{2}\right)}{2 \cos \left(\frac{3x + x}{2}\right) \cos \left(\frac{3x - x}{2}\right)}$

$= \frac{\cos 2x \sin x}{\cos 2x \cos x} = \frac{\sin x}{\cos x} = \tan x$

19. Verify $\frac{\cot 2\theta}{\tan 3\theta} = \frac{\cos 5\theta + \cos \theta}{\cos \theta - \cos 5\theta}$.

Work with the right side.

$\frac{\cos 5\theta + \cos \theta}{\cos \theta - \cos 5\theta}$

$= \frac{2 \cos \left(\frac{5\theta + \theta}{2}\right) \cos \left(\frac{5\theta - \theta}{2}\right)}{-2 \sin \left(\frac{\theta + 5\theta}{2}\right) \sin \left(\frac{\theta - 5\theta}{2}\right)}$

$= \frac{2 \cos 3\theta \cos 2\theta}{-2 \sin 3\theta \sin (-2\theta)}$

$= \frac{2 \cos 3\theta \cos 2\theta}{2 \sin 3\theta \sin 2\theta}$

$= \frac{1}{\frac{2 \sin 3\theta}{2 \cos 3\theta}}\left(\frac{\cos 2\theta}{\sin 2\theta}\right)$

$= \frac{1}{\tan 3\theta} \cot 2\theta = \frac{\cot 2\theta}{\tan 3\theta}$

21. Verify $\frac{1}{\tan 2s} = \frac{\sin 3s - \sin s}{\cos s - \cos 3s}$.

Work with the right side.

$\frac{\sin 3s - \sin s}{\cos s - \cos 3s}$

$= \frac{2 \cos \left(\frac{3s + s}{2}\right) \sin \left(\frac{3s - s}{2}\right)}{-2 \sin \left(\frac{s + 3s}{2}\right) \sin \left(\frac{s - 3s}{2}\right)}$

$$= \frac{2 \cos 2s \sin s}{-2 \sin 2s \sin (-s)}$$

$$= \frac{\cos 2s \sin s}{\sin 2s \sin s} = \frac{\cos 2s}{\sin 2s}$$

$$= \cot 2s = \frac{1}{\tan 2s}$$

23. Verify $\sin 6\theta \cos 4\theta - \sin 3\theta \cos 7\theta$
$= \sin 3\theta \cos \theta$.

$\sin 6\theta \cos 4\theta - \sin 3\theta \cos 7\theta$

$$= \frac{1}{2}[\sin (6\theta + 4\theta) + \sin (6\theta - 4\theta)]$$
$$- \frac{1}{2}[\sin (3\theta + 7\theta) + \sin (3\theta - 7\theta)]$$

$$= \frac{1}{2}[\sin 10\theta + \sin 2\theta$$
$$- \sin 10\theta - \sin (-4\theta)]$$

$$= \frac{1}{2}(\sin 2\theta + \sin 4\theta)$$

$$= \frac{1}{2}\left[2 \sin \left(\frac{2\theta + 4\theta}{2}\right) \cdot \cos \left(\frac{2\theta - 4\theta}{2}\right)\right]$$

$$= \sin 3\theta \cos (-\theta)$$

$$= \sin 3\theta \cos \theta$$

25. Verify $\sin^2 u - \sin^2 v$
$= \sin (u + v) \sin (u - v)$.

Work with the right side.

$\sin (u + v) \sin (u - v)$

$$= (\sin u \cos v + \sin v \cos u) \cdot (\sin u \cos v - \sin v \cos u)$$

$$= \sin^2 u \cos^2 v - \sin^2 v \cos^2 u$$

$$= \sin^2 u(1 - \sin^2 v) - \sin^2 v(1 - \sin^2 u)$$

$$= \sin^2 u - \sin^2 u \sin^2 v - \sin^2 v + \sin^2 v \sin^2 u$$

$$= \sin^2 u - \sin^2 v$$

27. $\sin s \cos t$

$$= \left(\frac{1}{2}\right)[\sin (s + t) + \sin (s - t)]$$

Let $s = t$ and multiply both sides by 2.

$2 \sin t \cos t$

$$= 2\left\{\frac{1}{2}[\sin (t + t) + \sin (t - t)]\right\}$$

$$= 2\left[\frac{1}{2}(\sin 2t + \sin 0)\right]$$

$$= 1 \sin 2t = \sin 2t$$

29. $\cos A \sin B = \frac{1}{2}[\sin (A + B) - \sin (A - B)]$

The product of the cosine of one angle and the sine of another angle is equal to one-half the difference of the sine of the sum of the angles and the sine of the difference of the angles.

$\cos A \cos B = \frac{1}{2}[\cos (A + B) + \cos (A - B)]$

The product of the cosine of one angle and the cosine of another angle is equal to one-half the sum of the cosine of the sum of the angles and the cosine of the difference of the angles.

$\sin A \sin B = \frac{1}{2}[\cos (A - B) - \cos (A + B)]$

The product of the sine of one angle and the sine of another angle is equal to one-half the difference of the cosine of the difference of the angles and the cosine of the sum of the angles.

Chapter 5 Review Exercises

1. $\cos x = \frac{3}{5}$, x is in quadrant IV.

$$\sin^2 x = 1 - \cos^2 x = 1 - \left(\frac{3}{5}\right)^2$$

$$= \frac{16}{25}$$

Since x is in quadrant IV, $\sin x < 0$.

$$\sin x = -\sqrt{\frac{16}{25}} = -\frac{4}{5}$$

$$\tan x = \frac{\sin x}{\cos x} = \frac{-\frac{4}{5}}{\frac{3}{5}} = -\frac{4}{3}$$

$$\sec x = \frac{1}{\cos x} = \frac{1}{\frac{3}{5}} = \frac{5}{3}$$

$$\csc x = \frac{1}{\sin x} = \frac{1}{-\frac{4}{5}} = -\frac{5}{4}$$

$$\cot x = \frac{1}{\tan x} = \frac{1}{-\frac{4}{3}} = -\frac{3}{4}$$

3. (a) $\sin \frac{\pi}{12} = \sin\left(\frac{\pi}{4} - \frac{\pi}{6}\right)$

$$= \sin\frac{\pi}{4}\cos\frac{\pi}{6} - \cos\frac{\pi}{4}\sin\frac{\pi}{6}$$

$$= \frac{\sqrt{2}}{2}\cdot\frac{\sqrt{3}}{2} - \frac{\sqrt{2}}{2}\cdot\frac{1}{2}$$

$$= \frac{\sqrt{6}}{4} - \frac{\sqrt{2}}{4} = \frac{\sqrt{6} - \sqrt{2}}{4}$$

$$\cos\frac{\pi}{12} = \cos\left(\frac{\pi}{4} - \frac{\pi}{6}\right)$$

$$= \cos\frac{\pi}{4}\cos\frac{\pi}{6} + \sin\frac{\pi}{4}\sin\frac{\pi}{6}$$

$$= \frac{\sqrt{2}}{2}\cdot\frac{\sqrt{3}}{2} + \frac{\sqrt{2}}{2}\cdot\frac{1}{2}$$

$$= \frac{\sqrt{6}}{4} + \frac{\sqrt{2}}{4} = \frac{\sqrt{6} + \sqrt{2}}{4}$$

$$\tan\frac{\pi}{12} = \tan\left(\frac{\pi}{4} - \frac{\pi}{6}\right)$$

$$= \frac{\tan\frac{\pi}{4} - \tan\frac{\pi}{6}}{1 + \tan\frac{\pi}{4}\tan\frac{\pi}{6}}$$

$$= \frac{1 - \frac{\sqrt{3}}{3}}{1 + (1)\frac{\sqrt{3}}{3}}$$

$$= \frac{3 - \sqrt{3}}{3 + \sqrt{3}}\cdot\frac{3 - \sqrt{3}}{3 - \sqrt{3}}$$

$$= \frac{12 - 6\sqrt{3}}{6} = 2 - \sqrt{3}$$

(b) In this exercise $\pi/12$ is in quadrant I, where all circular functions are positive.

$$\sin\frac{\pi}{12} = \sin\left(\frac{\frac{\pi}{6}}{2}\right)$$

$$= \sqrt{\frac{1 - \cos\frac{\pi}{6}}{2}}$$

$$= \sqrt{\frac{1 - \frac{\sqrt{3}}{2}}{2}}$$

$$= \sqrt{\frac{2 - \sqrt{3}}{4}}$$

$$= \frac{\sqrt{2 - \sqrt{3}}}{2}$$

$$\cos\frac{\pi}{12} = \cos\left(\frac{\frac{\pi}{6}}{2}\right)$$

$$= \sqrt{\frac{1 + \cos\frac{\pi}{6}}{2}}$$

$$= \sqrt{\frac{1 + \frac{\sqrt{3}}{2}}{2}}$$

$$= \sqrt{\frac{2 + \sqrt{3}}{4}}$$

$$= \frac{\sqrt{2 + \sqrt{3}}}{2}$$

$$\tan\frac{\pi}{12} = \tan\left(\frac{\frac{\pi}{6}}{2}\right)$$

$$= \frac{\sin\frac{\pi}{6}}{1 + \cos\frac{\pi}{6}}$$

$$= \frac{\frac{1}{2}}{1 + \frac{\sqrt{3}}{2}} = \frac{1}{2 + \sqrt{3}}$$

$$= \frac{1}{2+\sqrt{3}} \cdot \frac{2-\sqrt{3}}{2-\sqrt{3}}$$

$$= 2 - \sqrt{3}$$

5. $\cos 210° = \cos (150° + 60°)$
$= \cos 150° \cos 60° - \sin 150° \sin 60°$ (e)

7. $\tan (-35°) = \cot [90° - (-35°)]$
$= \cot 125°$ (j)

9. $\cos 35° = \cos (-35°)$ (i)

11. $\sin 75° = \sin (15° + 60°)$
$= \sin 15° \cos 60° + \cos 15° \sin 60°$ (h)

13. $\cos 300° = \cos 2(150°)$
$= \cos^2 150° - \sin^2 150°$ (g)

15. $\csc x = \dfrac{1}{\sin x}$ (a)

17. $\cot x = \dfrac{\cos x}{\sin x}$ (f)

19. $\tan^2 x + 1 = \sec^2 x$
$= \dfrac{1}{\cos^2 x}$ (e)

21. $\sec^2 \theta - \tan^2 \theta$

$$= \frac{1}{\cos^2 \theta} - \frac{\sin^2 \theta}{\cos^2 \theta}$$

$$= \frac{1 - \sin^2 \theta}{\cos^2 \theta}$$

$$= \frac{\cos^2 \theta}{\cos^2 \theta} = 1$$

23. $\tan^2 \theta\ (1 + \cot^2 \theta)$

$$= \frac{\sin^2 \theta}{\cos^2 \theta}\left(1 + \frac{\cos^2 \theta}{\sin^2 \theta}\right)$$

$$= \frac{\sin^2 \theta}{\cos^2 \theta}\left(\frac{\sin^2 \theta + \cos^2 \theta}{\sin^2 \theta}\right)$$

$$= \frac{\sin^2 \theta}{\cos^2 \theta}\left(\frac{1}{\sin^2 \theta}\right)$$

$$= \frac{1}{\cos^2 \theta} \text{ or } \sec^2 \theta$$

25. $\csc^2 \theta + \sec^2 \theta$

$$= \frac{1}{\sin^2 \theta} + \frac{1}{\cos^2 \theta}$$

$$= \frac{\cos^2 \theta + \sin^2 \theta}{\sin^2 \theta \cos^2 \theta}$$

$$= \frac{1}{\sin^2 \theta \cos^2 \theta}$$

27. Find $\sin (x + y)$, $\cos (x - y)$, and $\tan (x + y)$, given $\sin x = -1/4$, $\cos y = -4/5$, x and y are in quadrant III.

Since x and y are in quadrant III, cos x and cos y are negative.

$$\cos x = -\sqrt{1 - \sin^2 x}$$

$$= -\sqrt{1 - \frac{1}{16}}$$

$$= -\sqrt{\frac{15}{16}}$$

$$= -\frac{\sqrt{15}}{4}$$

$$\sin y = -\sqrt{1 - \cos^2 x}$$

$$= -\sqrt{1 - \left(-\frac{4}{5}\right)^2}$$

$$= -\sqrt{\frac{25 - 16}{25}} = -\frac{3}{5}$$

$\sin (x + y)$

$$= \sin x \cos y + \cos x \sin y$$

$$= \left(-\frac{1}{4}\right)\left(-\frac{4}{5}\right) + \left(-\frac{\sqrt{15}}{4}\right)\left(-\frac{3}{5}\right)$$

$$= \frac{4}{20} + \frac{3\sqrt{15}}{20} = \frac{4 + 3\sqrt{15}}{20}$$

cos (x − y)

$= \cos x \cos y + \sin x \sin y$

$= \left(-\frac{\sqrt{15}}{4}\right)\left(-\frac{4}{5}\right) + \left(-\frac{1}{4}\right)\left(-\frac{3}{5}\right)$

$= \frac{\sqrt{15}}{5} + \frac{3}{20} = \frac{4\sqrt{15}}{20} + \frac{3}{20}$

$= \frac{4\sqrt{15} + 3}{20}$

tan (x + y)

$= \frac{\sin (x + y)}{\cos (x + y)}$

$= \frac{\frac{4 + 3\sqrt{15}}{20}}{\frac{4\sqrt{15} - 3}{20}}$

$= \frac{4 + 3\sqrt{15}}{4\sqrt{15} - 3} \cdot \frac{4\sqrt{15} + 3}{4\sqrt{15} + 3}$

$= \frac{16\sqrt{15} + 12 + 12(15) + 9\sqrt{15}}{16(15) - 9}$

$= \frac{192 + 25\sqrt{15}}{231}$

To find the quadrant of x + y, notice that sin (x + y) > 0, which implies x + y is in quadrant I or II. Also tan (x + y) > 0, which implies that x + y is in quadrant I or III. Therefore, x + y is in quadrant I.

29. Find sin (x + y), cos (x − y), and tan (x + y), given sin x = 1/10, cos y = 4/5, x is in quadrant I, y is in quadrant IV.

Since x is in quadrant I, cos x > 0.

$\cos x = \sqrt{1 - \sin^2 x}$

$= \sqrt{1 - \frac{1}{100}}$

$= \sqrt{\frac{99}{100}} = \frac{3\sqrt{11}}{10}$

Since y is in quadrant IV, sin y < 0.

$\sin y = -\sqrt{1 - \cos^2 x}$

$= -\sqrt{1 - \frac{16}{25}}$

$= -\sqrt{\frac{25 - 16}{25}} = -\frac{3}{5}$

sin (x + y)

$= \sin x \cos y + \cos x \sin y$

$= \left(\frac{1}{10}\right)\left(\frac{4}{5}\right) + \frac{3\sqrt{11}}{10}\left(-\frac{3}{5}\right)$

$= \frac{4}{50} - \frac{9\sqrt{11}}{50} = \frac{4 - 9\sqrt{11}}{50}$

cos (x − y)

$= \cos x \cos y + \sin x \sin y$

$= \left(\frac{3\sqrt{11}}{10}\right)\left(\frac{4}{5}\right) + \left(\frac{1}{10}\right)\left(-\frac{3}{5}\right)$

$= \frac{12\sqrt{11}}{50} - \frac{3}{50} = \frac{12\sqrt{11} - 3}{50}$

tan (x + y)

$= \frac{\sin (x + y)}{\cos (x + y)}$

$= \frac{\frac{4 - 9\sqrt{11}}{50}}{\frac{12\sqrt{11} + 3}{50}}$

$= \frac{4 - 9\sqrt{11}}{12\sqrt{11} + 3} \cdot \frac{12\sqrt{11} - 3}{12\sqrt{11} - 3}$

$= \frac{48\sqrt{11} - 12 - 11(108) + 27\sqrt{11}}{144(11) - 9}$

$= \frac{75\sqrt{11} - 1200}{1575}$

$= \frac{75(\sqrt{11}) - 75(16)}{75(21)}$

$= \frac{\sqrt{11} - 16}{21}$

To find the quadrant of x + y, notice that sin (x + y) < 0, which implies x + y is in quadrant III or IV. Notice also that tan (x + y) < 0, which implies that x + y is in quadrant II or IV. Therefore, x + y is in quadrant IV.

31. Find sin θ and cos θ, given cos $2\theta = -\frac{3}{4}$, $90° < 2\theta < 180°$.

Since 2θ is in quadrant II, θ is in quadrant I.

$$\cos 2\theta = 1 - 2\sin^2\theta$$

$$-\frac{3}{4} = 1 - 2\sin^2\theta$$

$$-\frac{7}{4} = -2\sin^2\theta$$

$$\frac{7}{8} = \sin^2\theta$$

Since θ is in quadrant I, $\sin\theta > 0$.

$$\sqrt{\frac{7}{8}} = \sin\theta$$

$$\sin\theta = \frac{\sqrt{14}}{4}$$

$$\cos\theta = \sqrt{1 - \sin^2\theta}$$

$$= \sqrt{1 - \frac{7}{8}} = \sqrt{\frac{1}{8}} = \frac{\sqrt{2}}{4}$$

33. Find sin 2x and cos 2x, given tan x = 3, sin x < 0.

If tan x = 3 > 0 and sin x < 0, then x is in quadrant III and 2x is in quadrant I or II.

$$\tan 2x = \frac{2\tan x}{1 - \tan^2 x}$$

$$= \frac{2(3)}{1 - 3^2}$$

$$= -\frac{6}{8} = -\frac{3}{4}$$

Since tan 2x < 0, 2x is in quadrant II. Thus, sec 2x < 0 and sin 2x > 0.

$$\sec 2x = -\sqrt{1 + \tan^2 2x}$$

$$= -\sqrt{1 + \left(-\frac{3}{4}\right)^2}$$

$$= -\sqrt{\frac{25}{16}} = -\frac{5}{4}$$

$$\cos 2x = \frac{1}{\sec 2x}$$

$$= -\frac{4}{5}$$

$$\sin 2x = \sqrt{1 - \left(-\frac{4}{5}\right)^2}$$

$$= \frac{3}{5}$$

35. Find cos $\theta/2$, given cos $\theta = -1/2$, $90° < \theta < 180°$.

Since θ is in quadrant II, $\theta/2$ is in quadrant I, so cos $\theta/2 > 0$.

$$\cos\frac{\theta}{2} = \sqrt{\frac{1 + \left(-\frac{1}{2}\right)}{2}}$$

$$= \sqrt{\frac{\frac{1}{2}}{2}} = \sqrt{\frac{1}{4}} = \frac{1}{2}$$

37. Find tan x, given tan 2x = 2, $\pi < x < 3\pi/2$.

$$\tan 2x = \frac{2\tan x}{1 - \tan^2 x} = 2$$

$$2\tan x = 2 - 2\tan^2 x$$

$$\tan^2 x + \tan x - 1 = 0$$

$$\tan x = \frac{-1 \pm \sqrt{1 + 4}}{2}$$

Since x is in quadrant III, tan > 0.

$$\tan x = \frac{-1 + \sqrt{5}}{2}$$

39. Verify $\sin^2 x - \sin^2 y = \cos^2 y - \cos^2 x$.

$$\begin{aligned}\sin^2 x - \sin^2 y &= (1 - \cos^2 x) - (1 - \cos^2 y)\\ &= 1 - \cos^2 x - 1 + \cos^2 y\\ &= \cos^2 y - \cos^2 x\end{aligned}$$

41. Verify $-\cot \frac{x}{2} = \frac{\sin 2x + \sin x}{\cos 2x - \cos x}$.

Work with the right side.

$$\begin{aligned}\frac{\sin 2x + \sin x}{\cos 2x - \cos x} &= \frac{2 \sin \left(\frac{2x + x}{2}\right) \cos \left(\frac{2x - x}{2}\right)}{-2 \sin \left(\frac{2x + x}{2}\right) \sin \left(\frac{2x - x}{2}\right)}\\ &= \frac{\cos \frac{x}{2}}{-\sin \frac{x}{2}}\\ &= -\cot \frac{x}{2}\end{aligned}$$

43. Verify $\frac{\sin 2x}{\sin x} = \frac{2}{\sec x}$.

$$\begin{aligned}\frac{\sin 2x}{\sin x} &= \frac{2 \sin x \cos x}{\sin x}\\ &= 2 \cos 2x\\ &= \frac{2}{\frac{1}{\cos x}}\\ &= \frac{2}{\sec x}\end{aligned}$$

45. Verify $\frac{2 \tan B}{\sin 2B} = \sec^2 B$.

$$\begin{aligned}\frac{2 \tan B}{\sin 2B} &= \frac{\frac{\sin B}{\cos B}}{2 \sin B \cos B}\\ &= \frac{1}{\cos^2 B}\\ &= \sec^2 B\end{aligned}$$

47. Verify $1 + \tan^2 \alpha = 2 \tan \alpha \csc 2\alpha$.

Work with the right side.

$$\begin{aligned}2 \tan \alpha \csc 2\alpha &= \frac{2 \tan \alpha}{\sin 2\alpha}\\ &= \frac{2 \frac{\sin \alpha}{\cos \alpha}}{2 \sin \alpha \cos \alpha}\\ &= \frac{2 \sin \alpha}{2 \sin \alpha \cos^2 \alpha}\\ &= \frac{1}{\cos^2 \alpha} = \sec^2 \alpha\\ &= 1 + \tan^2 \alpha\end{aligned}$$

49. Verify $\frac{\sin t}{1 - \cos t} = \cot \frac{t}{2}$.

$$\begin{aligned}\frac{\sin t}{1 - \cos t} &= \frac{2 \sin \frac{t}{2} \cos \frac{t}{2}}{1 - \left(1 - 2 \sin^2 \frac{t}{2}\right)}\\ &= \frac{\sin \frac{t}{2} \cos \frac{t}{2}}{\sin^2 \frac{t}{2}}\\ &= \frac{\cos \frac{t}{2}}{\sin \frac{t}{2}} = \cot \frac{t}{2}\end{aligned}$$

51. Verify $\frac{2 \cot x}{\tan 2x} = \csc^2 x - 2$.

$$\begin{aligned}\frac{2 \cot x}{\tan 2x} &= \frac{2}{\tan x \left(\frac{2 \tan x}{1 - \tan^2 x}\right)}\\ &= \frac{1 - \tan^2 x}{\tan^2 x}\\ &= \frac{1 - \frac{\sin^2 x}{\cos^2 x}}{\frac{\sin^2 x}{\cos^2 x}}\\ &= \frac{\cos^2 x - \sin^2 x}{\sin^2 x}\end{aligned}$$

$$= \frac{1 - 2\sin^2 x}{\sin^2 x}$$

$$= \csc^2 x - 2$$

53. Verify $\tan\theta \sin 2\theta = 2 - 2\cos^2\theta$.

$$\tan\theta \sin 2\theta$$

$$= \tan\theta\,(2\sin\theta\cos\theta)$$

$$= \frac{\sin\theta}{\cos\theta}(2\sin\theta\cos\theta)$$

$$= 2\sin^2\theta$$

$$= 2(1 - \cos^2\theta)$$

$$= 2 - 2\cos^2\theta$$

55. Verify $2\tan x \csc 2x - \tan^2 x = 1$.

$$2\tan x \csc 2x - \tan^2 x$$

$$= 2\tan x \frac{1}{\sin 2x} - \tan^2 x$$

$$= 2\tan x \frac{1}{2\sin x\cos x} - \tan^2 x$$

$$= \frac{\sin x}{\sin x\cos^2 x} - \frac{\sin^2 x}{\cos^2 x}$$

$$= \frac{1}{\cos^2 x} - \frac{\sin^2 x}{\cos^2 x} = \frac{\cos^2 x}{\cos^2 x} = 1$$

57. Verify $\tan\theta\cos^2\theta = \frac{2\tan\theta\cos^2\theta - \tan\theta}{1 - \tan^2\theta}$.

Work with the right side.

$$\frac{2\tan\theta\cos^2\theta - \tan\theta}{1 - \tan^2\theta}$$

$$= \frac{\tan\theta\,(2\cos^2\theta - 1)}{1 - \frac{\sin^2\theta}{\cos^2\theta}}$$

$$= \frac{\tan\theta\,(2\cos^2\theta - 1)}{\frac{\cos^2\theta - \sin^2\theta}{\cos^2\theta}}$$

$$= \frac{\cos^2\theta\tan\theta(2\cos^2\theta - 1)}{2\cos^2\theta - 1}$$

$$= \cos^2\theta\tan\theta$$

59. Verify $2\cos^3 x - \cos x = \frac{\cos^2 x - \sin^2 x}{\sec x}$.

Work with the right side.

$$\frac{\cos^2 x - \sin^2 x}{\sec x}$$

$$= \frac{2\cos^2 x - 1}{\frac{1}{\cos x}}$$

$$= (2\cos^2 x - 1)\cos x$$

$$= 2\cos^3 x - \cos x$$

61. Verify $\cos^4\theta = \frac{3}{8} + \frac{1}{2}\cos 2\theta + \frac{1}{8}\cos 4\theta$.

Work with the right side.

$$\frac{3}{8} + \frac{1}{2}\cos 2\theta + \frac{1}{8}\cos 4\theta$$

$$= \frac{3}{8} + \frac{2\cos^2\theta - 1}{2} + \frac{1}{8}(2\cos^2 2\theta - 1)$$

$$= \frac{3}{8} + \frac{2\cos^2\theta - 1}{2} + \frac{1}{8}[2(2\cos^2\theta - 1)^2] - \frac{1}{8}$$

$$= \frac{3}{8} + \frac{2\cos^2\theta - 1}{2} + \frac{1}{4}(4\cos^4\theta - 4\cos^2\theta + 1) - \frac{1}{8}$$

$$= \frac{3}{8} - \frac{1}{2} + \frac{1}{4} - \frac{1}{8} + \cos^2\theta - \cos^2\theta + \cos^4\theta$$

$$= \cos^4\theta$$

63. Verify $\sec^2\alpha - 1 = \frac{\sec 2\alpha - 1}{\sec 2\alpha + 1}$.

Work with the right side.

$$\frac{\sec 2\alpha - 1}{\sec 2\alpha + 1}$$

$$= \frac{\frac{1}{\cos 2\alpha} - 1}{\frac{1}{\cos 2\alpha} + 1}$$

$$= \frac{\frac{1}{\cos^2\alpha - \sin^2\alpha} - 1}{\frac{1}{\cos^2\alpha - \sin^2\alpha} + 1}$$

$$= \frac{1 - \cos^2 \alpha + \sin^2 \alpha}{1 + \cos^2 \alpha - \sin^2 \alpha}$$

$$= \frac{2 \sin^2 \alpha}{2 \cos^2 \alpha} = \tan^2 \alpha$$

$$= \sec^2 \alpha - 1$$

65. Verify $\tan 4\theta = \dfrac{2 \tan 2\theta}{2 - \sec^2 2\theta}$.

$\tan 4\theta$

$$= \frac{2 \tan 2\theta}{2 - \sec^2 2\theta} = \frac{2 \tan 2\theta}{1 - (\sec^2 2\theta - 1)}$$

$$= \frac{2 \tan 2\theta}{1 - \sec^2 2\theta + 1} = \frac{2 \tan 2\theta}{2 - \sec^2 2\theta}$$

67. Verify $2 \cos^2 \frac{x}{2} \tan x = \tan x + \sin x$.

Work with the right side.

$\tan x + \sin x$

$$= \frac{\sin x}{\cos x} + \sin x$$

$$= \sin x\left(1 + \frac{1}{\cos x}\right)$$

$$= \sin x\left(1 + \frac{1}{2 \cos^2 \frac{x}{2} - 1}\right)$$

$$= \sin x \left(\frac{2 \cos^2 \frac{x}{2} - 1 + 1}{2 \cos^2 \frac{x}{2} - 1}\right)$$

$$= \frac{2 \sin x \cos^2 \frac{x}{2}}{2 \cos^2 \frac{x}{2} - 1}$$

$$= 2 \cos^2 \frac{x}{2} \cdot \frac{\sin x}{\cos x}$$

$$= 2 \tan x \cos^2 \frac{x}{2}$$

69. Verify $\tan\left(\frac{x}{2} + \frac{\pi}{4}\right) = \sec x + \tan x$.

Work with the left side.

$\tan\left(\frac{x}{2} + \frac{\pi}{4}\right)$

$$= \frac{\tan \frac{x}{2} + \tan \frac{\pi}{4}}{1 - \tan \frac{x}{2} \tan \frac{\pi}{4}}$$

$$= \frac{\tan \frac{x}{2} + 1}{1 - \tan \frac{x}{2}}$$

Work with the right side.

$\sec x + \tan x$

$$= \frac{1}{\cos x} + \frac{\sin x}{\cos x}$$

$$= \frac{\cos^2 \frac{x}{2} + \sin^2 \frac{x}{2} + 2 \sin \frac{x}{2} \cos \frac{x}{2}}{\cos^2 \frac{x}{2} - \sin^2 \frac{x}{2}}$$

$$= \frac{\left(\cos \frac{x}{2} + \sin \frac{x}{2}\right)^2}{\left(\cos \frac{x}{2} - \sin \frac{x}{2}\right)\left(\cos \frac{x}{2} + \sin \frac{x}{2}\right)}$$

$$= \frac{\cos \frac{x}{2} + \sin \frac{x}{2}}{\cos \frac{x}{2} - \sin \frac{x}{2}}$$

$$= \frac{\dfrac{\cos \frac{x}{2}}{\cos \frac{x}{2}} + \dfrac{\sin \frac{x}{2}}{\cos \frac{x}{2}}}{\dfrac{\cos \frac{x}{2}}{\cos \frac{x}{2}} - \dfrac{\sin \frac{x}{2}}{\cos \frac{x}{2}}} = \frac{1 + \tan \frac{x}{2}}{1 - \tan \frac{x}{2}}$$

For Exercises 71 and 73 see figure in the textbook for labels of lines and angles.

71. Given angle A = 60°, angle B = 30°, and AB = 2

(a) Prove that triangle ABD is isosceles. AB and BD are radii of the same circle, so

AB = BD

and triangle ABD is isosceles.

(b) Prove angle ABD is 150°.

$$\text{Angle ABD} = 180° - 30° = 150°$$

(c) Prove angle DAB and angle ADB are each 15°.

In triangle ADB,

$$\text{angle DAB} + \text{angle ADB} + \text{angle ABD} = 180°$$
$$\text{angle DAB} + \text{angle ADB} + 150° = 180°$$
$$\text{angle DAB} + \text{angle ADB} = 30°.$$

Since triangle ABD is isosceles,

$$\text{angle DAB} = \text{angle ADB} = \frac{1}{2}(30°) = 15°.$$

(d) Prove DC has length $2 + \sqrt{3}$.

Since triangle ABC is a 30°-60° right triangle and AB = 2, the side opposite the 30° angle (that is, AC) is 1.

By the Pythagorean theorem,

$$BC = \sqrt{2^2 - 1^2} = \sqrt{3}.$$

DB, like AB, is a radius of the circle, so DB = AB = 2.

$$DC = DB + BC = 2 + \sqrt{3}$$

(e) $AD = \sqrt{1^2 + (2 + \sqrt{3})^2}$

$$= \sqrt{1 + 4 + 4\sqrt{3} + 3}$$
$$= \sqrt{8 + 4\sqrt{3}}$$

Since

$$(\sqrt{6} + \sqrt{2})^2 = 6 + 2\sqrt{12} + 2$$
$$= 8 + 2\sqrt{12}$$
$$= 8 + 4\sqrt{3},$$

then

$$\sqrt{6} + \sqrt{2} = \sqrt{8 + 4\sqrt{3}}.$$

73. Show that the length of AE is $\sqrt{6} - \sqrt{2}$ and then find sin 15°.

Since triangle AED is a right triangle,

$$AE = \sqrt{1^2 + (2 - \sqrt{3})^2}$$
$$= \sqrt{1 + 4 - 4\sqrt{3} + 3}$$
$$= \sqrt{8 - 4\sqrt{3}}.$$

Since

$$(\sqrt{6} - \sqrt{2})^2 = 6 - 2\sqrt{12} + 2$$
$$= 8 - 2\sqrt{12}$$
$$= 8 - 4\sqrt{3},$$

then

$$AE = \sqrt{6} - \sqrt{2}.$$

DE is a diameter so its length is twice the radius, or 2(2) = 4. In right triangle AED, the sine of angle ADE is

$$\sin ADE = \frac{AE}{DE}.$$

From Exercise 71 (c) angle ADE = 15°, so

$$\sin 15° = \frac{\sqrt{6} - \sqrt{2}}{4}.$$

CHAPTER 5 TEST

[5.1] Let $\tan x = -2/3$, with x in quadrant II. Find each of the following.

1. $\sin x$

2. $\cot x$

3. $\sec x$

Let $\sin s = 1/4$ with s in quadrant II and let $\cos t = -2/3$ with t in quadrant III. Find each of the following.

[5.3] 4. $\cos (s - t)$

[5.4] 5. $\sin (s + t)$

[5.5] 6. $\sin 2s$

[5.6] 7. $\cos \frac{1}{2}t$

Suppose $x = -\pi/12$. Use identities to find the exact value of each of the following.

[5.3] 8. $\cos x$

[5.4] 9. $\sin x$

Answer *true* or *false* for each of the following.

[5.6] 10. $\cos 17° = 2 \cos^2 34° - 1$

[5.1] 11. $\csc (-57°) = -\csc 57°$

[5.4] 12. $\sin 91° = \sin 34° \cos 57° + \cos 34° \sin 57°$

[5.4] 13. $\tan 23° = \dfrac{\tan 70° + \tan 47°}{1 - \tan 70° \tan 47°}$

[5.2] 14. A student claims that

$$\frac{\cos\theta}{\sin\theta\cot\theta} = 1$$

cannot be verified as an identity, since by letting $\theta = 90°$ we get $0/0$, an undefined expression. Comment on this student's reasoning.

[5.2] **Use identities to express each of the following in terms of sin θ and cos θ, and simplify.**

15. $\dfrac{\tan\theta}{\sec\theta}$

16. $\tan^2\theta\,(1 + \cot^2\theta)$

Verify that each of the following is an identity.

[5.2] 17. $\dfrac{\cos^2\beta}{\sin\beta} = \csc\beta - \sin\beta$

[5.5] 18. $\dfrac{\sin 2\alpha}{\tan\alpha} = 2 - \dfrac{2}{\csc^2\alpha}$

[5.6] 19. $\dfrac{\sec A}{\csc A}\tan\dfrac{A}{2} = \sec A - 1$

[5.7] 20. $\dfrac{\cos 3\theta + \cos\theta}{\sin 3\theta + \sin\theta} = \cot 2\theta$

CHAPTER 5 TEST ANSWERS

1. $\frac{2\sqrt{13}}{13}$ 2. $-\frac{3}{2}$ 3. $-\frac{\sqrt{13}}{3}$ 4. $\frac{2\sqrt{15}-\sqrt{5}}{12}$ 5. $\frac{-2+5\sqrt{3}}{12}$

6. $-\frac{\sqrt{15}}{8}$ 7. $-\frac{\sqrt{6}}{6}$ 8. $\frac{\sqrt{2}+\sqrt{6}}{4}$ 9. $\frac{\sqrt{2}-\sqrt{6}}{4}$ 10. False

11. True 12. True 13. False 14. See student's comment.

15. $\sin\theta$ 16. $\frac{1}{\cos^2\theta}$

17. $\text{RHS} = \csc\beta - \sin\beta = \frac{1}{\sin\beta} - \sin\beta = \frac{1-\sin^2\beta}{\sin\beta} = \frac{\cos^2\beta}{\sin\beta} = \text{LHS}$

18. $\text{LHS} = \frac{\sin 2\alpha}{\tan\alpha} = \frac{2\sin\alpha\cos\alpha}{\frac{\sin\alpha}{\cos\alpha}} = 2\cos^2\alpha = 2(1-\sin^2\alpha)$

$= 2 - \frac{2}{\csc^2\alpha} = \text{RHS}$

19. $\text{LHS} = \frac{\sec A}{\csc A}\tan\frac{A}{2} = \frac{\frac{1}{\cos A}}{\frac{1}{\sin A}}\left(\frac{1-\cos A}{\sin A}\right) = \frac{\sin A}{\cos A}\left(\frac{1-\cos A}{\sin A}\right)$

$= \frac{1-\cos A}{\cos A} = \frac{1}{\cos A} - \frac{\cos A}{\cos A} = \sec A - 1 = \text{RHS}$

20. $\text{LHS} = \frac{\cos 3\theta + \cos\theta}{\sin 3\theta + \sin\theta} = \frac{2\cos\left(\frac{3\theta+\theta}{2}\right)\cos\left(\frac{3\theta-\theta}{2}\right)}{2\sin\left(\frac{3\theta+\theta}{2}\right)\cos\left(\frac{3\theta-\theta}{2}\right)}$

$= \frac{\cos\left(\frac{4\theta}{2}\right)}{\sin\left(\frac{4\theta}{2}\right)} = \frac{\cos 2\theta}{\sin 2\theta} = \cot 2\theta = \text{RHS}$

CHAPTER 6 INVERSE TRIGONOMETRIC FUNCTIONS AND TRIGONOMETRIC EQUATIONS

Section 6.1

1. $y = \text{Arcsin}\left(-\frac{1}{2}\right)$

$\text{Sin } y = -\frac{1}{2}, -\frac{\pi}{2} \le y \le \frac{\pi}{2}$

y is in quadrant IV. The reference angle is $\pi/6$.

$y = -\frac{\pi}{6}$

3. $y = \text{Tan}^{-1} 1$

$\text{Tan } y = 1, -\frac{\pi}{2} < y < \frac{\pi}{2}$

y is an quadrant I.

$y = \frac{\pi}{4}$

5. $y = \text{Arcsin } 0$

$\text{Sin } y = 0, -\frac{\pi}{2} \le y \le \frac{\pi}{2}$

$y = 0$

7. $y = \text{Cos}^{-1} \frac{1}{2}$

$\text{Cos } y = \frac{1}{2}, 0 \le y \le \pi$

y is in quadrant I.

$y = \frac{\pi}{3}$

9. $y = \text{Sec}^{-1} (-\sqrt{2})$

$\text{Sec } y = -\sqrt{2}, 0 \le y \le \pi, y \ne \frac{\pi}{2}$

y is in quadrant II. The reference angle is $\pi/4$.

$y = \frac{3\pi}{4}$

11. $y = \text{Arccot } (-\sqrt{3})$

$\text{Cot } y = (-\sqrt{3}), 0 < y < \pi$

y is in quadrant II. The reference angle is $\pi/6$.

$y = \frac{5\pi}{6}$

13. $\theta = \text{Arctan } (-1)$

$\text{Tan } \theta = -1, -90° < \theta < 90°$

θ is in quadrant IV. The reference angle is 45°.

Thus, $\theta = -45°$.

15. $\theta = \text{Arcsin}\left(-\frac{\sqrt{3}}{2}\right)$

$\text{Sin } \theta = -\frac{\sqrt{3}}{2}, -90° \le \theta \le 90°$

θ is in quadrant IV. The reference angle is 60°.

$\theta = -60°$

17. $\theta = \text{Cot}^{-1}\left(-\frac{\sqrt{3}}{3}\right)$

$\text{Cot } \theta = -\frac{\sqrt{3}}{3}$ and $0° < \theta < 180°$

θ is in quadrant II. The reference angle is 60°.

$\theta = 180° - 60° = 120°$

19. $\theta = \text{Csc}^{-1} (-2)$

$\text{Csc } \theta = -2$ and $-90° < \theta < 90°$, $\theta \ne 0°$

θ is in quadrant IV. The reference angle is 30°.

$\theta = -30°$.

For Exercises 21–25, be sure that your calculator is in radian mode.

21. Arctan 1.1111111

Tan θ = 1.1111111, $-\frac{\pi}{2} < \theta < \frac{\pi}{2}$

θ is in quadrant I.

With a calculator:

Enter: 1.1111111 [INV] [TAN]

Display: .83798122

Arctan 1.1111111 = .83798122

23. Cot^{-1} (-.92170128)

Cot θ = -.92170128, $0 < \theta < \pi$

θ is in quadrant II.

With a calculator:

Enter: -.92170128

[+/-] [1/x] [INV] [TAN]

The absolute value of this result is the reference angle of θ. To find θ in quadrant II, subtract the result from π. Since the value is negative, we can simply add π.

Continuing with calculator:

Enter: [+] [π]

Display: 2.3154725

Cot^{-1} (-.92170128) = 2.3154725

25. Arcsin .92837781

Sin θ = .92837781, $-\frac{\pi}{2} \le \theta \le \frac{\pi}{2}$

θ is in quadrant I.

With a calculator:

Enter: .92837781 [INV] [SIN]

Display: 1.1900238

Arcsin .92837781 = 1.1900238

In Exercises 27–31, be sure that your calculator is in degree mode.

27. Sin^{-1} (-.13349122)

Sin θ = -.13349122, $-\frac{\pi}{2} \le \theta \le \frac{\pi}{2}$

In degrees, the range is

$$-90° \le \theta \le 90°.$$

θ is in quadrant IV.

With a calculator:

Enter: -.13349122 [+/-] [INV] [SIN]

Display: -7.6713835

θ = -7.6713835°

29. Arccos (-.39876459)

Cos θ = -.39876449, $< \theta < \pi$, or

$0° < \theta < 180°$.

θ is in quadrant II.

With a calculator:

Enter: -.39876459 [+/-] [INV] [COS]

Display: 113.50097

Arccos -.39876459 = 113.50097°

31. Csc^{-1} 1.9422833

Csc θ = 1.9422833, $-\frac{\pi}{2} \le \theta \le \frac{\pi}{2}$, or

$0° < \theta < 180°$

θ is in quadrant I.

With a calculator:

Enter: 1.9422833 [1/x] [INV] [SIN]

Display: 30.987961

Csc^{-1} 1.9422833 = 30.987961°

33. $y = \text{Cot}^{-1} x$

$y = \text{Cot}^{-1} x$ means $\text{Cot } y = x$.

Domain: $(-\infty, \infty)$

Range: $(0, \pi)$

See the answer graph at the back of the textbook.

35. $y = \text{Arcsec } x$

$y = \text{Arcsec } x$ means $\text{Sec } y = x$.

Domain: $(-\infty, -1] \cup [1, \infty)$

Range: $[0, \frac{\pi}{2}) \cup (\frac{\pi}{2}, \pi]$

See the answer graph at the back of the textbook.

For Exercises 37 and 39, be sure that your calculator is in radian mode.

37. $\text{Sin}^{-1} 1.003$ produces an error message because 1.003 is not in the domain of the Arcsin function, which is $[-1, 1]$.

39. 1.74283 is in quadrant II.

$$\sin 1.74283 = .9852387$$

$\text{Sin}^{-1} .9852387 = \theta$ means

$\text{Sin } \theta = .9852387$ and $-\frac{\pi}{2} \le \theta \le \frac{\pi}{2}$.

So $\text{Sin}^{-1} .9852387$ gives an angle in quadrant I or IV whose sine is .9852387.

$\text{Sin}^{-1} .9852387 = 1.3987627$, which equals $\pi - 1.74283$ and is the reference angle for 1.74283.

41. $\tan\left(\text{Arccos } \frac{3}{4}\right)$

Let $\omega = \text{Arccos } \frac{3}{4}$, $0 \le \omega \le \pi$.

$\text{Cos } \omega = \frac{3}{4}$; ω is in quadrant I.

Sketch ω and label a triangle with the side opposite ω equal to

$$\sqrt{4^2 - 3^2} = \sqrt{7}.$$

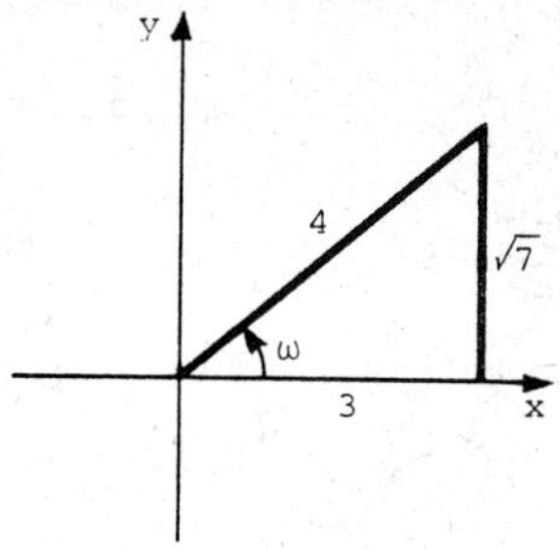

$$\tan\left(\text{Arccos } \frac{3}{4}\right) = \tan \omega = \frac{\sqrt{7}}{3}$$

43. $\cos(\text{Tan}^{-1}(-2))$

Let $\omega = \text{Tan}^{-1}(-2)$, $-\frac{\pi}{2} < \omega < \frac{\pi}{2}$.

$\text{Tan } \omega = -2$; ω is in quadrant IV.

Sketch ω and label a triangle with the hypotenuse equal to

$$\sqrt{(-2)^2 + 1} = \sqrt{5}.$$

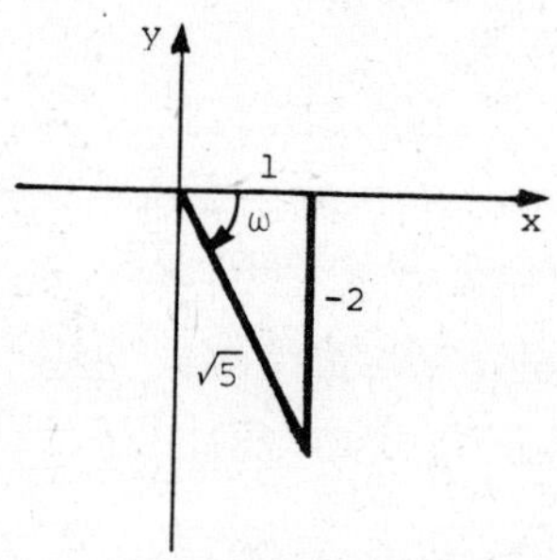

$\cos\ (\text{Tan}^{-1}\ (-2)) = \cos\ \omega$

$= \frac{1}{\sqrt{5}} = \frac{\sqrt{5}}{5}$

45. $\cot\ \left(\text{Arcsin}\ \left(-\frac{2}{3}\right)\right)$

Let $\omega = \text{Arcsin}\ \left(-\frac{2}{3}\right),\ -\frac{\pi}{2} \le \omega \le \frac{\pi}{2}$.

$\text{Sin}\ \omega = -\frac{2}{3}$, ω is in quadrant IV.

Sketch ω and label a triangle with the side adjacent to ω equal to

$\sqrt{3^2 - (-2)^2} = \sqrt{5}$.

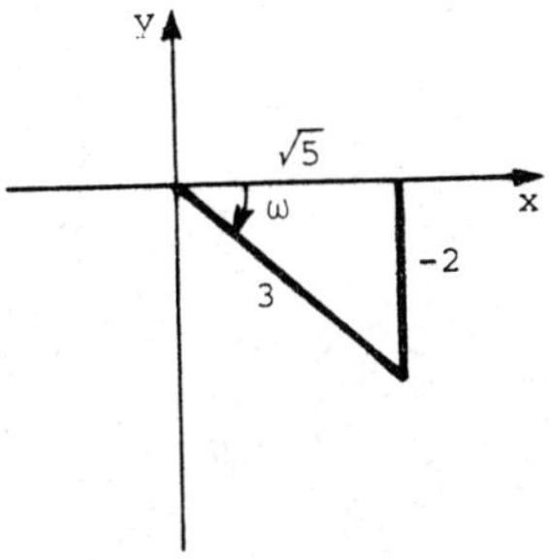

$\cot\ \left(\text{Arcsin}\ \left(-\frac{2}{3}\right)\right) = \cot\ \omega$

$= -\frac{\sqrt{5}}{2}$

47. $\sec\ \left(\text{Arccot}\ \frac{3}{5}\right)$

Let $\omega = \text{Arccot}\ \frac{3}{5},\ 0 < \omega < \pi$.

$\text{Cot}\ \omega = \frac{3}{5}$, ω is in quadrant I.

Sketch ω and label a triangle with the hypotenuse equal to

$\sqrt{3^2 + 5^2} = \sqrt{34}$.

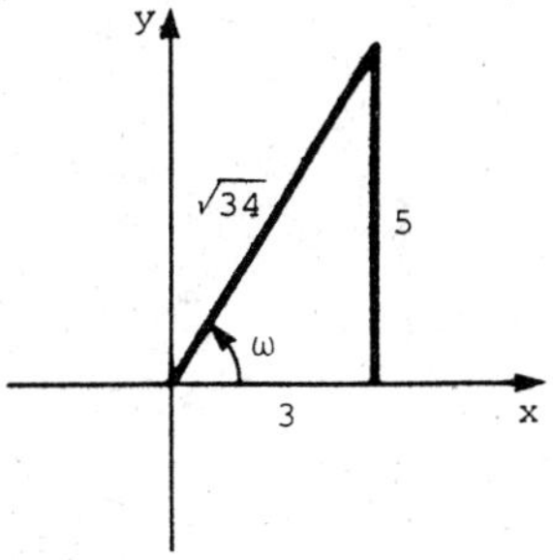

$\sec\ \left(\text{Arccot}\ \frac{3}{5}\right) = \sec\ \omega = \frac{\sqrt{34}}{3}$

49. $\cos\ \left(\text{Arccos}\ \frac{1}{2}\right)$

Let $\theta = \text{Arccos}\ \frac{1}{2},\ 0 \le \theta \le \pi$.

$\text{Cos}\ \theta = \frac{1}{2}$, θ is in quadrant I.

$\cos\ \left(\text{Arccos}\ \frac{1}{2}\right) = \cos\ \theta = \frac{1}{2}$

51. $\tan\ (\text{Tan}^{-1}\ (-1))$

Let $\theta = \text{Tan}^{-1}\ (-1),\ -\frac{\pi}{2} < \theta < \frac{\pi}{2}$.

$\text{Tan}\ \theta = -1$, θ is in quadrant IV.

$\tan\ (\text{Tan}^{-1}\ (-1)) = \tan\ \theta = -1$

53. $\sec\ (\text{Sec}^{-1}\ 2)$

Let $\theta = \text{Sec}^{-1}\ 2,\ 0 \le \theta \le \pi$.

$\text{Sec}\ \theta = 2$, θ is in quadrant I.

$\sec\ (\text{Sec}^{-1}\ 2) = \sec\ \theta = 2$

55. $\text{Arccos}\ \left(\cos\ \frac{\pi}{4}\right)$

Let $\text{Arccos}\ \left(\cos\ \frac{\pi}{4}\right) = \theta$.

$\text{Cos}\ \theta = \cos\ \frac{\pi}{4}$, and $0 \le \theta \le \pi$.

$\theta = \frac{\pi}{4}$

57. Arcsin $\left(\sin \frac{\pi}{3}\right)$

Let Arcsin $\left(\sin \frac{\pi}{3}\right) = \theta$.

Sin $\theta = \sin \frac{\pi}{3}$ and $-\frac{\pi}{2} \le \theta \le \frac{\pi}{2}$.

$$\theta = \frac{\pi}{3}$$

59. $\sin\left(2 \text{ Tan}^{-1} \frac{12}{5}\right)$

Let $\omega = \text{Tan}^{-1} \frac{12}{5}$, $-\frac{\pi}{2} < \omega < \frac{\pi}{2}$.

Tan $\omega = \frac{12}{5}$

Sketch ω in quadrant I and a right triangle with a hypotenuse equal to $\sqrt{12^2 + 5^2} = 13$.

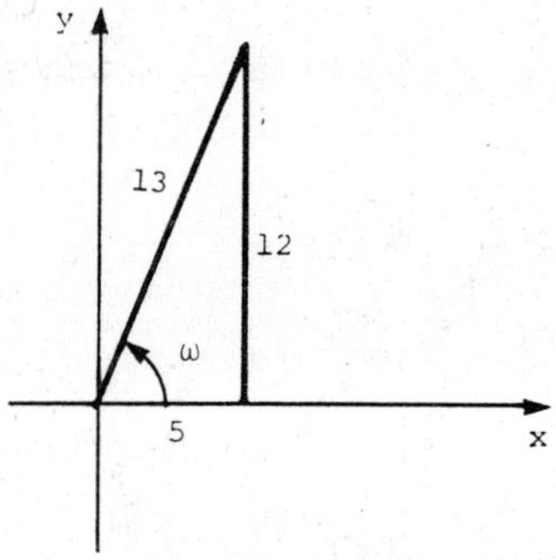

$$\sin \omega = \frac{12}{13}$$

$$\cos \omega = \frac{5}{13}$$

$$\begin{aligned}\sin\left(2 \text{ Tan}^{-1} \tfrac{12}{5}\right) &= \sin(2\omega)\\ &= 2 \sin \omega \cos \omega\\ &= 2\left(\tfrac{12}{13}\right)\left(\tfrac{5}{13}\right)\\ &= \tfrac{120}{169}\end{aligned}$$

61. $\cos\left(2 \text{ Arctan } \frac{4}{3}\right)$

Let $\omega = \text{Arctan } \frac{4}{3}$, $-\frac{\pi}{2} < \omega < \frac{\pi}{2}$.

Tan $\omega = \frac{4}{3}$, ω is in quadrant I.

Sketch ω in quadrant I and a right triangle with a hypotenuse equal to

$$\sqrt{4^2 + 3^2} = 5.$$

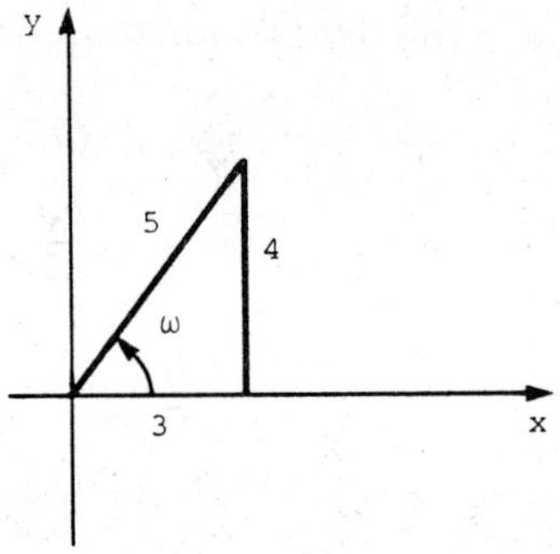

$$\cos \omega = \frac{3}{5}$$

$$\sin \omega = \frac{4}{5}$$

$$\begin{aligned}\cos\left(2 \text{ Arctan } \tfrac{4}{3}\right) &= \cos(2\omega)\\ &= \cos^2 \omega - \sin^2 \omega\\ &= \left(\tfrac{3}{5}\right)^2 - \left(\tfrac{4}{5}\right)^2\\ &= \tfrac{9}{25} - \tfrac{16}{25} = -\tfrac{7}{25}\end{aligned}$$

63. $\sin\left(2 \text{ Cos}^{-1} \frac{1}{5}\right)$

Let $\omega = \text{Cos}^{-1} \frac{1}{5}$, $0 \le \omega \le \pi$.

Cos $\omega = \frac{1}{5}$, ω is in quadrant I.

The side opposite ω is

$$\sqrt{5^2 - 1^2} = \sqrt{24} = 2\sqrt{6}.$$

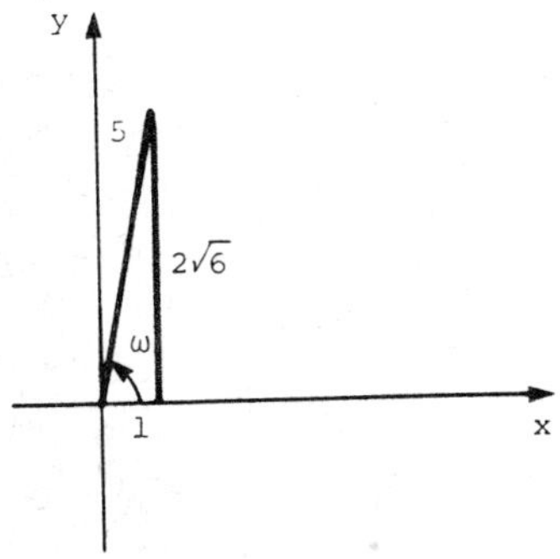

$$\sin \omega = \frac{2\sqrt{6}}{5}$$

$$\cos \omega = \frac{1}{5}$$

$$\sin\left(2 \cos^{-1} \frac{1}{5}\right) = \sin 2\omega$$
$$= 2 \sin \omega \cos \omega$$
$$= 2\left(\frac{2\sqrt{6}}{5}\right)\left(\frac{1}{5}\right)$$
$$= \frac{4\sqrt{6}}{25}$$

65. $\tan\left(2 \text{ Arcsin } \left(-\frac{3}{5}\right)\right)$

Let $\omega = \text{Arcsin } \left(-\frac{3}{5}\right)$, $-\frac{\pi}{2} \leq \omega \leq \frac{\pi}{2}$.

$\text{Sin } \omega = -\frac{3}{5}$, ω is in quadrant IV.

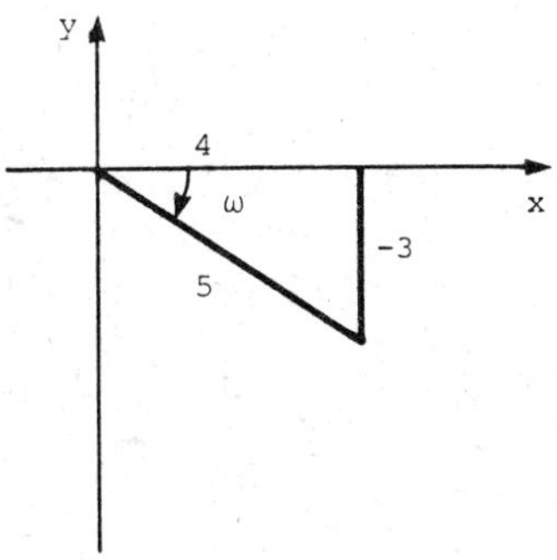

$$\tan \omega = -\frac{3}{4}$$

$$\tan\left(2 \text{ Arcsin } \left(-\frac{3}{5}\right)\right) = \tan (2\omega)$$
$$= \frac{2 \tan \omega}{1 - \tan^2 \omega}$$
$$= \frac{2\left(-\frac{3}{4}\right)}{1 - \left(-\frac{3}{4}\right)^2}$$
$$= \frac{-\frac{3}{2}}{1 - \frac{9}{16}} \cdot \frac{16}{16}$$
$$= \frac{-24}{16 - 9} = -\frac{24}{7}$$

67. $\sin\left(\sin^{-1} \frac{1}{2} + \tan^{-1} (-3)\right)$

Let $\omega_1 = \sin^{-1} \frac{1}{2}$, $-\frac{\pi}{2} \leq \omega_1 \leq \frac{\pi}{2}$.

$\text{Sin } \omega_1 = \frac{1}{2}$; ω_1 is in quadrant I.

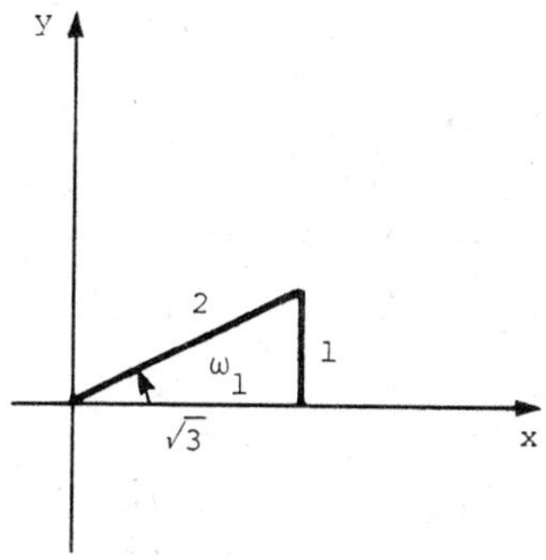

$\cos \omega_1 = \frac{\sqrt{3}}{2}$, $\sin \omega_1 = \frac{1}{2}$

Let $\omega_2 = \text{Tan}^{-1} (-3)$, $-\frac{\pi}{2} < \omega_2 < \frac{\pi}{2}$.

$\text{Tan } \omega_2 = -3$

$= -\frac{3}{1}$, ω_2 is in quadrant IV.

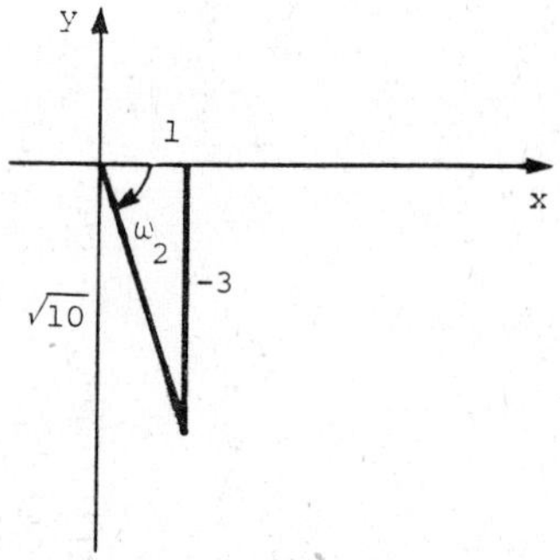

$$\cos \omega_2 = \frac{1}{\sqrt{10}} = \frac{\sqrt{10}}{10}$$

$$\sin \omega_2 = \frac{-3}{\sqrt{10}} = -\frac{3\sqrt{10}}{10}$$

$$\sin \left(\text{Sin}^{-1} \frac{1}{2} + \text{Tan}^{-1} (-3)\right)$$

$$= \sin (\omega_1 + \omega_2)$$

$$= \sin \omega_1 \cos \omega_2 + \cos \omega_1 \sin \omega_2$$

$$= \left(\frac{1}{2}\right)\left(\frac{\sqrt{10}}{10}\right) + \left(\frac{\sqrt{3}}{2}\right)\left(\frac{-3\sqrt{10}}{10}\right)$$

$$= \frac{\sqrt{10}}{20} - \frac{3\sqrt{30}}{20}$$

$$= \frac{\sqrt{10} - 3\sqrt{30}}{20}$$

69. $\cos \left(\text{Arcsin } \frac{3}{5} + \text{Arccos } \frac{5}{13}\right)$

Let $\omega_1 = \text{Arcsin } \frac{3}{5}, \ -\frac{\pi}{2} \le \omega_1 \le \frac{\pi}{2}$.

$\text{Sin } \omega_1 = \frac{3}{5}$, ω_1 is in quadrant I.

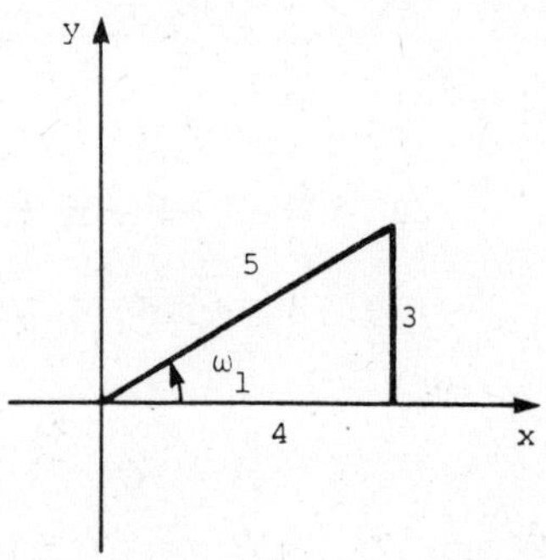

$$\cos \omega_1 = \frac{x}{5} = \frac{4}{5},$$

$$\sin \omega_1 = \frac{3}{5}$$

Let $\omega_2 = \text{Arccos } \frac{5}{13}, \ 0 \le \omega \le \pi$.

$\text{Cos } \omega_2 = \frac{5}{13}$, ω_2 is in quadrant I.

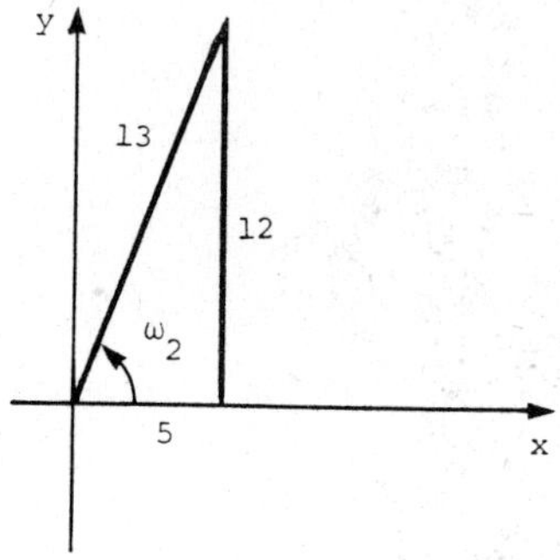

$$\sin \omega_2 = \frac{12}{13}$$

$$\cos \omega_2 = \frac{5}{13}$$

$$\cos \left(\text{Arcsin } \frac{3}{5} + \text{Arccos } \frac{5}{13}\right)$$

$$= \cos (\omega_1 + \omega_2)$$

$$= \cos \omega_1 \cos \omega_2 - \sin \omega_1 \sin \omega_2$$

$$= \left(\frac{4}{5}\right)\left(\frac{5}{13}\right) - \left(\frac{3}{5}\right)\left(\frac{12}{13}\right)$$

$$= \frac{20}{65} - \frac{36}{65} = -\frac{16}{65}$$

71. $\cos (\text{Tan}^{-1} .5)$

Enter: .5 [INV] [TAN] [COS]

$\cos (\text{Tan}^{-1} .5) = .89442719$

73. tan (Arcsin .12251014)

Enter: .12251014 [INV] [SIN] [TAN]

tan (Arcsin .12251014) = .12343998

75. sin (Arccos u)

Let $\omega = \text{Arccos } u, \ 0 < \omega < \pi$.

$\text{Cos } \omega = u = \frac{u}{1}$

If $u > 0$, then ω is in quadrant I.

If $u < 0$, then ω is in quadrant II.

In either quadrant, $\sin \omega > 0$.

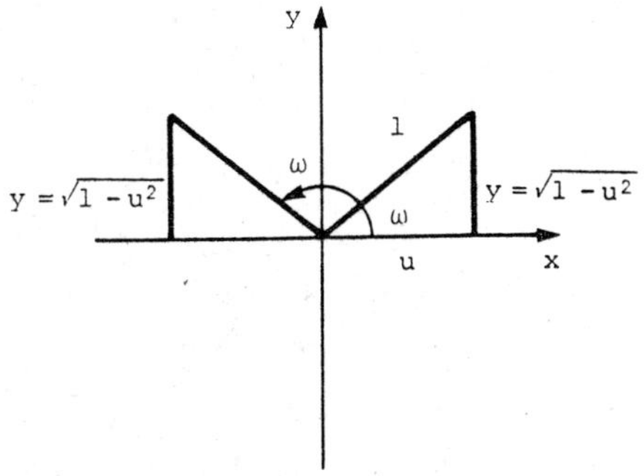

Since $y > 0$, from the Pythagorean theorem,

$$y = \sqrt{1 - u^2}.$$

$$\sin \omega = \frac{\sqrt{1 - u^2}}{1}$$

$$= \sqrt{1 - u^2}$$

Then $\sin(\text{Arccos } u) = \sin \omega$

$$= \sqrt{1 - u^2}.$$

77. $\sec(\text{Cot}^{-1} u)$

Let $\omega = \text{Cot}^{-1} u$, $0 < \omega < \pi$.

Then $\text{Cot } \omega = u = \frac{u}{1}$.

If $u > 0$, then ω is in quadrant I and $\sec \omega > 0$.

If $u < 0$, then ω is in quadrant II and $\sec \omega < 0$.

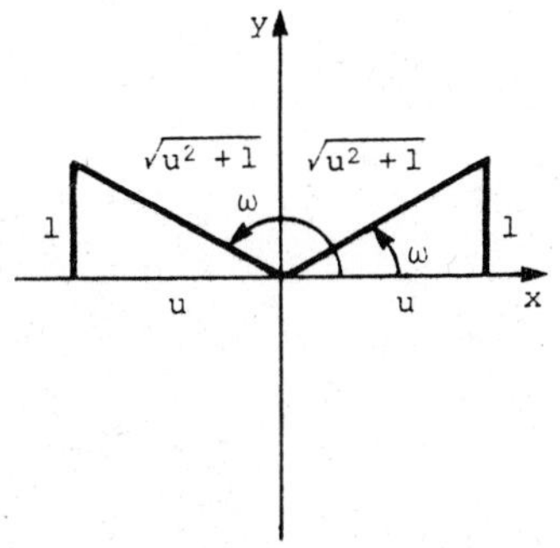

$$\sec \omega = \frac{\sqrt{u^2 + 1}}{u}$$

Then,

$$\sec(\text{Cot}^{-1} u) = \sec \omega$$

$$= \frac{\sqrt{u^2 + 1}}{u}.$$

79. $\cot(\text{Arcsin } u)$

Let $\omega = \text{Arcsin } u$, $-\frac{\pi}{2} \le \omega \le \frac{\pi}{2}$.

Then $\text{Sin } \omega = u = \frac{u}{1}$.

If $u > 0$, then ω is in quadrant I and $\cot \omega > 0$.

If $u < 0$, then ω is in quadrant IV and $\cot \omega < 0$.

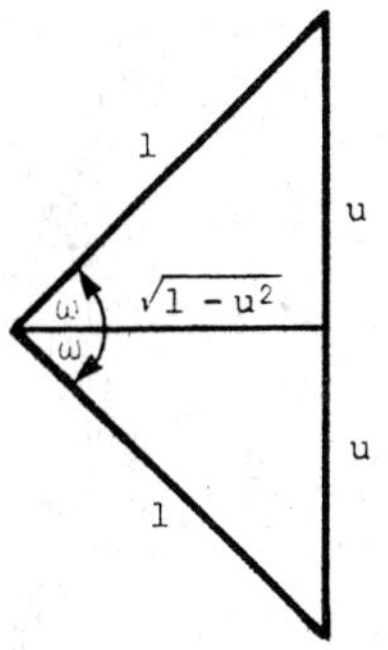

$$\cot \omega = \frac{\sqrt{1 - u^2}}{u}$$

Then

$$\cot(\text{Arcsin } u) = \cot \omega$$

$$= \frac{\sqrt{1 - u^2}}{u}.$$

81. $\sin\left(\text{Sec}^{-1} \frac{u}{2}\right)$

Let $\omega = \text{Sec}^{-1} \frac{u}{2}$, $0 \le \omega \le \pi$, $\omega \ne \frac{\pi}{2}$.

Since ω is in quadrant I or II, $\sin \omega > 0$.

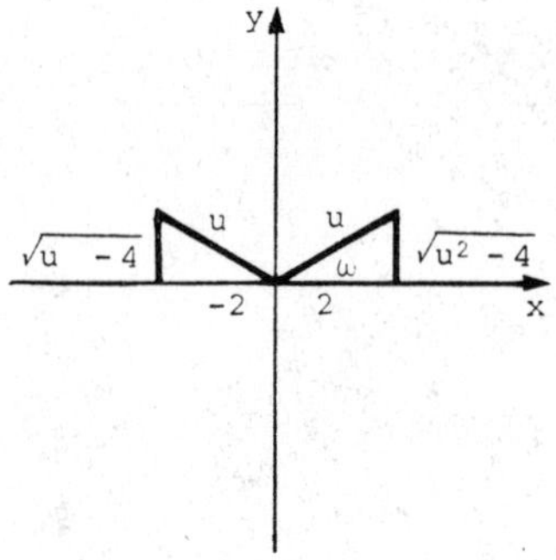

$$\sin \omega = \frac{\sqrt{u^2 - 4}}{u}$$

Then

$$\sin\left(\text{Sec}^{-1} \frac{u}{2}\right) = \sin \omega$$

$$= \frac{\sqrt{u^2 - 4}}{u}.$$

83. $\tan\left(\text{Arcsin} \dfrac{u}{\sqrt{u^2 + 2}}\right)$

Let $\omega = \text{Arcsin} \dfrac{u}{\sqrt{u^2 + 2}}$, $-\dfrac{\pi}{2} \le \omega \le \dfrac{\pi}{2}$.

Then $\text{Sin } \omega = \dfrac{u}{\sqrt{u^2 + 2}}$.

If $u > 0$, then ω is in quadrant I and $\tan \omega > 0$.

If $u < 0$, then ω is in quadrant IV and $\omega < 0$.

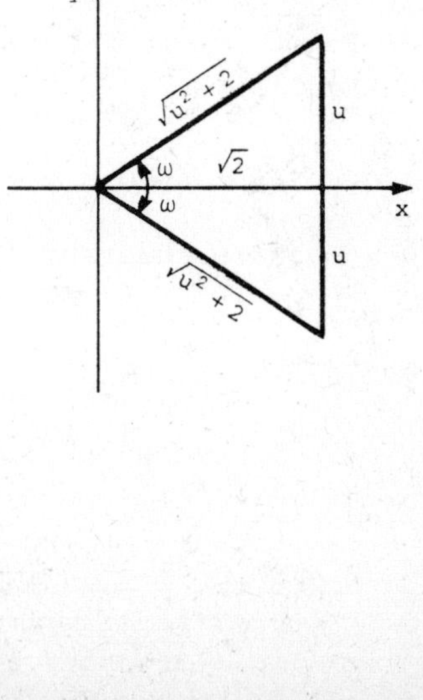

$$\tan \omega = \frac{u}{\sqrt{2}}$$

$$= \frac{u\sqrt{2}}{2}$$

$$\tan\left(\text{Arcsin} \frac{u}{\sqrt{u^2 + 2}}\right) = \tan \omega$$

$$= \frac{u}{\sqrt{2}} = \frac{u\sqrt{2}}{2}$$

85. $\sec\left(\text{Arccot} \dfrac{\sqrt{4 - u^2}}{u}\right)$

Let $\omega = \text{Arccot} \dfrac{\sqrt{4 - u^2}}{u}$, $0 < \omega < \pi$.

Then $\text{Cot } \omega = \dfrac{\sqrt{4 - u^2}}{u}$.

If u is negative (when ω is in quadrant II) the side adjacent to ω' is $-\sqrt{4 - u^2}$.

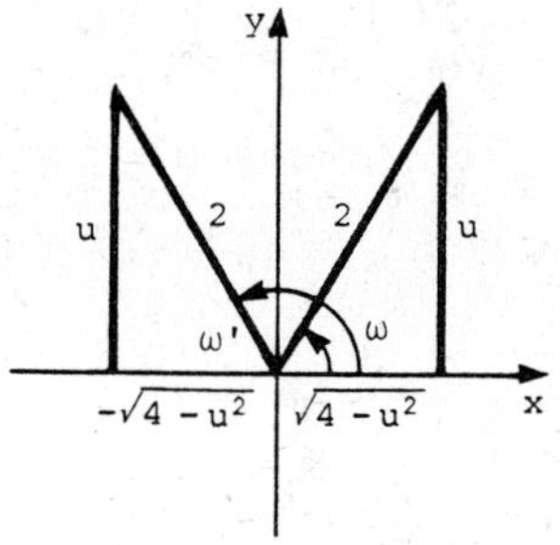

$$\sec \omega = \frac{2}{\pm\sqrt{4 - u^2}}$$

$$\sec\left(\text{Arccot} \frac{\sqrt{4 - u^2}}{u}\right) = \frac{2}{\pm\sqrt{4 - u^2}}$$

$$= \frac{\pm 2\sqrt{4 - u^2}}{4 - u^2}$$

87. (a) $m = 1.2$,

$$\alpha = 2 \text{ Arcsin } \left(\frac{1}{m}\right)$$

$$\alpha = 2 \text{ Arcsin } \left(\frac{1}{1.2}\right) = 113°$$

(b) $m = 1.5$,

$$\alpha = 2 \text{ Arcsin } \left(\frac{1}{1.5}\right) = 84°$$

(c) $m = 2$,

$$\alpha = 2 \text{ Arcsin } \left(\frac{1}{2}\right) = 60°$$

(d) $m = 2.5$,

$$\alpha = 2 \text{ Arcsin } \left(\frac{1}{2.5}\right) = 47°$$

Section 6.2

1. $2 \cot x + 1 = -1$

$$2 \cot x = -2$$

$$\cot x = -1$$

$$x = \frac{3\pi}{4}, \frac{7\pi}{4}$$

3. $2 \sin x + 3 = 4$

$$2 \sin x = 1$$

$$\sin x = \frac{1}{2}$$

$$x = \frac{\pi}{6}, \frac{5\pi}{6}$$

5. $\tan^2 x - 3 = 0$

$$\tan^2 x = 3$$

$$\tan x = \pm\sqrt{3}$$

If $\tan x = \sqrt{3}$,

$$x = \frac{\pi}{3}, \frac{4\pi}{3}.$$

If $x = -\sqrt{3}$, $x = \frac{2\pi}{3}, \frac{5\pi}{3}$.

$$x = \frac{\pi}{3}, \frac{2\pi}{3}, \frac{4\pi}{3}, \frac{5\pi}{3}$$

7. $(\cot x - \sqrt{3})(2 \sin x + \sqrt{3}) = 0$

$$\cot x - \sqrt{3} = 0$$

$$\cot x = \sqrt{3}$$

$$x = \frac{\pi}{6}, \frac{7\pi}{6}$$

or

$$2 \sin x + \sqrt{3} = 0$$

$$\sin x = -\frac{\sqrt{3}}{2}$$

$$x = \frac{4\pi}{3}, \frac{5\pi}{3}$$

$$x = \frac{\pi}{6}, \frac{7\pi}{6}, \frac{4\pi}{3}, \frac{5\pi}{3}$$

9. $(\cot x - 1)(\sqrt{3} \cot x + 1) = 0$

$$\cot x - 1 = 0$$

$$\cot x = 1$$

$$x = \frac{\pi}{4}, \frac{5\pi}{4}$$

or

$$\sqrt{3} \cot x + 1 = 0$$

$$\cot x = \frac{-1}{\sqrt{3}} = -\frac{\sqrt{3}}{3}$$

$$x = \frac{2\pi}{3}, \frac{5\pi}{3}$$

$$x = \frac{\pi}{4}, \frac{2\pi}{3}, \frac{5\pi}{4}, \frac{5\pi}{3}$$

11. $\cos^2 x + 2 \cos x + 1 = 0$

$$(\cos x + 1)^2 = 0$$

$$\cos x + 1 = 0$$

$$\cos x = -1$$

$$x = \pi$$

13. $-2 \sin^2 x = 3 \sin x + 1$

$$2 \sin^2 x + 3 \sin x + 1 = 0$$
$$(2 \sin x + 1)(\sin x + 1) = 0$$
$$2 \sin x + 1 = 0$$
$$\sin x = -\frac{1}{2}$$

$$x = \frac{7\pi}{6}, \frac{11\pi}{6}$$

or

$$\sin x + 1 = 0$$
$$\sin x = -1$$
$$x = \frac{3\pi}{2}$$

$$x = \frac{3\pi}{2}$$

$$x = \frac{7\pi}{6}, \frac{3\pi}{2}, \frac{11\pi}{6}$$

15. $2 \sin \theta - 1 = \csc \theta$

Use the identity

$$\csc \theta = \frac{1}{\sin \theta}, \ \sin \theta \neq 0.$$

$$2 \sin \theta - 1 = \frac{1}{\sin \theta}$$
$$2 \sin^2 \theta - \sin \theta = 1$$
$$2 \sin^2 \theta - \sin \theta - 1 = 0$$
$$(2 \sin \theta + 1)(\sin \theta - 1) = 0$$
$$2 \sin \theta + 1 = 0$$
$$\sin \theta = -\frac{1}{2}$$

$$\theta = 210°, 330°$$

or

$$\sin \theta - 1 = 0$$
$$\sin \theta = 1$$
$$\theta = 90°$$

$$\theta = 90°, 210°, 330°$$

17. $\tan \theta - \cot \theta = 0$

Use the identity

$$\cot \theta = \frac{1}{\tan \theta}, \ \tan \theta \neq 0.$$

$$\tan \theta - \frac{1}{\tan \theta} = 0$$
$$\tan^2 \theta - 1 = 0$$
$$\tan^2 \theta = 1$$
$$\tan \theta = \pm 1$$

If $\tan \theta = 1$,

$$\theta = 45°, 225°.$$

If $\tan \theta = -1$,

$$\theta = 135°, 315°.$$

$$\theta = 45°, 135°, 225°, 315°$$

19.

$$\csc^2 \theta - 2 \cot \theta = 0$$
$$(1 + \cot^2 \theta) - 2 \cot \theta = 0$$
$$\cot^2 \theta - 2 \cot \theta + 1 = 0$$
$$(\cot \theta - 1)^2 = 0$$
$$\cot \theta = 1$$

$$\theta = 45°, 225°$$

21.

$$2 \cos^4 \theta = \cos^2 \theta$$
$$2 \cos^4 \theta - \cos^2 \theta = 0$$
$$\cos^2 \theta (2 \cos^2 \theta - 1) = 0$$
$$\cos^2 \theta = 0 \quad \text{or} \quad 2 \cos^2 \theta = 1$$
$$\cos \theta = 0 \quad \text{or} \quad \cos \theta = \pm\frac{\sqrt{2}}{2}$$

If $\cos \theta = 0$,

$$\theta = 90°, 270°.$$

If $\cos \theta = \pm\frac{\sqrt{2}}{2}$,

$$\theta = 45°, 135°, 225°, 315°.$$

$$\theta = 45°, 90°, 135°, 225°, 270°, 315°$$

23. $2 \tan^2 \theta \sin \theta - \tan^2 \theta = 0$

$\tan^2 \theta(2 \sin \theta - 1) = 0$

$\tan^2 \theta = 0$

$\tan \theta = 0$

$\theta = 0°, 180°$

or

$2 \sin \theta - 1 = 0$

$\sin \theta = \frac{1}{2}$

$\theta = 30°, 150°$

$\theta = 0°, 30°, 150°, 180°$

25. $\sec^2 \theta \tan \theta = 2 \tan \theta$

$\sec^2 \theta \tan \theta - 2 \tan \theta = 0$

$\tan \theta(\sec^2 \theta - 2) = 0$

$\tan \theta = 0$ or $\sec^2 \theta = 2$

$\sec \theta = \pm\sqrt{2}$

$\theta = 0°, 180°$

or

$\theta = 45°, 135°, 225°, 315°$

$\theta = 0°, 45°, 135°, 180°, 225°, 315°$

27. $\sin \theta + \cos \theta = 1$

$(\sin \theta + \cos \theta)^2 = 1^2$

$\sin^2 \theta + 2 \sin \theta \cos \theta + \cos^2 \theta = 1$

$1 + 2 \sin \theta \cos \theta = 1$

$2 \sin \theta \cos \theta = 0$

$\sin \theta \cos \theta = 0$

$\sin \theta = 0$

$\theta = 0°, 180°$

or

$\cos \theta = 0$

$\theta = 90°, 270°$

Check these solutions since both sides were squared.

$\sin 180° + \cos 180° = -1$ and

$\sin 270° + \cos 270° = -1$ so

180° and 270° are not solutions.

$\theta = 0°, 90°$

29. $3 \sin^2 x - \sin x = 2$

$3 \sin^2 x - \sin x - 2 = 0$

$(3 \sin x + 2)(\sin x - 1) = 0$

$\sin x = -\frac{2}{3}$ or $\sin x = 1$

$x = 221.8°, 318.2°$, $x = 90°$

$x = 90°, 221.8°, 318.2°$

31. $\sec^2 \theta = 2 \tan \theta + 4$

$\tan^2 \theta + 1 = 2 \tan \theta + 4$

$\tan^2 \theta - 2 \tan \theta - 3 = 0$

$(\tan \theta - 3)(\tan \theta + 1) = 0$

$\tan \theta = 3$ or $\tan \theta = -1$

$\theta = 71.6°, 251.6°$ or $\theta = 135°, 315°$

33. $3 \cot^2 \theta = \cot \theta$

$3 \cot^2 \theta - \cot \theta = 0$

$\cot \theta(3 \cot \theta - 1) = 0$

$\cot \theta = 0$

$\theta = 90°, 270°$

or

$\cot \theta = \frac{1}{3}$

$\theta = 71.6°, 251.6°$

$\theta = 71.6°, 90°, 251.6°, 270°$

35. $9 \sin^2 x - 6 \sin x = 1$

$9 \sin^2 x - 6 \sin x - 1 = 0$

Use the quadratic formula with $a = 9$, $b = -6$, $c = -1$.

$$\sin x = \frac{6 \pm \sqrt{36 - [4 \cdot 9(-1)]}}{2 \cdot 9}$$

$$= \frac{6 \pm \sqrt{72}}{18}$$

$$= \frac{1 \pm \sqrt{2}}{3}$$

If $\sin x = \frac{1 + \sqrt{2}}{3}$

$$= \frac{1 + 1.4142136}{3}$$

$$= .80473787,$$

$$x = 53.6°, 126.4°.$$

If $\sin x = \frac{1 - \sqrt{2}}{3}$

$$= \frac{1 - 1.4142136}{3}$$

$$= -.13807119,$$

$$x = -7.9°.$$

Since this solution is not in the interval [0°, 360°), we must use it as a reference angle to find angles in the interval.

$$x = 180° + (-7.9°) = 187.9°$$

$$\text{or} \quad (-7.9°) + 360° = 352.1°$$

$x = 53.6°, 126.4°, 187.9°, 352.1°$

37. $\tan^2 x + 4 \tan x + 2 = 0$

$a = 1,\ b = 4,\ c = 2$

$$\tan x = \frac{-4 \pm \sqrt{16 - 4 \cdot 1 \cdot 2}}{2 \cdot 1}$$

$$= \frac{-4 \pm \sqrt{8}}{2} = \frac{-4 \pm 2\sqrt{2}}{2}$$

$$= -2 \pm \sqrt{2}$$

If $\tan x = -2 + \sqrt{2}$

$$= -2 + 1.4142136$$

$$= -.58578644,$$

$x = 149.6°, 329.6°.$

If $\tan x = -2 - \sqrt{2}$

$$= -2 - 1.4142136$$

$$= -3.4142136,$$

$$x = 106.3°, 286.3°.$$

$x = 106.3°, 149.6°, 286.3°, 329.6°$

39. $\sin^2 x - 2 \sin x + 3 = 0$

$a = 1,\ b = -2,\ c = 3$

$$\sin x = \frac{3 \pm \sqrt{4 - (4 \cdot 1 \cdot 3)}}{2 \cdot 1}$$

$$= \frac{2 \pm \sqrt{-8}}{2 \cdot 1}$$

Since $\sqrt{-8}$ is not a real number, the equation has no solution.

41. $\cot x + 2 \csc x = 3$

$$\frac{\cos x}{\sin x} + \frac{2}{\sin x} = 3$$

Multiply both sides by $\sin x$.

$$\cos x + 2 = 3 \sin x$$

Square both sides. Squaring both sides may introduce extraneous solutions, so all solutions must be checked in the original equation.

$$(\cos x + 2)^2 = (3 \sin x)^2$$

$$\cos^2 x + 4 \cos x + 4 = 9 \sin^2 x$$

$$\cos^2 x + 4 \cos x + 4 = 9(1 - \cos^2 x)$$

$$\cos^2 x + 4 \cos x + 4 = 9 - 9 \cos^2 x$$

$$10 \cos^2 x + 4 \cos x - 5 = 0$$

Use the quadratic formula with $a = 10$, $b = 4$, and $c = -5$.

$$\cos x = \frac{-4 \pm \sqrt{4^2 - 4(10)(-5)}}{2(10)}$$

$$\cos x = \frac{-4 \pm \sqrt{216}}{20}$$

$$\cos x = \frac{-4 \pm 6\sqrt{6}}{20}$$

$$\cos x = \frac{-2 \pm 3\sqrt{6}}{10}$$

$$\cos x = \frac{-2 + 3\sqrt{6}}{10} \text{ or } \cos x = \frac{-2 - 3\sqrt{6}}{10}$$

$\cos x = .53484692$ or $\cos x = -.93484692$

$x = 57.7°$ or $302.3°$ or $x \approx 159.2°$ or $200.8°$

These four angles must be checked in the original equation, $\cot x + 2 \csc x = 3$. We see that 57.7° and 159.2° check, but that 302.3° and 200.8° do not check.

43. From Example 1,

$$\cos \theta = -\frac{1}{2} \quad \text{or} \quad \cos \theta = 1$$

$\theta = 120° + 360° \cdot n$ or $240° + 360° \cdot n$ or $\theta = 0° + 360° \cdot n$ where n is any integer.

Thus,

$\theta = 120° + 360° \cdot n$, $240° + 360° \cdot n$, $360° \cdot n$.

45. The error occurs in going from the step

$$\sin^2 x - \sin x = 0$$

to the step

$$\sin x - 1 = 0.$$

We cannot divide both sides by sin x because sin x may equal zero and division by zero is undefined.

The equation should be solved by factoring.

$$\sin^2 x - \sin x = 0$$

$$\sin x (\sin x - 1) = 0$$

$$\sin x = 0 \quad \text{or} \quad \sin x = 1$$

$$x = 0 \quad \text{or} \quad x = \frac{\pi}{2}$$

47. $V = \cos 2\pi t,\ 0 \le t \le \frac{1}{2}$

(a) $V = 0$, $\cos 2\pi t = 0$

$$2\pi t = \text{Cos}^{-1}\ 0$$

$$2\pi t = \frac{\pi}{2}$$

$$t = \frac{\frac{\pi}{2}}{2\pi} = \frac{1}{4} \text{ sec}$$

(b) $V = .5$, $\cos 2\pi t = .5$

$$2\pi t = \text{Cos}^{-1}\ (.5)$$

$$2\pi t = \frac{\pi}{3}$$

$$t = \frac{\frac{\pi}{3}}{2\pi} = \frac{1}{6} \text{ sec}$$

(c) $V = .25$, $\cos 2\pi t = .25$

$$2\pi t = \text{Cos}^{-1}\ (.25)$$

$$2\pi t = 1.3181161$$

$$t = \frac{1.3181161}{2\pi} = .21 \text{ sec}$$

49. $.342D \cos \theta + h \cos^2 \theta = \frac{16D^2}{V_0{}^2}$

$V_0 = 60$, $D = 80$, $h = 2$

$$.342(80)\cos \theta + 2 \cos^2 \theta = \frac{16(80)^2}{60^2}$$

$$2\cos^2\theta + 27.36\cos\theta - \frac{256}{9} = 0$$

$$\cos^2\theta + 13.68\cos\theta - \frac{128}{9} = 0$$

$a = 1,\ b = 13.68,\ c = -128/9$

$$\cos\theta = \frac{-13.68 \pm \sqrt{(13.68)^2 - 4(-128/9)}}{2}$$

$$= \frac{-13.68 \pm 15.6215}{2}$$

$\cos\theta = .97075$ or $\cos\theta = -14.65075$

-14.65075 does not lead to a solution because it is not in the range of $\cos\theta$.

If $\cos\theta = .97075$,

$\theta = 14°$.

Section 6.3

1. $\cos 2x = \frac{\sqrt{3}}{2}$

Since $0 \le x < 2\pi$,

$0 \le 2x < 4\pi$.

$$2x = \frac{\pi}{6}, \frac{11\pi}{6}, \frac{13\pi}{6}, \frac{23\pi}{6}$$

$$x = \frac{\pi}{12}, \frac{11\pi}{12}, \frac{13\pi}{12}, \frac{23\pi}{12}$$

3. $\sin 3x = -1$,

Since $0 \le x < 2\pi$,

$0 \le 3x < 6\pi$.

$$3x = \frac{3\pi}{2}, \frac{7\pi}{2}, \frac{11\pi}{2}$$

$$x = \frac{\pi}{2}, \frac{7\pi}{6}, \frac{11\pi}{6}$$

5. $3\tan 3x = \sqrt{3},\ 0 \le 3x < 6\pi$

$$\tan 3x = \frac{\sqrt{3}}{3}$$

$$3x = \frac{\pi}{6}, \frac{7\pi}{6}, \frac{13\pi}{6}, \frac{19\pi}{6}, \frac{25\pi}{6}, \frac{31\pi}{6}$$

$$x = \frac{\pi}{18}, \frac{7\pi}{18}, \frac{13\pi}{18}, \frac{19\pi}{18}, \frac{25\pi}{18}, \frac{31\pi}{18}$$

7. $\sqrt{2}\cos 2x = -1,\ 0 \le 2x < 4\pi$

$$\cos 2x = \frac{-1}{\sqrt{2}} = -\frac{\sqrt{2}}{2}$$

$$2x = \frac{3\pi}{4}, \frac{5\pi}{4}, \frac{11\pi}{4}, \frac{13\pi}{4}$$

$$x = \frac{3\pi}{8}, \frac{5\pi}{8}, \frac{11\pi}{8}, \frac{13\pi}{8}$$

9. $\sin\frac{x}{2} = \sqrt{2} - \sin\frac{x}{2},\ 0 \le \frac{x}{2} < \pi$

$$\sin\frac{x}{2} + \sin\frac{x}{2} = \sqrt{2}$$

$$2\sin\frac{x}{2} = \sqrt{2}$$

$$\sin\frac{x}{2} = \frac{\sqrt{2}}{2}$$

$$\frac{x}{2} = \frac{\pi}{4}, \frac{3\pi}{4}$$

$$x = \frac{\pi}{2}, \frac{3\pi}{2}$$

11. $\tan 4x = 0,\ 0 \le 4x < 8\pi$

$4x = 0, \pi, 2\pi, 3\pi, 4\pi, 5\pi, 6\pi, 7\pi$

$$x = 0, \frac{\pi}{4}, \frac{\pi}{2}, \frac{3\pi}{4}, \pi, \frac{5\pi}{4}, \frac{3\pi}{2}, \frac{7\pi}{4}$$

13. $\sec^4 2x = 4,\ 0 \le 2x < 4\pi$

$\sec 2x = \pm\sqrt[4]{4} = \pm\sqrt{2}$

If $\sec 2x = \sqrt{2}$,

$$2x = \frac{\pi}{4}, \frac{7\pi}{4}, \frac{9\pi}{4}, \frac{15\pi}{4}.$$

If $\sec 2x = -\sqrt{2}$,

$$2x = \frac{3\pi}{4}, \frac{5\pi}{4}, \frac{11\pi}{4}, \frac{13\pi}{4}.$$

$$x = \frac{\pi}{8}, \frac{3\pi}{8}, \frac{5\pi}{8}, \frac{7\pi}{8}, \frac{9\pi}{8}, \frac{11\pi}{8}, \frac{13\pi}{8}, \frac{15\pi}{8}$$

15. $\sin \frac{x}{2} = \cos \frac{x}{2}, \ 0 \le \frac{x}{2} < \pi$

$$\sin^2 \frac{x}{2} = \cos^2 \frac{x}{2}$$

Use the identity

$$\sin^2 \theta + \cos^2 \theta = 1$$
$$\cos^2 \theta = 1 - \sin^2 \theta.$$

$$\sin^2 \frac{x}{2} = 1 - \sin^2 \frac{x}{2}$$

$$2 \sin^2 \frac{x}{2} = 1$$

$$\sin^2 \frac{x}{2} = \frac{1}{2}$$

$$\sin \frac{x}{2} = \pm\sqrt{\frac{1}{2}} = \pm\frac{\sqrt{2}}{2}$$

If $\sin \frac{x}{2} = \frac{\sqrt{2}}{2}$,

$$\frac{x}{2} = \frac{\pi}{4}, \frac{3\pi}{4}.$$

$\sin \frac{x}{2} = -\frac{\sqrt{2}}{2}$ has no solution in the interval $[0, \pi)$.

$$x = \frac{\pi}{2}, \frac{3\pi}{2}$$

Check solutions, since both sides were squared.

$x = \frac{\pi}{2}$:

$$\sin \frac{x}{2} = \sin \frac{\pi}{4} = \frac{\sqrt{2}}{2}$$

$$\cos \frac{x}{2} = \cos \frac{\pi}{2} = \frac{\sqrt{2}}{2}$$

$x = \frac{3\pi}{2}$:

$$\sin \frac{x}{2} = \sin \frac{3\pi}{4} = \frac{\sqrt{2}}{2}$$

$$\cos \frac{x}{2} = \cos \frac{3\pi}{4} = -\frac{\sqrt{2}}{2}$$

The only solution is $\pi/2$.

17. $\cos 2x + \cos x = 0, \ 0 \le x < 2\pi$

Use the identity

$\cos 2x = 2 \cos^2 x - 1.$

$$2 \cos^2 x - 1 + \cos x = 0$$
$$2 \cos^2 x + \cos x - 1 = 0$$
$$(2 \cos x - 1)(\cos x + 1) = 0$$

$$2 \cos x - 1 = 0$$
$$\cos x = \frac{1}{2}$$

$$x = \frac{\pi}{3}, \frac{5\pi}{3}$$

or

$$\cos x + 1 = 0$$
$$\cos x = -1$$
$$x = \pi$$

$$x = \frac{\pi}{3}, \pi, \frac{5\pi}{3}$$

19. $\sqrt{2} \sin 3\theta - 1 = 0$

$0° \le 3\theta < 3(360°) = 1080°$

$\sqrt{2} \sin 3\theta = 1$

$$\sin 3\theta = \frac{1}{\sqrt{2}} = \frac{\sqrt{2}}{2}$$

In quadrants I and II, sine is positive.

Therefore,

$3\theta = 45°, 135°, 405°, 495°, 765°, 855°$

$\theta = 15°, 45°, 135°, 165°, 255°, 285°.$

21. $\cos \frac{\theta}{2} = 1, \ 0 \le \frac{\theta}{2} < 180°$

$\frac{\theta}{2} = 0°$

$\theta = 0°$

23. $2\sqrt{3} \sin \frac{\theta}{2} = 3, \ 0° \le \frac{\theta}{2} < 180°$

$\sin \frac{\theta}{2} = \frac{3}{2\sqrt{3}}$

$= \frac{3\sqrt{3}}{6} = \frac{\sqrt{3}}{2}$

$\frac{\theta}{2} = 60°, 120°$

$\theta = 120°, 240°$

25. $2 \sin \theta = 2 \cos 2\theta$

$\sin \theta = \cos 2\theta$

Use the identity

$\cos 2\theta = 1 - 2 \sin^2 \theta$.

$\sin \theta = 1 - 2 \sin^2 \theta$

$2 \sin^2 \theta + \sin \theta - 1 = 0$

$(2 \sin \theta - 1)(\sin \theta + 1) = 0$

$2 \sin \theta - 1 = 0$

$\sin \theta = \frac{1}{2}$

$\theta = 30°, 150°$

or

$\sin \theta + 1 = 0$

$\sin \theta = -1$

$\theta = 270°$

$\theta = 30°, 150°, 270°$

27. $1 - \sin \theta = \cos 2\theta$

Use the identity

$\cos 2\theta = 1 - 2 \sin^2 \theta$.

$1 - \sin \theta = 1 - 2 \sin^2 \theta$

$2 \sin^2 \theta - \sin \theta = 0$

$\sin \theta (2 \sin \theta - 1) = 0$

$\sin \theta = 0$

$\theta = 0°, 180°$

or

$2 \sin \theta - 1 = 0$

$\sin \theta = \frac{1}{2}$

$\theta = 30°, 150°$

$\theta = 0°, 30°, 150°, 180°$

29. $\csc^2 \frac{\theta}{2} = 2 \sec \theta,$

Use the identities

$\csc x = \frac{1}{\sin x}$ and $\sec x = \frac{1}{\cos x}$.

$\frac{1}{\sin^2 \frac{\theta}{2}} = \frac{2}{\cos \theta}$

$2 \sin^2 \frac{\theta}{2} = \cos \theta$

Use the identity

$\sin^2 \frac{x}{2} = \frac{1 - \cos x}{2}$, and substitute.

$2\left(\frac{1 - \cos \theta}{2}\right) = \cos \theta$

$1 - \cos \theta = \cos \theta$

$1 = 2 \cos \theta$

$\frac{1}{2} = \cos \theta$

$\theta = 60°, 300°$

31. $2 - \sin 2\theta = 4 \sin 2\theta,$
$0° \le 2\theta < 720°$

$2 = 5 \sin 2\theta$

$\sin 2\theta = \frac{2}{5} = .4$

$2\theta = 23.6°, 156.4°, 383.6°, 516.4°$

$\theta = 11.8°, 78.2°, 191.8°, 258.2°$

33. $2\cos^2 2\theta = 1 - \cos 2\theta,$

$0° \le 2\theta < 720°$

$2\cos^2 2\theta + \cos 2\theta - 1 = 0$

$(2\cos 2\theta - 1)(\cos 2\theta + 1) = 0$

$2\cos 2\theta - 1 = 0$

$\cos 2\theta = \frac{1}{2}$

$2\theta = 60°, 300°, 420°, 660°$

$\theta = 30°, 150°, 210°, 330°$

or

$\cos 2\theta + 1 = 0$

$\cos 2\theta = -1$

$2\theta = 180°, 540°$

$\theta = 90°, 270°$

$\theta = 30°, 90°, 150°, 210°, 270°,$

$330°$

35. $\tan 3\theta + \sec 3\theta = 1,$

$0° \le 3\theta < 1080°$

$\tan 3\theta = 1 - \sec 3\theta$

$(\tan 3\theta)^2 = (1 - \sec 3\theta)^2$

$\tan^2 3\theta = 1 - 2\sec 3\theta + \sec^2 3\theta$

Use the identity

$\tan^2 x + 1 = \sec^2 x$

$\tan^2 x = \sec^2 x - 1.$

$\sec^2 3\theta - 1 = 1 - 2\sec 3\theta + \sec^2 3\theta$

$2\sec 3\theta = 2$

$\sec 3\theta = 1$

$3\theta = 0°, 360°, 720°$

$\theta = 0°, 120°, 240°$

A check shows that all these values are solutions.

37. $\sin x + \sin 3x = \cos x$

Since $0 \le x < 2\pi$, $0 \le 2x < 4\pi$.

Use the identity

$\sin a + \sin b$

$= 2\sin\left(\frac{a+b}{2}\right)\cos\left(\frac{a-b}{2}\right).$

$2\sin\left(\frac{x+3x}{2}\right)\cos\left(\frac{x-3x}{2}\right) = \cos x$

$2\sin 2x \cos(-x) = \cos x$

Use the identity $\cos x = \cos(-x)$.

$2\sin 2x \cos x = \cos x$

$2\sin 2x \cos x - \cos x = 0$

$\cos x(2\sin 2x - 1) = 0$

$\cos x = 0$

$x = \frac{\pi}{2}, \frac{3\pi}{2}$

or

$2\sin 2x - 1 = 0$

$\sin 2x = \frac{1}{2}$

$2x = \frac{\pi}{6}, \frac{5\pi}{6}, \frac{13\pi}{6}, \frac{17\pi}{6}$

$x = \frac{\pi}{12}, \frac{5\pi}{12}, \frac{13\pi}{12}, \frac{17\pi}{12}$

$x = \frac{\pi}{12}, \frac{5\pi}{12}, \frac{\pi}{2}, \frac{13\pi}{12}, \frac{17\pi}{12}, \frac{3\pi}{2}$

39. $\sin 3x - \sin x = 0$

Since $0 \le x < 2\pi$, $0 \le 2x < 4\pi$.

Use the identity

$\sin a - \sin b$

$= 2\cos\left(\frac{a+b}{2}\right)\sin\left(\frac{a-b}{2}\right).$

$2\cos\left(\frac{3x+x}{2}\right)\sin\left(\frac{3x-x}{2}\right) = 0$

$2\cos 2x \sin x = 0$

$\cos 2x \sin x = 0$

$\cos 2x = 0$

$2x = \frac{\pi}{2}, \frac{3\pi}{2}, \frac{5\pi}{2}, \frac{7\pi}{2}$

$x = \frac{\pi}{4}, \frac{3\pi}{4}, \frac{5\pi}{4}, \frac{7\pi}{4}$

or

$\sin x = 0$

$x = 0, \pi$

$x = 0, \frac{\pi}{4}, \frac{3\pi}{4}, \pi, \frac{5\pi}{4}, \frac{7\pi}{4}$

41. $\sin 4x + \sin 2x = 2 \cos x,$

$0 \le x < 2\pi,\ 0 \le 3x < 6\pi$

Use the identity

$\sin a + \sin b$

$= 2 \sin \left(\frac{a + b}{2}\right) \cos \left(\frac{a - b}{2}\right).$

$$2 \sin \left(\frac{4x + 2x}{2}\right) \cos \left(\frac{4x - 2x}{2}\right) = 2 \cos x$$

$$2 \sin 3x \cos x = 2 \cos x$$

$$2 \sin 3x \cos x - 2 \cos x = 0$$

$$2 \cos x(\sin 3x - 1) = 0$$

$\cos x = 0$

$x = \frac{\pi}{2}, \frac{3\pi}{2}$

or

$\sin 3x - 1 = 0$

$\sin 3x = 1$

$3x = \frac{\pi}{2}, \frac{5\pi}{2}, \frac{9\pi}{2}$

$x = \frac{\pi}{6}, \frac{5\pi}{6}, \frac{3\pi}{2}$

$x = \frac{\pi}{6}, \frac{\pi}{2}, \frac{5\pi}{6}, \frac{3\pi}{2}$

43. If $\sin \frac{1}{2}x = -\frac{1}{2},$

we must write

$\frac{1}{2}x = 210° + n \cdot 360°$

or

$\frac{1}{2}x = 330° + n \cdot 360°,$

where n is any integer. Then multiplying by 2, we have

$x = 420° + n \cdot 720°$

or

$x = 660° + n \cdot 720°.$

Since $n \in \{\ldots, -3, -2, -1, 0, 1, 2, 3, \ldots\}$, there is no value of n which will give values of x in the interval $[0°, 360°)$.

45. $h = \frac{35}{3} + \frac{7}{3} \sin \frac{2\pi x}{365}$

(a) Find x such that $h = 14$.

$$14 = \frac{35}{3} + \frac{7}{3} \sin \frac{2\pi x}{365}$$

$$14 - \frac{35}{3} = \frac{7}{3} \sin \frac{2\pi x}{365}$$

$$\frac{7}{3} = \frac{7}{3} \sin \frac{2\pi x}{365}$$

$$\sin \frac{2\pi x}{365} = 1$$

$\frac{2\pi x}{365} = \frac{\pi}{2} + 2\pi n$, n is an integer.

$$x = \left(\frac{\pi}{2} + 2\pi n\right)\left(\frac{365}{2\pi}\right)$$

$$= \frac{365}{4} \pm 365n$$

$$= 91.25 \pm 365n$$

x = 91.25 means about 91.3 days after March 21, on June 20.

(b) h assumes its least value when $\sin \frac{2\pi x}{365}$ takes on its least value, which is −1.

$$\sin \frac{2\pi x}{365} = -1$$
$$\frac{2\pi x}{365} = \frac{3\pi}{2} \pm 2\pi n, \text{ n is an integer.}$$
$$x = \left(\frac{3\pi}{2} \pm 2\pi n\right)\left(\frac{365}{2\pi}\right)$$
$$= \frac{3(365)}{4} \pm 365n$$
$$= 273.75 \pm 365$$

x = 273.75 means about 273.8 days after March 21, on December 19.

(c) Let h = 10.

$$10 = \frac{35}{3} + \frac{7}{3} \sin \frac{2\pi x}{365}$$
$$30 = 35 + 7 \sin \frac{2\pi x}{365}$$
$$-5 = 7 \sin \frac{2\pi x}{365}$$
$$-\frac{5}{7} = \sin \frac{2\pi x}{365}$$
$$-.71428571 = \sin \left(\frac{2\pi x}{365}\right)$$

In quadrants III and IV, sine is negative.

In quadrant III,

$$\frac{2\pi x}{365} = \pi + .79560295$$
$$= 3.9371956$$
$$x = \frac{365}{2\pi}(3.9371956)$$
$$= 228.7$$

x = 228.7 means 228.7 days after March 21, on November 4.

In quadrant IV,

$$\frac{2\pi x}{365} = 2\pi - .79560295$$
$$= 5.4875823$$
$$x = \frac{365}{2\pi}(5.4878263)$$
$$= 318.8$$

x = 318.8 means about 318.8 days after March 21, on February 2.

47. $i = I_{max} \sin 2\pi ft$

Let $i = 40$, $I_{max} = 100$, $f = 60$.

$$40 = 100 \sin 2\pi(60)(t)$$
$$40 = 100 \sin 120\pi t$$
$$.4 = \sin 120\pi t$$

From a calculator,

$$120\pi t = .4115168$$
$$t = \frac{.4115168}{120\pi}$$
$$t = .0010916$$
$$t \approx .001 \text{ sec}$$

49. $i = I_{max} \sin 2\pi ft$

Let $i = I_{max}$, $f = 60$.

$$I_{max} = I_{max} \sin 2\pi(60)t$$
$$1 = \sin 120\pi t$$
$$120\pi t = \frac{\pi}{2}$$
$$120t = \frac{1}{2}$$
$$t = \frac{1}{240} \approx .004 \text{ sec}$$

Section 6.4

1. $y = 5 \cos x$

$\frac{y}{5} = \cos x$

$x = \text{Arccos } \frac{y}{5}$

3. $2y = \cot 3x$

$3x = \text{Arccot } 2y$

$x = \frac{1}{3} \text{Arccot } 2y$

5. $y = 3 \tan 2x$

$\frac{y}{3} = \tan 2x$

$2x = \text{Arctan } \frac{y}{3}$

$x = \frac{1}{2} \text{Arctan } \frac{y}{3}$

7. $y = 6 \cos \frac{x}{4}$

$\frac{y}{6} = \cos \frac{x}{4}$

$\frac{x}{4} = \text{Arccos } \frac{y}{6}$

$x = 4 \text{ Arccos } \frac{y}{6}$

9. $y = -2 \cos 5x$

$-\frac{y}{2} = \cos 5x$

$5x = \text{Arccos } (-\frac{y}{2})$

$x = \frac{1}{5} \text{Arccos } (-\frac{y}{2})$

11. $y = \cos (x + 3)$

$x + 3 = \text{Arccos } y$

$x = -3 + \text{Arccos } y$

13. $y = \sin x - 2$

$y + 2 = \sin x$

$x = \text{Arcsin } (y + 2)$

15. $y = 2 \sin x - 4$

$y + 4 = 2 \sin x$

$\frac{y + 4}{2} = \sin x$

$x = \text{Arcsin } (\frac{y + 4}{2})$

17. The expression sin x - 2 means (sin x) - 2 which is not equivalent to the expression sin (x - 2).

19. $\frac{4}{3} \text{Sin}^{-1} \frac{y}{4} = \pi$

$\text{Sin}^{-1} \frac{y}{4} = \frac{3\pi}{4}$

$\frac{y}{4} = \text{Sin } \frac{3\pi}{4}$

$y = 4 \text{ Sin } \frac{3\pi}{4}$

$y = 4(\frac{\sqrt{2}}{2})$

$y = 2\sqrt{2}$

21. $2 \text{ Arccos } (\frac{y - \pi}{3}) = 2\pi$

$\text{Arccos } (\frac{y - \pi}{3}) = \pi$

$\frac{y - \pi}{3} = \text{Cos } \pi$

$\frac{y - \pi}{3} = -1$

$y - \pi = -3$

$y = \pi - 3$

23. Arcsin x = Arctan $\frac{3}{4}$

Let Arctan $\frac{3}{4}$ = u, so Tan u = $\frac{3}{4}$, u is in quadrant I.

Sketch a triangle and label it. The hypotenuse is $\sqrt{3^2 + 4^2} = 5$.

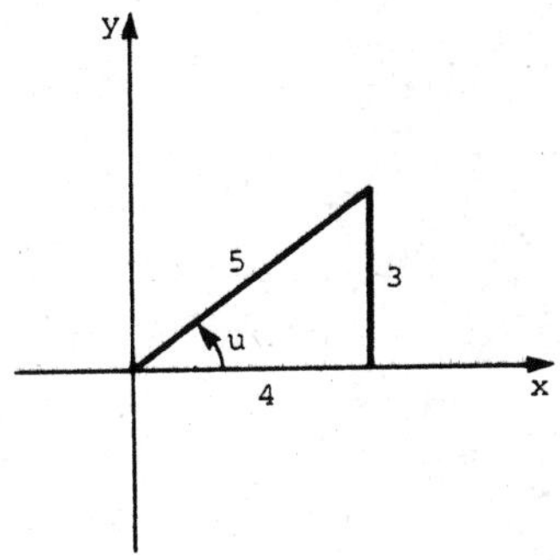

$$\sin u = \frac{3}{r} = \frac{3}{5}$$

The equation becomes

Arcsin x = u, or x = Sin u.

$$x = \frac{3}{5}$$

25. $\text{Cos}^{-1} x = \text{Sin}^{-1} \frac{3}{5}$

Let $\text{Sin}^{-1} \frac{3}{5} = u$, so Sin u = $\frac{3}{5}$.

u is in quadrant I.

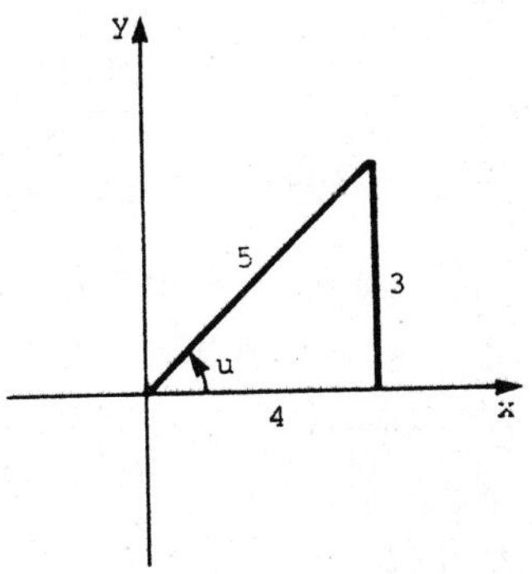

$$\cos u = \frac{4}{5}$$

The equation becomes

$\text{Cos}^{-1} x = u$, or x = Cos u.

$$x = \frac{4}{5}$$

27. $\text{Sin}^{-1} x - \text{Tan}^{-1} 1 = -\frac{\pi}{4}$

$$\text{Sin}^{-1} x = \text{Tan}^{-1} 1 - \frac{\pi}{4}$$

$$\text{Sin}^{-1} x = \frac{\pi}{4} - \frac{\pi}{4}$$

$$\text{Sin}^{-1} x = 0$$

$$\text{Sin } 0 = x$$

$$x = 0$$

29. Arccos x + 2 Arcsin $\frac{\sqrt{3}}{2} = \pi$

$$\text{Arccos } x = \pi - 2 \text{ Arcsin } \frac{\sqrt{3}}{2}$$

$$\text{Arccos } x = \pi - 2\left(\frac{\pi}{3}\right)$$

$$\text{Arccos } x = \pi - \frac{2\pi}{3}$$

$$\text{Arccos } x = \frac{\pi}{3}$$

$$x = \text{Cos } \frac{\pi}{3}$$

$$x = \frac{1}{2}$$

31. Arcsin 2x + Arccos x = $\frac{\pi}{6}$

$$\text{Arcsin } 2x = \frac{\pi}{6} - \text{Arccos } x$$

$$2x = \sin\left(\frac{\pi}{6} - \text{Arccos } x\right)$$

Use identity for sin (A − B).

$$2x = \sin \frac{\pi}{6} \cos (\text{Arccos } x) - \cos \frac{\pi}{6} \sin (\text{Arccos } x)$$

Let u = Arccos x.

$\text{Cos } u = x = \frac{x}{1}$

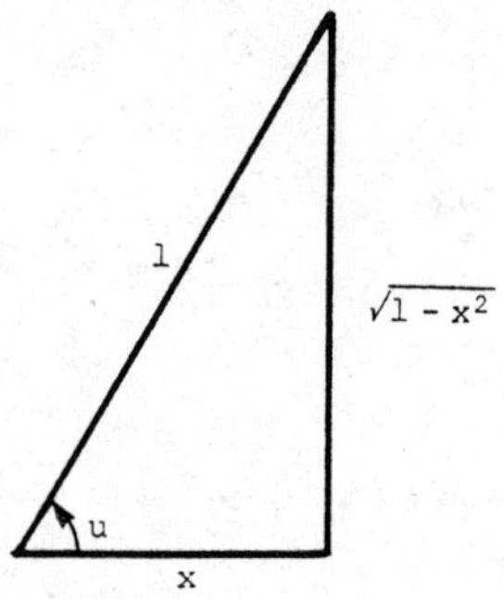

$\sin u = \sqrt{1 - x^2}$

$$2x = \sin\frac{\pi}{6} \cdot \cos u - \cos\frac{\pi}{6}\sin u$$

$$2x = \frac{1}{2}x - \frac{\sqrt{3}}{2}(\sqrt{1-x^2})$$

$$4x = x - \sqrt{3}\sqrt{1-x^2}$$

$$3x = -\sqrt{3}\sqrt{1-x^2}$$

$$(3x)^2 = (-\sqrt{3}\sqrt{1-x^2})^2$$

$$9x^2 = 3(1-x^2)$$

$$9x^2 = 3 - 3x^2$$

$$12x^2 = 3$$

$$x^2 = \frac{3}{12} = \frac{1}{4}$$

$$x = \pm\frac{1}{2}$$

Check these solutions since both sides were squared.

$x = \frac{1}{2}$:

$$\text{Arcsin }\left(2 \cdot \frac{1}{2}\right) + \text{Arccos }\left(\frac{1}{2}\right)$$

$$= \frac{\pi}{2} + \frac{\pi}{3} \neq \frac{\pi}{6}$$

so 1/2 is not a solution.

$x = -\frac{1}{2}$:

$$\text{Arcsin }\left(2\left(-\frac{1}{2}\right)\right) + \text{Arccos }\left(-\frac{1}{2}\right)$$

$$= -\frac{\pi}{2} + \frac{2\pi}{3} = \frac{\pi}{6}$$

The only solution is −1/2.

33. $\text{Cos}^{-1} x + \text{Tan}^{-1} x = \frac{\pi}{2}$

$$\text{Cos}^{-1} x = \frac{\pi}{2} - \text{Tan}^{-1} x$$

$$x = \text{Cos }\left(\frac{\pi}{2} - \text{Tan}^{-1} x\right)$$

Use identity for cos (A − B).

$$x = \cos\frac{\pi}{2}\cos(\text{Tan}^{-1} x) + \sin\frac{\pi}{2}\sin(\text{Tan}^{-1} x)$$

Let $u = \text{Tan}^{-1} x$.

Tan u = x

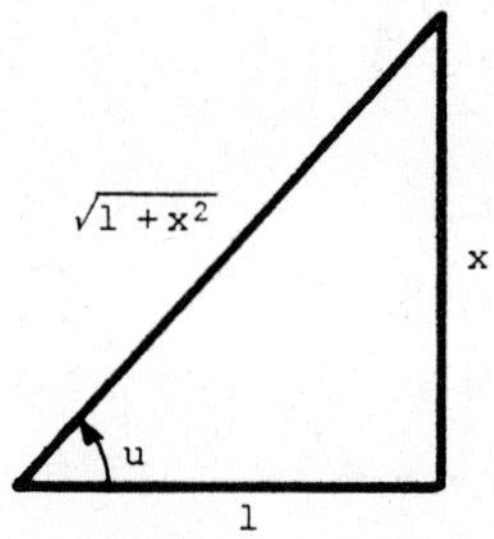

$$\cos u = \frac{1}{\sqrt{1+x^2}}$$

$$\sin u = \frac{x}{\sqrt{1+x^2}}$$

$$x = \cos\frac{\pi}{2} \cdot \cos u + \sin\frac{\pi}{2} \cdot \sin u$$

$$x = 0\left(\frac{1}{\sqrt{1+x^2}}\right) + 1 \cdot \frac{x}{\sqrt{1+x^2}}$$

$$x = \frac{x}{\sqrt{1+x^2}}$$

$$x - \frac{x}{\sqrt{1 + x^2}} = 0$$

$$x\left(1 - \frac{1}{\sqrt{1 + x^2}}\right) = 0$$

$x = 0,$

or

$$1 - \frac{1}{\sqrt{1 + x^2}} = 0$$

$$\frac{1}{\sqrt{1 + x^2}} = 1$$

$$\sqrt{1 + x^2} = 1$$

$$1 + x^2 = 1$$

$$x^2 = 0$$

$$x = 0$$

35. $d = 550 + 450 \cos \frac{\pi}{50}t$

$$d - 550 = 450 \cos \frac{\pi}{50}t$$

$$\frac{d - 550}{450} = \cos \frac{\pi}{50}t$$

$$\frac{\pi}{50} \cdot t = \text{Arccos}\left(\frac{d - 550}{450}\right)$$

$$t = \frac{50}{\pi} \text{Arccos}\left(\frac{d - 550}{450}\right)$$

37. (a) $e = E_{max} \sin 2\pi ft$

$$\frac{e}{E_{max}} = \sin 2\pi ft$$

$$2\pi ft = \text{Arcsin} \frac{e}{E_{max}}$$

$$t = \frac{1}{2\pi f} \text{Arcsin} \frac{e}{E_{max}}$$

(b) Let $E_{max} = 12$, $e = 5$, and $f = 100$.

$$t = \frac{1}{2\pi(100)} \text{Arcsin} \frac{5}{12}$$

$$t = \frac{1}{200\pi} \text{Arcsin } .41666667$$

$$= .00068 \text{ sec}$$

39. (a) $u = \text{Arcsin } x$

$x = \text{Sin } u, \; -\frac{\pi}{2} \le u \le \frac{\pi}{2}$

(b) See the answer graph at the back of the textbook.

(c) $\tan u = \dfrac{x}{\sqrt{1 - x^2}}$

$$= \frac{x\sqrt{1 - x^2}}{1 - x^2}$$

(d) $u = \text{Arctan } \dfrac{x\sqrt{1 - x^2}}{1 - x^2}$

Chapter 6 Review Exercises

1. $y = \text{Sin}^{-1} \frac{\sqrt{2}}{2}$

$\text{Sin } y = \frac{\sqrt{2}}{2}, \; -\frac{\pi}{2} \le y \le \frac{\pi}{2}$

$y = \frac{\pi}{4}$

3. $y = \text{Tan}^{-1} (-\sqrt{3})$

$\text{Tan } y = -\sqrt{3}, \; -\frac{\pi}{2} \le y \le \frac{\pi}{2}$

$y = -\frac{\pi}{3}$

5. $y = \text{Cos}^{-1} \left(-\frac{\sqrt{2}}{2}\right)$

$\text{Cos } y = -\frac{\sqrt{2}}{2}, \; 0 \le y \le \pi$

$y = \frac{3\pi}{4}$

7. $y = \text{Sec}^{-1} (-2)$

$\text{Sec } y = -2, \; 0 \le y \le \pi, \; y \ne \frac{\pi}{2}$

$y = \frac{2\pi}{3}$

9. $y = \text{Arccot}\ (-1)$

$\text{Cot}\ y = -1,\ 0 < y < \pi$

$y = \frac{3\pi}{4}$

11. $\theta = \text{Arcsin}\ \left(-\frac{\sqrt{3}}{2}\right)$

$\text{Sin}\ \theta = -\frac{\sqrt{3}}{2},\ -90° \le \theta \le 90°$

$\theta = -60°$

13. $\theta = \text{Arctan}\ 1.7804675$

Enter: 1.7804675 [INV] [TAN]

Display: 60.679245

$\theta = 60.679245°$

15. $\theta = \text{Cos}^{-1}\ .80396577$

Enter: .80396577 [INV] [COS]

Display: 36.489508

$\theta = 36.489508°$

17. $\theta = \text{Arcsec}\ 3.4723155$

Enter: 3.4723155 [1/x] [INV] [COS]

Display: 73.262206

$\theta = 73.262206°$

19. If $\theta = \text{Sin}^{-1}\ 3$, then $\text{Sin}\ \theta = 3$ and $-90° \le \theta \le 90°$. But the sine function can never have a value greater than 1, so $\text{Sin}\ \theta = 3$ is impossible and $\sin^{-1} 3$ cannot be defined.

21. The domain of the Arccot function is $(-\infty, \infty)$ since the range of the cotangent function is $(-\infty, \infty)$.

23. $\sin\left(\text{Sin}^{-1}\ \frac{1}{2}\right)$

Let $\omega = \text{Sin}^{-1}\ \frac{1}{2}$.

Then

$$\sin \omega = \frac{1}{2},\ -\frac{\pi}{2} \le \omega \le \frac{\pi}{2}.$$

$\sin\left(\text{Sin}^{-1}\ \frac{1}{2}\right) = \sin \omega = \frac{1}{2}$

25. $\cos\ (\text{Arccos}\ (-1))$

Let $\omega = \text{Arccos}\ (-1)$

Then $\cos \omega = -1,\ 0 \le \omega \le \pi$.

$\cos\ (\text{Arccos}\ (-1)) = \cos \omega = -1$

27. $\text{Arccos}\left(\cos \frac{3\pi}{4}\right) = x$

$\cos x = \cos \frac{3\pi}{4},\ 0 \le x \le \pi,$

so $x = \frac{3\pi}{4}$.

$\text{Arccos}\left(\cos \frac{3\pi}{4}\right) = \frac{3\pi}{4}$

29. $\text{Tan}^{-1}\left(\tan \frac{\pi}{4}\right) = x$

$\tan x = \tan \frac{\pi}{4},\ -\frac{\pi}{2} < x < \frac{\pi}{2},$

so $x = \frac{\pi}{4}$.

$\text{Tan}^{-1}\left(\tan \frac{\pi}{4}\right) = \frac{\pi}{4}$

31. $\sin\left(\text{Arccos}\ \frac{3}{4}\right)$

Let $\omega = \text{Arccos}\ \frac{3}{4}$, so $\text{Cos}\ \omega = \frac{3}{4}$,

$0 \le \omega \le \pi$.

Sketch a triangle and label it using $\text{Cos}\ \omega = 3/4$. The side opposite $\omega = \sqrt{4^2 - 3^2} = \sqrt{7}$.

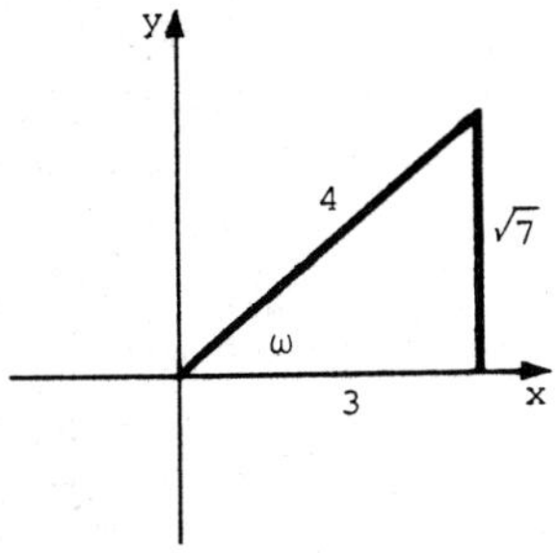

$\sin \omega = \frac{\sqrt{7}}{4}$

$\sin\left(\text{Arccos } \frac{3}{4}\right) = \sin \omega = \frac{\sqrt{7}}{4}$

33. $\cos(\text{Csc}^{-1}(-2))$

Let $\omega = \text{Csc}^{-1}(-2)$, $-\frac{\pi}{2} \le \omega < \frac{\pi}{2}$,

$\omega \ne 0$,

$\csc \omega = -2$, ω is in quadrant IV.

The side adjacent to ω is

$$\sqrt{2^2 - (-1)^2} = \sqrt{3}.$$

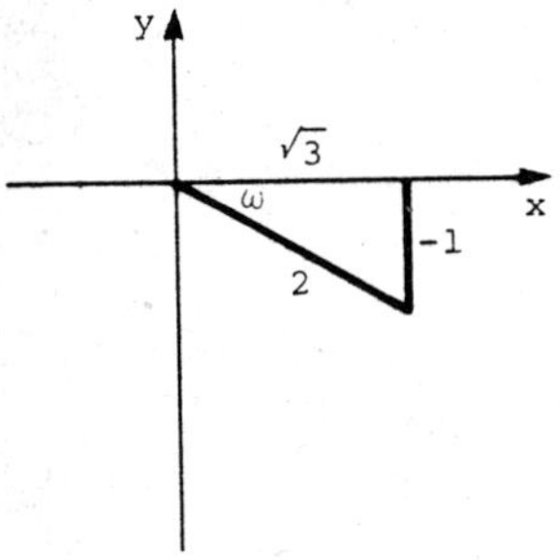

$\cos \omega = \frac{\sqrt{3}}{2}$

$\cos(\text{Csc}^{-1}(-2)) = \frac{\sqrt{3}}{2}$

35. $\tan\left(\text{Arcsin } \frac{3}{5} + \text{Arccos } \frac{5}{7}\right)$

Let $\omega_1 = \text{Arcsin } \frac{3}{5}$, $-\frac{\pi}{2} \le \omega_1 \le \frac{\pi}{2}$.

$\sin \omega_1 = \frac{3}{5}$ The side adjacent to ω_1 is

$$\sqrt{5^2 - 3^2} = 4.$$

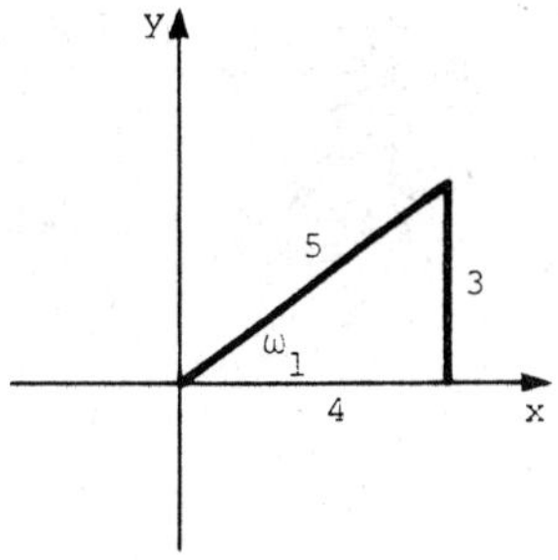

$\tan \omega_1 = \frac{3}{4}$

Let $\omega_2 = \text{Arccos } \frac{5}{7}$, $0 \le \omega_2 \le \pi$.

$\cos \omega_2 = \frac{5}{7}$ The side opposite ω_2 is

$$\sqrt{7^2 - 5^2} = \sqrt{24} = 2\sqrt{6}.$$

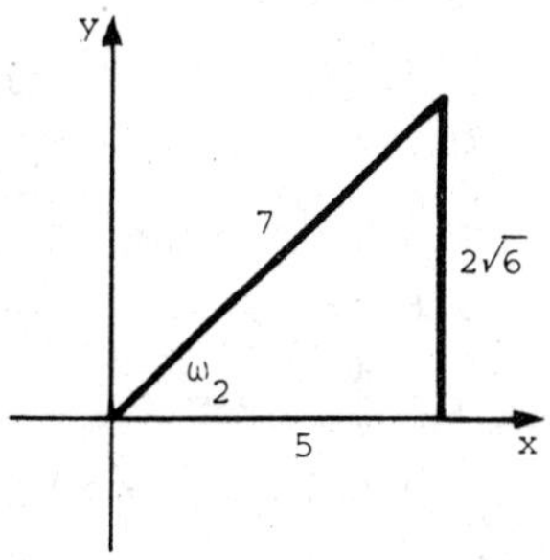

$\tan \omega_2 = \frac{2\sqrt{6}}{5}$

Use $\tan(\omega_1 + \omega_2)$

$$= \frac{\tan \omega_1 + \tan \omega_2}{1 - \tan \omega_1 \tan \omega_2}.$$

$$\tan\left(\text{Arcsin } \frac{3}{5} + \text{Arccos } \frac{5}{7}\right)$$

$$= \frac{\frac{3}{4} + \frac{2\sqrt{6}}{5}}{1 - \left(\frac{3}{4}\right)\left(\frac{2\sqrt{6}}{5}\right)} = \frac{\frac{15 + 8\sqrt{6}}{20}}{\frac{20 - 6\sqrt{6}}{20}}$$

$$= \frac{15 + 8\sqrt{6}}{20 - 6\sqrt{6}} \cdot \frac{20 + 6\sqrt{6}}{20 + 6\sqrt{6}}$$

$$= \frac{588 + 250\sqrt{6}}{184}$$

$$= \frac{294 + 125\sqrt{6}}{92}$$

37. $\cos\left(\text{Arctan } \frac{u}{\sqrt{1 - u^2}}\right)$

Let $\omega = \text{Arctan } \frac{u}{\sqrt{1 - u^2}}$, $-\frac{\pi}{2} < \omega < \frac{\pi}{2}$,

$\tan \omega = \frac{u}{\sqrt{1 - u^2}}$ The hypotenuse is

$\sqrt{(1 - u^2) + u^2} = 1.$

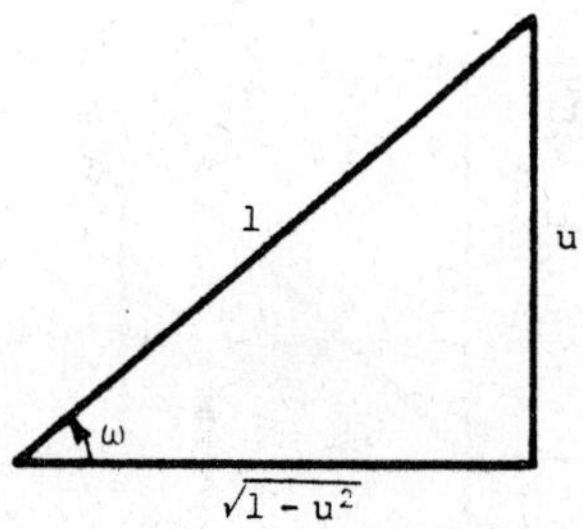

$\cos\left(\text{Arctan } \frac{u}{\sqrt{1 - u^2}}\right)$

$= \cos \omega = \sqrt{1 - u^2}$

39. $y = \text{Sin}^{-1} x$

Domain: $[-1, 1]$

Range: $\left[-\frac{\pi}{2}, \frac{\pi}{2}\right]$

See the answer graph at the back of the textbook.

41. $y = \text{Arccot } x$

Domain: $(-\infty, \infty)$

Range: $(0, \pi)$

See the answer graph at the back of the textbook.

43. $2 \tan x - 1 = 0$

$2 \tan x = 1$

$\tan x = \frac{1}{2}$

$x = .46364761$ or $x = 3.6052403$

45. $\tan x = \cot x$, $0 \le x < 2\pi$

Use the identity $\cot x = \frac{1}{\tan x}$.

$\tan x = \frac{1}{\tan x}$, $\tan \ne 0$

$\tan^2 x = 1$

$\tan x = \pm 1$

If $\tan x = 1$,

$x = \frac{\pi}{4}, \frac{5\pi}{4}$.

If $\tan x = -1$,

$x = \frac{3\pi}{4}, \frac{7\pi}{4}$.

$x = \frac{\pi}{4}, \frac{3\pi}{4}, \frac{5\pi}{4}, \frac{7\pi}{4}$

47. $\tan^2 2x - 1 = 0,\ 0 \le 2x < 4\pi$

$\tan^2 2x = 1$

$\tan 2x = \pm 1$

If $\tan 2x = 1$,

$$2x = \frac{\pi}{4}, \frac{5\pi}{4}, \frac{9\pi}{4}, \frac{13\pi}{4}$$

$$x = \frac{\pi}{8}, \frac{5\pi}{8}, \frac{9\pi}{8}, \frac{13\pi}{8}.$$

If $\tan 2x = -1$,

$$2x = \frac{3\pi}{4}, \frac{7\pi}{4}, \frac{11\pi}{4}, \frac{15\pi}{4}$$

$$x = \frac{3\pi}{8}, \frac{7\pi}{8}, \frac{11\pi}{8}, \frac{15\pi}{8}.$$

$$x = \frac{\pi}{8}, \frac{3\pi}{8}, \frac{5\pi}{8}, \frac{7\pi}{8}, \frac{9\pi}{8}, \frac{11\pi}{8}, \frac{13\pi}{8}, \frac{15\pi}{8}$$

49. $\cos 2x + \cos x = 0,\ 0 \le x < 2\pi$

Use the identity

$\cos 2x = 2\cos^2 x - 1.$

$$2\cos^2 x - 1 + \cos x = 0$$

$$2\cos^2 x + \cos x - 1 = 0$$

$$(2\cos x - 1)(\cos x + 1) = 0$$

$$2\cos x - 1 = 0$$

$$\cos x = \frac{1}{2}$$

$$x = \frac{\pi}{3}, \frac{5\pi}{3}$$

or

$$\cos x + 1 = 0$$

$$\cos x = -1$$

$$x = \pi$$

$$x = \frac{\pi}{3}, \pi, \frac{5\pi}{3}$$

51. $\sin^2\theta + 3\sin\theta + 2 = 0,$
$0^\circ \le \theta < 360^\circ$

$$(\sin\theta + 2)(\sin\theta + 1) = 0$$

$$\sin\theta + 2 = 0$$

$$\sin\theta = -2$$

-2 is not in the range of $\sin\theta$ so there is no solution.

On the other hand,

$$\sin\theta + 1 = 0$$

$$\sin\theta = -1$$

$$\theta = 270^\circ.$$

53. $\sin 2\theta = \cos 2\theta + 1,\ 0^\circ \le 2\theta < 720^\circ$

$$(\sin 2\theta)^2 = (\cos 2\theta + 1)^2$$

$$\sin^2 2\theta = \cos^2 2\theta + 2\cos 2\theta + 1$$

Use the identity

$\sin^2 x + \cos^2 x = 1$

$\sin^2 x = 1 - \cos^2 x.$

$$1 - \cos^2 2\theta = \cos^2 2\theta + 2\cos 2\theta + 1$$

$$2\cos^2 2\theta + 2\cos 2\theta = 0$$

$$\cos^2 2\theta + \cos 2\theta = 0$$

$$\cos 2\theta(\cos 2\theta + 1) = 0$$

$$\cos 2\theta = 0$$

$$2\theta = 90^\circ, 270^\circ, 450^\circ, 630^\circ$$

or

$$\cos 2\theta + 1 = 0$$

$$\cos 2\theta = -1$$

$$2\theta = 180^\circ, 540^\circ$$

Possible values for θ are

$\theta = 45^\circ, 135^\circ, 225^\circ, 315^\circ, 90^\circ, 270^\circ.$

Check these solutions since both sides were squared. A value for θ will be a solution if

$\sin 2\theta - \cos 2\theta = 1.$

$\theta = 45°,\ 2\theta = 90°$

$$\sin 90° - \cos 90° = 1 - 0 = 1$$

$\theta = 90°,\ 2\theta = 180°$

$$\sin 180° - \cos 180° = 0 - (-1) = 1$$

$\theta = 135°,\ 2\theta = 270°$

$$\sin 270° - \cos 270° = -1 - 0 \neq 1$$

$\theta = 225°,\ 2\theta = 450°$

$$\sin 450° - \cos 450° = 1 - 0 = 1$$

$\theta = 270°,\ 2\theta = 540°$

$$\sin 540° - \cos 540° = 0 - (-1) = 1$$

$\theta = 315°,\ 2\theta = 630°$

$$\sin 630° - \cos 630° = -1 - 0 \neq 1$$

Solutions are 45°, 90°, 225°, 270°.

55. $3 \cos^2 \theta + 2 \cos \theta - 1 = 0$

$(3 \cos \theta - 1)(\cos \theta + 1) = 0$

$3 \cos \theta - 1 = 0$ or $\cos \theta + 1 = 0$

$\cos \theta = \frac{1}{3}$ or $\cos \theta = -1$

$\theta = 70.5°, 289.5°$, or $\theta = 180°$

57. $\sin 2\theta + \sin 4\theta = 0,$

$0° \leq 2\theta < 720°$

$\sin 2\theta + \sin [2(2\theta)] = 0$

Use the identity

$$\sin 2x = 2 \sin x \cos x.$$

$\sin 2\theta + 2 \sin 2\theta \cos 2\theta = 0$

$\sin 2\theta(1 + 2 \cos 2\theta) = 0$

$\sin 2\theta = 0$

$2\theta = 0°, 180°, 360°, 540°$

$\theta = 0°, 90°, 180°, 270°$

or

$1 + 2 \cos 2\theta = 0$

$\cos 2\theta = -\frac{1}{2}$

$2\theta = 120°, 240°, 480°, 600°$

$\theta = 60°, 120°, 240°, 300°$

$\theta = 0°, 60°, 90°, 120°, 180°, 240°, 270°, 300°$

59. $4y = 2 \sin x$

$2y = \sin x$

$x = \text{Arcsin } 2y$

61. $2y = \tan (3x + 2)$

$3x + 2 = \text{Arctan } 2y$

$3x = \text{Arctan } 2y - 2$

$x = \frac{1}{3}(\text{Arctan } 2y - 2)$

$x = \left(\frac{1}{3} \text{ Arctan } 2y\right) - \frac{2}{3}$

63. $\frac{4}{3}$ Arctan $\frac{x}{2} = \pi$

Arctan $\frac{x}{2} = \frac{3\pi}{4}$

But, by definition, the range of Arctan is $(-\pi/2, \pi/2)$. So this equation has no solution.

65. Arccos x + Arctan 1 $= \frac{11\pi}{12}$

Arccos x $= \frac{11\pi}{12}$ − Arctan 1

Arccos x $= \frac{11\pi}{12} - \frac{\pi}{4}$

Arccos x $= \frac{11\pi}{12} - \frac{3\pi}{12} = \frac{8\pi}{12}$

Arccos x $= \frac{2\pi}{3}$

$\cos \frac{2\pi}{3} = x$

$x = -\frac{1}{2}$

67. $\frac{c_1}{c_2} = \frac{\sin \theta_1}{\sin \theta_2}$

$.752 = \frac{\sin \theta_1}{\sin \theta_2}$

If $\theta_2 = 90°$,

$.752 = \frac{\sin \theta_1}{\sin 90°}$

$= \frac{\sin \theta_1}{1}$

$= \sin \theta_1$

$\theta_1 = 48.8°$

CHAPTER 6 TEST

Do not use a calculator for this test.

[6.1] For each of the following, give the exact real number value of y.

1. $y = \text{Arccos } \frac{1}{2}$
2. $y = \text{Csc}^{-1} (-1)$
3. $y = \text{Arctan } (-\sqrt{3})$
4. $y = \text{Sec}^{-1} \sqrt{2}$

[6.1] Find each of the following.

5. $\cos \left(\text{Tan}^{-1} \frac{3}{4}\right)$
6. $\text{Sin}^{-1} \left(\cos \frac{\pi}{6}\right)$
7. $\sin \left(\text{Arcsin } \frac{2}{3} - \text{Arccos } \frac{1}{4}\right)$
8. $\text{Cot}^{-1} \left[\cot \left(-\frac{\pi}{3}\right)\right]$ Explain why the answer is *not* $-\pi/3$.

[6.2] Find all solutions for the following equations in the interval [0°, 360°).

9. $\sec \theta - \csc \theta = 0$
10. $\tan^2 \theta = \tan \theta$
11. $\sec^4 \theta = 2 \sec^2 \theta$
12. $\sin^2 \theta = 2 \sin \theta - 1$

Find all solutions for the following equations in the interval [0, 2π).

[6.2] 13. $\sec^2 x = 2\sqrt{3} \tan x - 2$

[6.3] 14. $2 \cos 2x = 1$

[6.3] 15. $\tan \frac{\theta}{2} = 1$

[6.3] 16. $\sin 2x = \cos x$

[6.4] **Solve each equation for x.**

17. $y = 3 \cos 2x$

18. $3y = 2 \tan x - 1$

[6.4] **Solve each equation.**

19. $\operatorname{Csc}^{-1} 2x = -\frac{\pi}{6}$

20. $\operatorname{Arcsin} x = \operatorname{Arctan} \frac{4}{3}$

[3.3] **Find the value of s in the interval $[0, \pi/2]$ that makes each of the following true.**

21. $\cot s = 1.3821596$

22. $\cos s = .92846507$

23. $\sec s = 3.8421657$

[3.4] 24. Find the linear velocity of a point on the tip of the minute hand of a clock, if the hand is 5 cm long.

[3.4] 25. Explain how you change radians per second to radians per hour.

CHAPTER 6 TEST ANSWERS

1. $\pi/3$ 2. $-\pi/2$ 3. $-\pi/3$ 4. $\pi/4$ 5. $4/5$ 6. $\pi/3$

7. $\frac{2 - 5\sqrt{3}}{12}$ 8. $2\pi/3$; $-\pi/3$ is not in the range of $\mathrm{Cot}^{-1}\, x$.

9. 45°, 225° 10. 0°, 45°, 225° 11. 45°, 135°, 225°, 315°

12. 90° 13. $\pi/3$, $4\pi/3$ 14. $\pi/6$, $5\pi/6$, $7\pi/6$, $11\pi/6$ 15. $\pi/2$

16. $\pi/6$, $\pi/2$, $5\pi/6$, $3\pi/2$ 17. $x = \frac{1}{2} \text{ Arccos } \frac{y}{3}$ 18. $x = \text{Arctan } \frac{3y + 1}{2}$

19. -1 20. $4/5$

CHAPTER 7 APPLICATIONS OF TRIGONOMETRY AND VECTORS

Section 7.1

1. $A = 37°$, $B = 48°$, $c = 18$ m

$C = 180° - A - B$

$C = 180° - 37° - 48°$

$C = 95°$

$$\frac{b}{\sin B} = \frac{c}{\sin C}$$

$$b = \frac{18 \sin 48°}{\sin 95°}$$

$b = 13$ m

$$\frac{a}{\sin A} = \frac{c}{\sin C}$$

$$a = \frac{18 \sin 37°}{\sin 95°}$$

$a = 11$ m

3. $A = 46°\ 30'$, $B = 52°\ 50'$, $b = 87.3$ mm

$C = 180° - A - B$

$C = 180° - 46°\ 30' - 52°\ 50'$

$C = 80°\ 40'$

$$\frac{a}{\sin A} = \frac{b}{\sin B}$$

$$a = \frac{87.3 \sin 46°\ 30'}{\sin 52°\ 50'}$$

$a = 79.5$ mm

$$\frac{b}{\sin B} = \frac{c}{\sin C}$$

$$c = \frac{87.3 \sin 80°\ 40'}{\sin 52°\ 50'}$$

$c = 108$ mm

5. $A = 27.2°$, $C = 115.5°$, $c = 76.0$ ft

$B = 180° - A - C$

$B = 180° - 27.2° - 115.5°$

$B = 37.3°$

$$\frac{a}{\sin A} = \frac{c}{\sin C}$$

$$a = \frac{76.0 \sin 27.2°}{\sin 115.5°}$$

$a = 38.5$ ft

$$\frac{b}{\sin B} = \frac{c}{\sin C}$$

$$b = \frac{76.0 \sin 37.3°}{\sin 115.5°}$$

$b = 51.0$ ft

7. $A = 68.41°$, $B = 54.23°$, $a = 12.75$ ft

$C = 180° - A - B$

$C = 180° - 68.41° - 54.23°$

$C = 57.36°$

$$\frac{a}{\sin A} = \frac{b}{\sin B}$$

$$b = \frac{12.75 \sin 54.23°}{\sin 68.41°}$$

$b = 11.13$ ft

$$\frac{a}{\sin A} = \frac{c}{\sin C}$$

$$c = \frac{12.75 \sin 57.36°}{\sin 68.41°}$$

$c = 11.55$ ft

9. $A = 87.2°$, $b = 75.9$ yd, $C = 74.3°$

$B = 180° - A - C$

$B = 180° - 87.2° - 74.3°$

$B = 18.5°$

$$\frac{a}{\sin A} = \frac{b}{\sin B}$$

$$a = \frac{75.9 \sin 87.2°}{\sin 18.5°}$$

$a = 239$ yd

$$\frac{b}{\sin B} = \frac{c}{\sin C}$$

$$c = \frac{75.9 \sin 74.3°}{\sin 18.5°}$$

$$c = 230 \text{ yd}$$

11. $B = 20° \ 50'$, $AC = 132$ ft, $C = 103° \ 10'$

$$A = 180° - B - C$$

$$A = 180° - 20° \ 50' - 103° \ 10'$$

$$A = 56° \ 00'$$

$$\frac{AC}{\sin B} = \frac{AB}{\sin C}$$

$$AB = \frac{132 \sin 103° \ 10'}{\sin 20° \ 50'}$$

$$AB = 361 \text{ ft}$$

$$\frac{BC}{\sin A} = \frac{AC}{\sin B}$$

$$BC = \frac{132 \sin 56° \ 00'}{\sin 20° \ 50'}$$

$$BC = 308 \text{ ft}$$

13. $A = 39.70°$, $C = 30.35°$, $b = 39.74$ m

$$B = 180° - A - C$$

$$B = 180° - 39.70° - 30.35°$$

$$B = 110.0° \quad \text{(rounded)}$$

$$\frac{a}{\sin A} = \frac{b}{\sin B}$$

$$a = \frac{39.74 \sin 39.70°}{\sin 109.95°}$$

$$a = 27.01 \text{ m}$$

$$\frac{b}{\sin B} = \frac{c}{\sin C}$$

$$c = \frac{39.74 \sin 30.35°}{\sin 109.95°}$$

$$c = 21.36 \text{ m}$$

15. $B = 42.88°$, $C = 102.40°$, $b = 3974$ ft

$$A = 180° - B - C$$

$$A = 180° - 42.88° - 102.40°$$

$$A = 34.72°$$

$$\frac{a}{\sin A} = \frac{b}{\sin B}$$

$$a = \frac{3974 \sin 34.72°}{\sin 42.88°}$$

$$a = 3326 \text{ ft}$$

$$\frac{b}{\sin B} = \frac{c}{\sin C}$$

$$c = \frac{3974 \sin 102.40°}{\sin 42.88°}$$

$$c = 5704 \text{ ft}$$

17. $A = 39° \ 54'$, $a = 268.7$ m, $B = 42° \ 32'$

$$C = 180° - A - B$$

$$C = 180° - 39° \ 54' - 42° \ 32'$$

$$C = 97° \ 34'$$

$$\frac{a}{\sin A} = \frac{b}{\sin B}$$

$$b = \frac{268.7 \sin 42° \ 32'}{\sin 39° \ 54'}$$

$$b = 283.2 \text{ m}$$

$$\frac{a}{\sin A} = \frac{c}{\sin C}$$

$$c = \frac{268.7 \sin 97° \ 34'}{\sin 39° \ 54'}$$

$$c = 415.2 \text{ m}$$

19. The law of sines cannot be used to solve a triangle if we are given the lengths of three sides of a triangle because the law of sines states

$$\frac{a}{\sin A} = \frac{b}{\sin B} = \frac{c}{\sin C}.$$

If we know a, b, and c, there is no way to solve for A, B, or C because we would have one equation with two unknowns or two equations with three unknowns.

23.

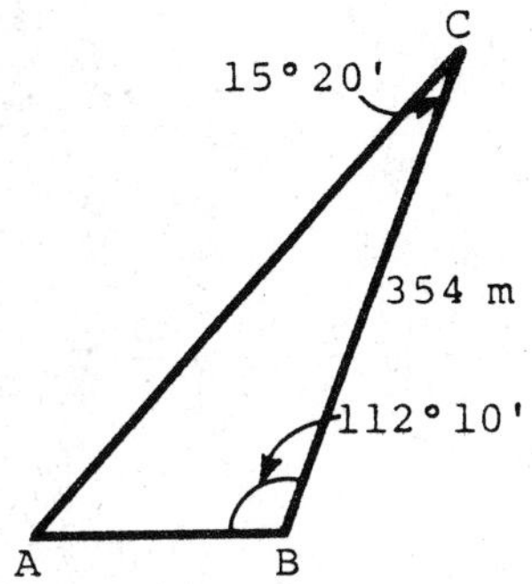

$A = 180° - B - C$

$A = 180° - 112°\ 10' - 15°\ 20'$

$A = 52°\ 30'$

$$\frac{BC}{\sin A} = \frac{AB}{\sin C}$$

$$AB = \frac{354 \sin 15°\ 20'}{\sin 52°\ 30'}$$

$AB = 118$ m

25. Let C = the transmitter.

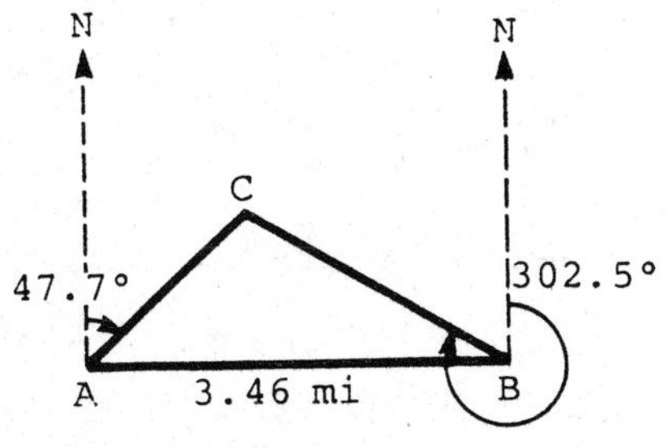

Since side AB is on an east-west line, the angle between it and any north-south line is 90°.

$A = 90° - 47.7° = 42.3°$

$B = 302.5° - 270° = 32.5°$

$C = 180° - A - B$

$= 180° - 42.3° - 32.5°$

$C = 105.2°$

$$\frac{AC}{\sin 32.5°} = \frac{3.46}{\sin 105.2°}$$

$$AC = \frac{3.46 \sin 32.5°}{\sin 105.2°}$$

$AC = 1.93$ mi

27. $$\frac{x}{\sin 54.8°} = \frac{12.0}{\sin 70.4°}$$

$$x = \frac{12.0 \sin 54.8°}{\sin 70.4°}$$

$x = 10.4$ in

29. Label α in the triangle as shown.

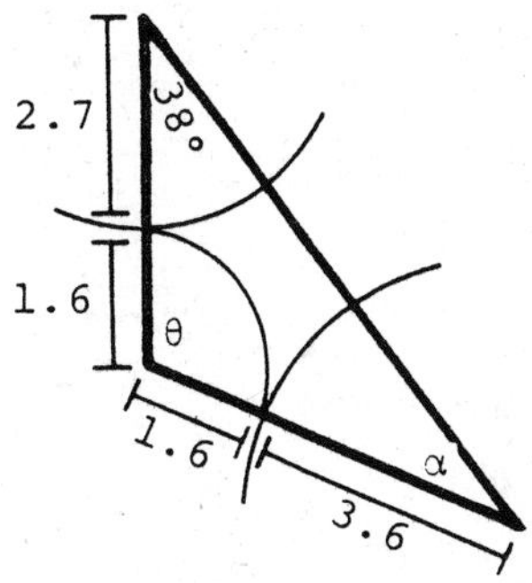

$$\frac{1.6 + 3.6}{\sin 38°} = \frac{1.6 + 2.7}{\sin \alpha}$$

$$\sin \alpha = \frac{(1.6 + 2.7) \sin 38°}{1.6 + 3.6}$$

$\alpha = 31°$

$\theta = 180° - 38° - \alpha$

$= 180° - 38° - 31°$

$\theta = 111°$

31. Let x = the distance to the lighthouse at bearing N 37° E

y = the distance to the lighthouse at bearing N 25° E.

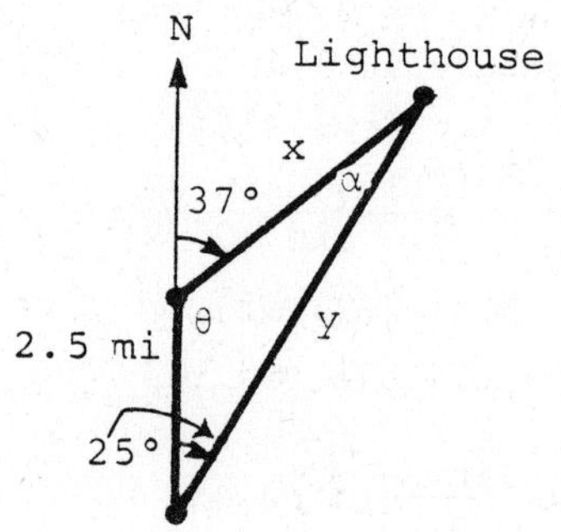

$\theta = 180° - 37° = 143°$

$\alpha = 180° - \theta - 25°$

$= 180° - 143° - 25° = 12°$

$$\frac{2.5}{\sin \alpha} = \frac{x}{\sin 25°}$$

$$x = \frac{2.5 \sin 25°}{\sin 12°}$$

$= 5.1$ mi

$$\frac{2.5}{\sin \alpha} = \frac{y}{\sin \theta}$$

$$y = \frac{2.5 \sin 143°}{\sin 12°}$$

$= 7.2$ mi

33. A = 42.5°, b = 13.6 m, c = 10.1 m

Angle A is included between sides b and c.

$$\text{Area} = \frac{1}{2} bc \sin A$$

$$= \frac{1}{2}(13.6)(10.1) \sin 42.5°$$

$= 46.4 \text{ m}^2$

35. B = 124.5°, a = 30.4 cm, c = 28.4 cm

Angle B is included between sides a and c.

$$\text{Area} = \frac{1}{2} ac \sin B$$

$$= \frac{1}{2}(30.4)(28.4) \sin 124.5°$$

$= 356 \text{ cm}^2$

37. A = 50.80°, b = 32.67 in, c = 52.89 in

Angle A is included between sides b and c.

$$\text{Area} = \frac{1}{2} bc \sin A$$

$$= \frac{1}{2}(32.67)(52.89) \sin 56.80°$$

$= 722.9 \text{ in}^2$

39. A = 24° 25′, B = 56° 20′, c = 78.40 cm

$A + B + C = 180°$

$C = 180° - A - B$

$C = 180° - 24° 25' - 56° 20'$

$= 99° 15'$

Use the law of sines to find b (or a).

$$\frac{b}{\sin B} = \frac{c}{\sin C}$$

$$b = \frac{78.40 \sin 56° 20'}{\sin 99° 15'}$$

$b = 66.11$

Angle A is included between sides b and c.

$$\text{Area} = \frac{1}{2}\,bc \sin A$$

$$= \frac{1}{2}(66.11)(78.40) \sin 24° \; 25'$$

$$= 1071 \text{ cm}^2$$

41. $\text{Area} = \frac{1}{2}\,ab \sin C$

$$= \frac{1}{2}(16.1)(15.2) \sin 125°$$

$$= 100 \text{ m}^2$$

43. Show that in any triangle, $\sin A + \sin B > \sin (A + B)$.

By the law of sines,

$$\frac{a}{\sin A} = \frac{b}{\sin B} = \frac{c}{\sin C}$$

in triangle ABC.

Let $k = \frac{a}{\sin A}$. Then $a = k \sin A$, $b = k \sin B$, and $c = k \sin C$.

$$C = 180° - (A + B).$$

$$\sin C = \sin (180° - (A + B))$$

$$= \sin (A + B) \text{ since}$$

$$\sin \theta = \sin (180° - \theta).$$

Then $a + b > c$ becomes

$$k \sin A + k \sin B > k \sin (A + B)$$

$$k (\sin A + \sin B) > k \sin (A + B).$$

Since $a > 0$ and $\sin A > 0$ for any angle A in a triangle because A is in quadrants I or II, then $k > 0$. Therefore, $1/k > 0$.

$$\sin A + \sin B > \sin (A + B)$$

Section 7.2

In Exercises 1-21, the number of possible triangles are found based on the given conditions. Remember that, for example, if angle B and sides a and b are given such that $a > b > a \sin B$, then two triangles are possible.

1. $A = 29.7°$, $b = 41.5$ ft, $a = 27.2$ ft

$b > a > b \sin A$ tells us that there are two possible triangles.

$$\frac{\sin B}{b} = \frac{\sin A}{a}$$

$$\sin B = \frac{41.5 \sin 29.7°}{27.2}$$

$$\sin B = .75593878$$

$$B_1 = 49.1°$$

$$C_1 = 180° - A - B_1$$

$$= 180° - 29.7° - 49.1°$$

$$= 101.2°$$

$$B_2 = 180° - B_1 = 130.9°$$

$$C_2 = 19.4°$$

3. $C = 41° \; 20'$, $b = 25.9$ m, $c = 38.4$ m

$c > b$ tells us that there is only one triangle.

$$\frac{\sin B}{b} = \frac{\sin C}{c}$$

$$\sin B = \frac{25.9 \sin 41° \; 20'}{38.4}$$

$$\sin B = .44545209$$

$$B = 26° \; 30'$$

Note that $B \neq 153° \; 30'$ because $153° \; 30' + C > 180°$.

$$A = 180° - 26° \; 30' - 41° \; 20'$$

$$A = 112° \; 10'$$

5. $B = 74.3°$, $a = 859$ m, $b = 783$ m

$$\frac{\sin A}{a} = \frac{\sin B}{b}$$

$$\sin A = \frac{859 \sin 74.3°}{783}$$

$\sin A = 1.0561331$

$\sin A > 1$ is impossible.

Note that $b < a \sin B$.

No such triangle exists.

7. $A = 142.13°$, $b = 5.432$ ft, $a = 7.297$ ft

$a > b$ tells us that there is only one triangle.

$$\frac{\sin B}{b} = \frac{\sin A}{a}$$

$$\sin B = \frac{5.432 \sin 142.13°}{7.297}$$

$\sin B = .45697581$

$B = 27.19°$

B must be acute since A is obtuse.

$C = 180° - 142.13° - 27.19°$

$C = 10.68°$

9. $C = 129°\ 18'$, $a = 372.9$ cm, $c = 416.7$ cm

$c > a$ tells us that there is only one triangle.

$$\frac{\sin A}{a} = \frac{\sin C}{c}$$

$$\sin A = \frac{372.9 \sin 129°\ 18'}{416.7}$$

$\sin A = .69250063$

$A = 43°\ 50'$

$B = 180° - 43°\ 50' - 129°\ 18'$

$B = 6°\ 52'$

11. $A = 42.5°$, $a = 15.6$ ft, $b = 8.14$ ft

$a > b$ tells us that there is only one triangle.

$$\frac{\sin B}{b} = \frac{\sin A}{a}$$

$$\sin B = \frac{b \sin A}{a} = \frac{8.14 \sin 42.5°}{15.6}$$

$\sin B = .35251951$

$B = 20.6°$

$B \neq 159.4°$ since $159.4° + A > 180°$.

$C = 180° - 42.5° - 20.6°$

$= 116.9°$

$$\frac{c}{\sin C} = \frac{a}{\sin A}$$

$$c = \frac{a \sin C}{\sin A} = \frac{15.6 \sin 116.9°}{\sin 42.5°}$$

$c = 20.6$ ft

13. $B = 72.2°$, $b = 78.3$ m, $c = 145$ m

$$\frac{\sin C}{c} = \frac{\sin B}{b}$$

$$\sin C = \frac{c \sin B}{b} = \frac{145 \sin 72.2°}{78.3}$$

$\sin C = 1.7632025$

$\sin C > 1$ is impossible. Note that $c \sin B > b$. No such triangle exists.

15. $A = 38°\ 40'$, $a = 9.72$ km, $b = 11.8$ km

$b > a > b \sin A$ tells us that there are two possible triangles.

$$\frac{\sin B}{b} = \frac{\sin A}{a}$$

$$\sin B = \frac{b \sin A}{a}$$

$$= \frac{11.8 \sin 38° 40'}{9.72}$$

$\sin B = .75848811$

$B_1 = 49° 20'$

$C_1 = 180° - 38° 40' - 49° 20'$

$C_1 = 92° 00'$

$$\frac{c_1}{\sin C_1} = \frac{a}{\sin A}$$

$$c_1 = \frac{a \sin C_1}{\sin A}$$

$$= \frac{9.72 \sin 92° 00'}{\sin 38° 40'}$$

$c_1 = 15.5$ km

$B_2 = 130° 40'$

$C_2 = 180° - 38° 40' - 130° 40'$

$C_2 = 10° 40'$

$$\frac{c_2}{\sin C_2} = \frac{a}{\sin A}$$

$$c_2 = \frac{a \sin C_2}{\sin A}$$

$$= \frac{9.72 \sin 10° 40'}{\sin 38° 40'}$$

$c_2 = 2.88$ km

17. $B = 32° 50'$, $a = 7540$ cm, $b = 5180$ cm

$a > b > a \sin B$ tells us that there are two possible triangles.

$$\frac{\sin A}{a} = \frac{\sin B}{b}$$

$$\sin A = \frac{a \sin B}{b}$$

$$= \frac{7540 \sin 32° 50'}{5180}$$

$\sin A = .78922132$

$A_1 = 52° 10'$

$C_1 = 180° - 52° 10' - 32° 50'$

$C_1 = 95° 00'$

$$\frac{b}{\sin B} = \frac{c_1}{\sin C_1}$$

$$c_1 = \frac{b \sin C_1}{\sin B}$$

$$= \frac{5180 \sin 95° 00'}{\sin 32° 50'}$$

$c_1 = 9520$ cm

$A_2 = 127° 50'$

$C_2 = 180 - 127° 50' - 32° 50'$

$C_2 = 19° 20'$

$$\frac{b}{\sin B} = \frac{c_2}{\sin C_2}$$

$$c_2 = \frac{b \sin C_2}{\sin B}$$

$$= \frac{5180 \sin 19° 20'}{\sin 32° 50'}$$

$c_2 = 3160$ cm

19. $A = 96.80°$, $b = 3.589$ ft, $a = 5.818$ ft

$a > b$ tells us that there is only one possible triangle.

$$\frac{\sin B}{b} = \frac{\sin A}{a}$$

$$\sin B = \frac{b \sin A}{a}$$

$$= \frac{3.589 \sin 96.80°}{5.818}$$

$\sin B = .61253922$

$B = 37.77°$

B must be acute since A is obtuse.

$C = 180° - 96.80° - 37.77°$

$C = 45.43°$

$$\frac{c}{\sin C} = \frac{a}{\sin A}$$

$$c = \frac{a \sin C}{\sin A}$$

$$= \frac{5.818 \sin 45.43°}{\sin 96.80°}$$

$$c = 4.174 \text{ ft}$$

21. B = 39.68°, a = 29.81 m, b = 23.76 m

a > b > a sin B tells us that there are two possible triangles.

$$\frac{\sin A}{a} = \frac{\sin B}{b}$$

$$\sin A = \frac{a \sin B}{b}$$

$$= \frac{29.81 \sin 39.68°}{23.76}$$

$$\sin A = .80108002$$

$$A_1 = 53.23°$$

$$C_1 = 180° - 53.23° - 39.68°$$

$$C_1 = 87.09°$$

$$\frac{b}{\sin B} = \frac{c_1}{\sin C_1}$$

$$c_1 = \frac{b \sin C_1}{\sin B}$$

$$= \frac{23.76 \sin 87.09°}{\sin 39.68°}$$

$$c_1 = 37.16 \text{ m}$$

$$A_2 = 126.77°$$

$$C_2 = 180° - 126.77° - 39.68°$$

$$C_2 = 13.55°$$

$$\frac{b}{\sin B} = \frac{c_2}{\sin C_2}$$

$$c_2 = \frac{b \sin C_2}{\sin B}$$

$$= \frac{23.76 \sin 13.55°}{\sin 39.68°}$$

$$c_2 = 8.719 \text{ m}$$

23. $a = \sqrt{5}$, $c = 2\sqrt{5}$, A = 30°

$$\frac{a}{\sin A} = \frac{c}{\sin C}$$

$$\sin C = \frac{2\sqrt{5} \sin 30°}{\sqrt{5}}$$

$$\sin C = 1$$

$$C = 90°$$

This is a right triangle.

25. If A = 103° 20′, a = 14.6 ft., b = 20,4 ft., since b > a, B should be larger than A. But A = 103° 20′, and A + B would be greater than 180°. Since the sum of the angles of a triangle is 180°, the given dimensions indicate that no triangle ABC exists.

27. Let A = 38° 50′, a = 21.9, b = 78.3.

$$b \sin A = 78.3 \sin 38° 50'$$

$$= 49.1$$

Thus, 21.9 < 49.1.
That is, a < b sin A.
The piece of property cannot exist with the given data.

29. Prove that $\frac{a + b}{b} = \frac{\sin A + \sin B}{\sin B}$.

Start with the law of sines.

$$\frac{a}{\sin A} = \frac{b}{\sin B}$$

$$\left(\frac{\sin A}{b}\right)\left(\frac{a}{\sin A}\right) = \left(\frac{\sin A}{b}\right)\left(\frac{b}{\sin B}\right)$$

$$\frac{a}{b} = \frac{\sin A}{\sin B}$$

$$\frac{a}{b} + 1 = \frac{\sin A}{\sin B} + 1$$

$$\frac{a}{b} + \frac{b}{b} = \frac{\sin A}{\sin B} + \frac{\sin B}{\sin B}$$

$$\frac{a + b}{b} = \frac{\sin A + \sin B}{\sin B}$$

Section 7.3

1. Find AB, or c, in the following triangle.

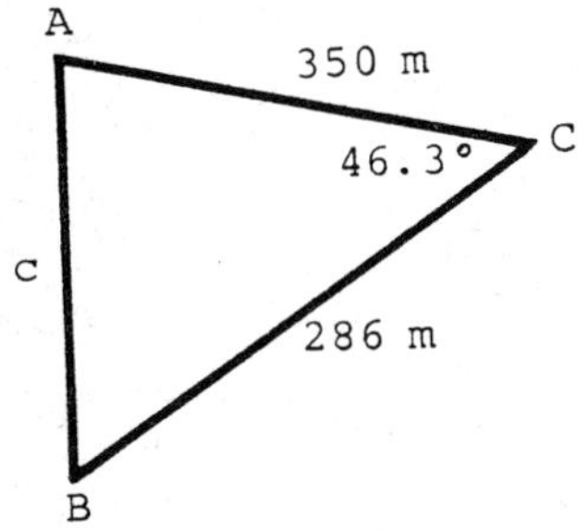

$$c^2 = a^2 + b^2 - 2ab \cos C$$

$$c = \sqrt{286^2 + 350^2 - 2(286)(350) \cos 46.3°}$$

$$c = 257$$

$$AB = 257 \text{ m}$$

3. Find AC, or b, in the following triangle.

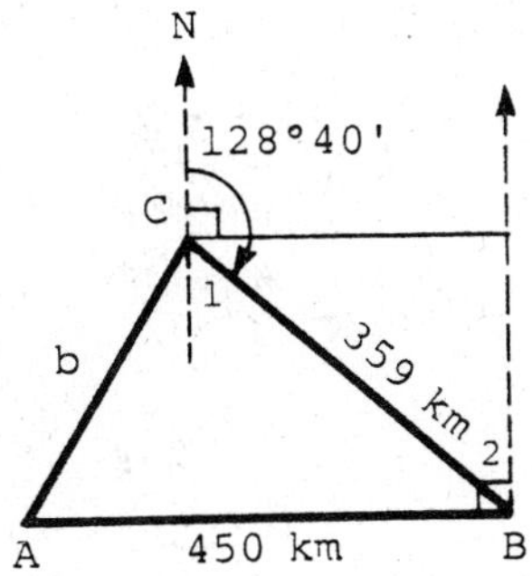

Angle 1 = 180° − 128° 40′ = 51° 20′

Angle 1 = Angle 2

Angle B = 90° − Angle 2 = 38° 40′

$$b^2 = a^2 + c^2 - 2ac \cos B$$

$$b = \sqrt{359^2 + 450^2 - 2(359)(450) \cos 38° 40'}$$

$$b = 281 \text{ km}$$

5. Find x.

$$x^2 = 25^2 + 25^2 - 2(25)(25) \cos 52°$$

$$= 480$$

$$x = 22 \text{ ft}$$

7. $AB^2 = 10^2 + 10^2 - 2(10)(10) \cos 128°$

$$AB^2 = 323$$

$$AB = 18 \text{ ft}$$

9. Let x = the distance from the tracking station to the satellite at 12:03 P.M.

Notice that the distance from the center of the earth to the satellite is

$$6400 + 1600 = 8000 \text{ km.}$$

Let θ = the angle made by the movement of the satellite from noon to 12:03 P.M.

$$\frac{2\pi \text{ radians}}{2 \text{ hours}} = \frac{\theta \text{ radians}}{3 \text{ minutes}}$$

$$\frac{2\pi}{2(60)} = \frac{\theta}{3}$$

$$\theta = \frac{6\pi}{120} = \frac{\pi}{20} \text{ radians}$$

$$x^2 = 6400^2 + 8000^2 - 2(6400)(8000) \cos \frac{\pi}{20}$$

$$x = 2000$$

The distance between the satellite and the tracking station is 2000 km.

11. Let θ = the angle opposite the diagonal.

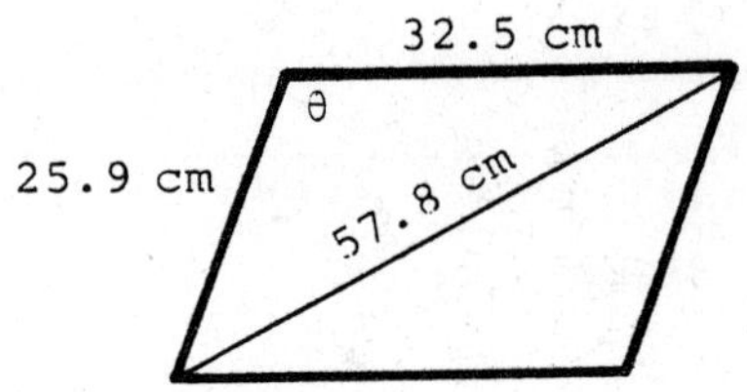

$$57.8^2 = 25.9^2 + 32.5^2 - 2(25.9)(32.5)\cos\theta$$
$$\theta = 163.5°$$

13. AB = 22.47928 mi, AC = 28.14276 mi,

A = 58.56989°

$BC^2 = AC^2 + AB^2 - 2(AC)(AB)\cos A$

Use a calculator and substitute.

$BC^2 = 637.5539346$

$BC = 25.24983$ mi

15. Applying the law of cosines when a = 3, b = 4, and c = 10 gives

$$c^2 = a^2 + b^2 - 2ab\cos C$$
$$10^2 = 3^2 + 4^2 - 2(3)(4)\cos C$$
$$75 = -24\cos C$$
$$-3.125 = \cos C$$

Since $-1 \le \cos C \le 1$, a triangle cannot have sides 3, 4, and 10.

17. C = 28.3°, b = 5.71 in, a = 4.21 in

$$c^2 = a^2 + b^2 - 2ab\cos C$$
$$c = \sqrt{(4.21)^2 + (5.71)^2 - 2(4.21)(5.71)\cos 28.3°}$$
$$c = 2.83 \text{ in}$$

Keep all digits of $\sqrt{c^2}$ in the calculator for use in the next calculation. If 2.83 is used, the answer will vary slightly due to round-off error.

Find angle A next, since it is the smaller angle and must be acute.

$$\sin A = \frac{a \sin C}{c} = \frac{4.21 \sin 28.3°}{\sqrt{c^2}}$$

$$\sin A = .70581857$$

$$A = 44.9°$$

$$B = 180° - 44.9° - 28.3°$$

$$B = 106.8°$$

19. $C = 45.6°$, $b = 8.94$ m, $a = 7.23$ m

$$c^2 = a^2 + b^2 - 2ab \cos C$$

$$c = \sqrt{(7.23)^2 + (8.94)^2 - 2(7.23)(8.94) \cos 45.6°}$$

$$c = 6.46 \text{ m}$$

Find angle A next, since it is the smaller angle and must be acute.

$$\sin A = \frac{a \sin C}{c} = \frac{7.23 \sin 45.6°}{\sqrt{c^2}}$$

$$\sin A = .79946437$$

$$A = 53.1°$$

$$B = 180° - 53.1° - 45.6°$$

$$B = 81.3°$$

21. $A = 80°\ 40'$, $b = 143$ cm, $c = 89.6$ cm

$$a^2 = b^2 + c^2 - 2bc \cos A$$

$$a = \sqrt{143^2 + (89.6)^2 - 2(143)(89.6) \cos 80°\ 40'}$$

$$a = 156 \text{ cm}$$

Find angle C next, since it is the smaller angle and must be acute.

$$\sin C = \frac{c \sin A}{a} = \frac{89.6 \sin 80°\ 40'}{\sqrt{a^2}}$$

$$\sin C = .56692713$$

$$C = 34°\ 30'$$

$$B = 180° - 80°\ 40' - 34°\ 30'$$

$$B = 64°\ 50'$$

23. $B = 74.80°$, $a = 8.919$ in, $c = 6.427$ in

$$b^2 = a^2 + c^2 - 2ac\cos B$$
$$b = \sqrt{(8.919)^2 + (6.427)^2 - 2(8.919)(6.427)\cos 74.80°}$$
$$b = 9.529 \text{ in}$$

Find angle C next, since it is the smaller angle and must be acute.

$$\sin C = \frac{c \sin B}{b} = \frac{6.427 \sin 74.80°}{\sqrt{b^2}}$$
$$\sin C = .65089219$$
$$C = 40.61°$$
$$A = 180° - 74.80° - 40.61°$$
$$A = 64.59°$$

25. $A = 112.8°$, $b = 6.28$ m, $c = 12.2$ m

$$a^2 = b^2 + c^2 - 2bc\cos A$$
$$a = \sqrt{(6.28)^2 + (12.2)^2 - 2(6.28)(12.2)\cos 112.8°}$$
$$a = 15.7 \text{ m}$$

Angle A is obtuse, so both B and C are acute. Find either angle next.

$$\sin B = \frac{b \sin A}{a} = \frac{6.28 \sin 112.8°}{\sqrt{a^2}}$$
$$\sin B = .36787456$$
$$B = 21.6°$$
$$C = 180° - 112.8° - 21.6°$$
$$C = 45.6°$$

27. $C = 24°\ 49'$, $a = 251.3$ m, $b = 318.7$ m

$$c^2 = a^2 + b^2 - 2ab\cos C$$
$$c = \sqrt{(251.3)^2 + (318.7)^2 - 2(251.3)(318.7)\cos 24°\ 49'}$$
$$c = 139.0 \text{ m}$$

Find angle A next, since it is the smaller angle and must be acute.

$\sin A = \dfrac{a \sin C}{c} = \dfrac{251.3 \sin 24° 49'}{\sqrt{c^2}}$

$\sin A = .75854779$

$A = 49° 20'$

$B = 180° - 49° 20' - 24° 49'$

$B = 105° 51'$

29. $a = 3.0$ ft, $b = 5.0$ ft, $c = 6.0$ ft

Angle C is the largest, so find it first.

$c^2 = a^2 + b^2 - 2ab \cos C$

$\cos C = \dfrac{3.0^2 + 5.0^2 - 6.0^2}{2(3.0)(5.0)}$

$\cos C = -.06666667$

$C = 94°$

$\sin B = \dfrac{b \sin C}{c}$

$\sin B = .83147942$

$B = 56°$

$A = 180° - 56° - 94°$

$A = 30°$

31. $a = 9.3$ cm, $b = 5.7$ cm, $c = 8.2$ cm

Angle A is the largest, so find it first.

$a^2 = b^2 + c^2 - 2bc \cos A$

$\cos A = \dfrac{5.7^2 + 8.2^2 - 9.3^2}{2(5.7)(8.2)}$

$\cos A = .14163457$

$A = 82°$

$\sin B = \dfrac{b \sin A}{a}$

$\sin B = .60672455$

$B = 37°$

$C = 180° - 82° - 37°$

$C = 61°$

33. $a = 42.9$ m, $b = 37.6$ m, $c = 62.7$ m

Angle C is the largest, so find it first.

$c^2 = a^2 + b^2 - 2ab \cos C$

$\cos C = \dfrac{42.9^2 + 37.6^2 - 62.7^2}{2(42.9)(37.6)}$

$\cos C = -.20988940$

$C = 102° 10'$

$\sin B = \dfrac{b \sin C}{c}$

$\sin B = .58632321$

$B = 35° 50'$

$A = 180° - B - C$

$A = 42° 00'$

35. $AB = 1240$ ft, $AC = 876$ ft, $BC = 918$ ft

$AB = c$, $AC = b$, $BC = a$

Angle C is the largest, so find it first.

$c^2 = a^2 + b^2 - 2ab \cos C$

$\cos C = \dfrac{a^2 + b^2 - c^2}{2ab}$

$\cos C = .04507765$

$C = 87° 20'$

$\sin B = \dfrac{b \sin C}{c}$

$\sin B = .70573350$

$B = 44° 50'$

$A = 180° - B - C$

$A = 47° 50'$

37. $a = 12$ m, $b = 16$ m, $c = 25$ m

$$s = \frac{1}{2}(12 + 16 + 25)$$

$$s = 26.5$$

$$\text{area} = \sqrt{s(s - a)(s - b)(s - c)}$$

$$= \sqrt{(26.5)(14.5)(10.5)(1.5)}$$

$$= 78 \text{ m}^2$$

39. $a = 154$ cm, $b = 179$ cm, $c = 183$ cm

$$s = \frac{1}{2}(154 + 179 + 183)$$

$$s = 258$$

$$\text{area} = \sqrt{s(s - a)(s - b)(s - c)}$$

$$\text{area} = \sqrt{(258)(104)(79)(75)}$$

$$= 12{,}600 \text{ cm}^2$$

41. $a = 76.3$ ft, $b = 109$ ft, $c = 98.8$ ft

$$s = \frac{1}{2}(76.3 + 109 + 98.8)$$

$$s = 142.05$$

$$\text{area} = \sqrt{s(s - a)(s - b)(s - c)}$$

$$= \sqrt{(142.05)(65.75)(33.05)(43.25)}$$

$$= 3650 \text{ ft}^2$$

43. $a = 74.14$ ft, $b = 89.99$ ft, $c = 51.82$ ft

$$s = \frac{1}{2}(74.14 + 89.99 + 51.82)$$

$$= 107.975$$

$$\text{area} = \sqrt{s(s - a)(s - b)(s - c)}$$

$$= \sqrt{(107.975)(33.835)(17.985)(56.155)}$$

$$= 1921 \text{ ft}^2$$

45. Find the area of the region.

$$s = \frac{1}{2}(75 + 68 + 85)$$

$$s = 114$$

$$\text{area} = \sqrt{(114)(39)(46)(29)}$$

$$= 2435.3571 \text{ m}^2$$

Number of cans needed

$= (\text{area in m}^2)/(\text{m}^2 \text{ per can})$

$$\frac{2435.3571}{75} = 32.471427 \text{ cans}$$

She will need to open 33 cans.

47. Find the area of the Bermuda triangle.

$$s = \frac{1}{2}(850 + 925 + 1300)$$

$$s = 1537.5$$

$$\text{area} = \sqrt{1537.5(687.5)(612.5)(237.5)}$$

$$= 392{,}000$$

The area of the Bermuda Triangle is 392,000 mi².

49. Verify

$$1 + \cos A = \frac{(b + c + a)(b + c - a)}{2bc}.$$

$$1 + \cos A = \frac{2bc}{2bc} + \frac{b^2 + c^2 - a^2}{2bc}$$

$$= \frac{b^2 + 2bc + c^2 - a^2}{2bc}$$

$$= \frac{(b + c)^2 - a^2}{2bc}$$

$$= \frac{(b + c + a)(b + c - a)}{2bc}$$

51. Verify

$$\cos \frac{A}{2} = \sqrt{\frac{s(s - a)}{bc}}.$$

$$\cos \frac{A}{2} = \sqrt{\frac{1 + \cos A}{2}}$$

Use the result of Exercise 49.

$$= \sqrt{\frac{(b + c + a)}{2} \cdot \frac{(b + c - a)}{2} \cdot \frac{1}{bc}}$$

Now use the fact that

$$2s = a + b + c$$

and

$$2s - 2a = b + c - a.$$

$$= \sqrt{\left(\frac{2s}{2}\right)\left(\frac{2s - 2a}{2}\right)\frac{1}{bc}}$$

$$\cos \frac{A}{2} = \sqrt{\frac{s(s - a)}{bc}}$$

53. Verify

$$\frac{1}{2} bc \sin A$$

$$= \sqrt{\frac{1}{2} bc(1 + \cos A) \cdot \frac{1}{2} bc(1 - \cos A)}.$$

$$\frac{1}{2} bc \sin A$$

$$= \frac{1}{2} bc\sqrt{\sin^2 A}$$

$$= \frac{1}{2} bc\sqrt{1 - \cos^2 A}$$

$$= \sqrt{\frac{1}{2} \cdot \frac{1}{2} \cdot b^2 \cdot c^2}\ \sqrt{1 - \cos^2 A}$$

$$= \sqrt{\frac{1}{2} \cdot \frac{1}{2} \cdot b^2 \cdot c^2 (1 + \cos A)(1 - \cos A)}$$

$$= \sqrt{\frac{1}{2} bc(1 + \cos A) \cdot \frac{1}{2} bc(1 - \cos A)}$$

55. Prove that if a and b are equal sides of an isosceles triangle, then $c^2 = 2a^2(1 - \cos C)$.

By the law of cosines,

$$c^2 = a^2 + b^2 - 2ab \cos C.$$

Since $a = b$,

$$c^2 = a^2 + a^2 - 2a^2 \cos C$$

$$c^2 = 2a^2 - 2a^2 \cos C$$

$$c^2 = 2a^2(1 - \cos C).$$

57. Prove that if angle C of triangle ABC is obtuse, then side c is longer than either side a or side b.

Since angle C is obtuse,

$$\cos C < 0.$$

Multiply by $-2ab$, a negative number.

$$-2ab \cos C > 0$$

Add $a^2 + b^2$.

$$a^2 + b^2 - 2ab \cos C > a^2 + b^2$$

Use the law of cosines.

$$c^2 > a^2 + b^2$$

Since $a > 0$, $b > 0$, and $c > 0$, it is apparent that $c > a$ and $c > b$.

59. Verify $\dfrac{\tan \frac{1}{2}(A - B)}{\tan \frac{1}{2}(A + B)} = \dfrac{a - b}{a + b}$.

$$\frac{\tan \frac{1}{2}(A - B)}{\tan \frac{1}{2}(A + B)}$$

$$= \frac{\dfrac{\sin (A - B)}{1 + \cos (A - B)}}{\dfrac{\sin (A + B)}{1 + \cos (A + B)}}$$

$$= \frac{\sin (A - B)[1 + \cos (A + B)]}{\sin (A + B)[1 + \cos (A - B)]}$$

$$= \frac{(\sin A \cos B - \cos A \sin B)(1 + \cos A \cos B - \sin A \sin B)}{(\sin A \cos B + \cos A \sin B)(1 + \cos A \cos B + \sin A \sin B)}$$

$$= \frac{\sin A \cos B + \sin A \cos A(\sin^2 B + \cos^2 B) - \sin B \cos B(\sin^2 A + \cos^2 A) - \cos A \sin B}{\sin A \cos B + \sin A \cos A(\sin^2 B + \cos^2 B) + \sin B \cos B(\sin^2 A + \cos^2 A) + \cos A \sin B}$$

$$= \frac{\sin A \cos B + \sin A \cos A - \sin B \cos B - \cos A \sin B}{\sin A \cos B + \sin A \cos A + \sin B \cos B + \cos A \sin B}$$

$$= \frac{\sin A(\cos A + \cos B) - \sin B(\cos A + \cos B)}{\sin A(\cos A + \cos B) + \sin B(\cos A + \cos B)}$$

$$= \frac{(\cos A + \cos B)(\sin A - \sin B)}{(\cos A + \cos B)(\sin A + \sin B)}$$

$$= \frac{\sin A - \sin B}{\sin A + \sin B}$$

$$= \frac{a - b}{a + b}$$

Section 7.4

1. Equal vectors have the same magnitude and direction:

$$\mathbf{m} = \mathbf{p}, \ \mathbf{n} = \mathbf{r}.$$

3. One vector is a positive scalar multiple of another if the two vectors point in the same direction; they may have different magnitudes.

$$\mathbf{m} = 1\mathbf{p}, \ \mathbf{m} = 2\mathbf{t}, \ \mathbf{n} = 1\mathbf{r}, \ \mathbf{p} = 2\mathbf{t},$$

$$\mathbf{p} = 1\mathbf{m}, \ \mathbf{t} = \frac{1}{2}\mathbf{m}, \ \mathbf{r} = 1\mathbf{n}, \ \mathbf{t} = \frac{1}{2}\mathbf{p}$$

In Exercises 5–21, see the answer sketches at the back of the textbook.

23. $\mathbf{a} + (\mathbf{b} + \mathbf{c}) = (\mathbf{a} + \mathbf{b}) + \mathbf{c}$

and

$\mathbf{h} + (\mathbf{e} + \mathbf{g}) = (\mathbf{h} + \mathbf{e}) + \mathbf{g}$

Yes, vector addition is associative.

In Exercises 25–29, sketches may vary slightly. See the sketches at the back of the textbook.

In Exercises 31–39, $\mathbf{x}$ is the horizontal component and $\mathbf{y}$ is the vertical component.

31. $\alpha = 38°$, $|\mathbf{v}| = 12$

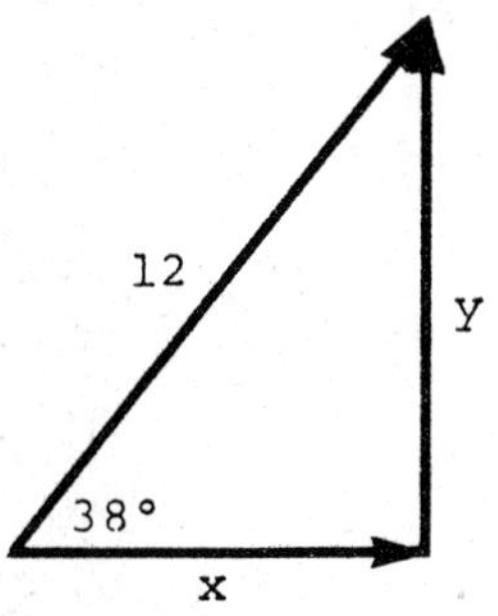

$$\cos 38° = \frac{|\mathbf{x}|}{12}$$

$$|\mathbf{x}| = 12 \cos 38° = 9.5$$

$$\sin 38° = \frac{|\mathbf{y}|}{12}$$

$$|\mathbf{y}| = 12 \sin 38° = 7.4$$

33. $\alpha = 50°$, $|\mathbf{v}| = 26$

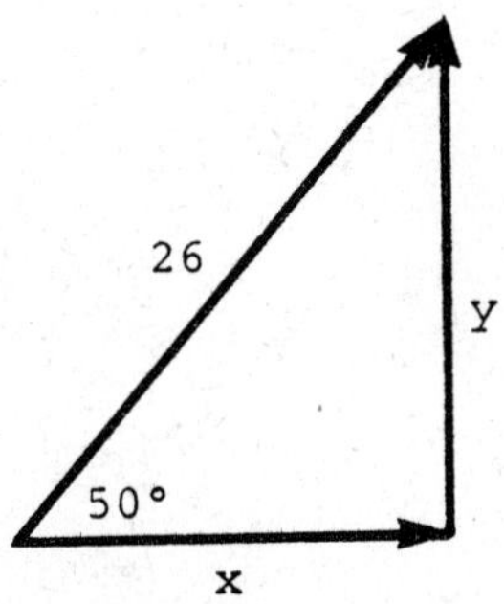

$$\cos 50° = \frac{|\mathbf{x}|}{26}$$

$$|\mathbf{x}| = 26 \cos 50° = 17$$

$$\sin 50° = \frac{|\mathbf{y}|}{26}$$

$$|\mathbf{y}| = 26 \sin 50° = 20$$

35. $\alpha = 27°\ 30'$, $|\mathbf{v}| = 15.4$

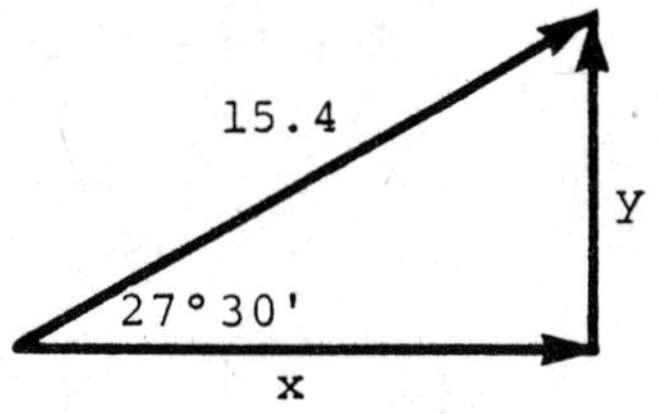

$$\cos 27°\ 30' = \frac{|\mathbf{x}|}{15.4}$$

$$|\mathbf{x}| = 15.4 \cos 27°\ 30' = 13.7$$

$$\sin 27°\ 30' = \frac{|\mathbf{y}|}{15.4}$$

$$|\mathbf{y}| = 15.4 \sin 27°\ 30' = 7.11$$

37. $\alpha = 128.5°$, $|\mathbf{v}| = 198$

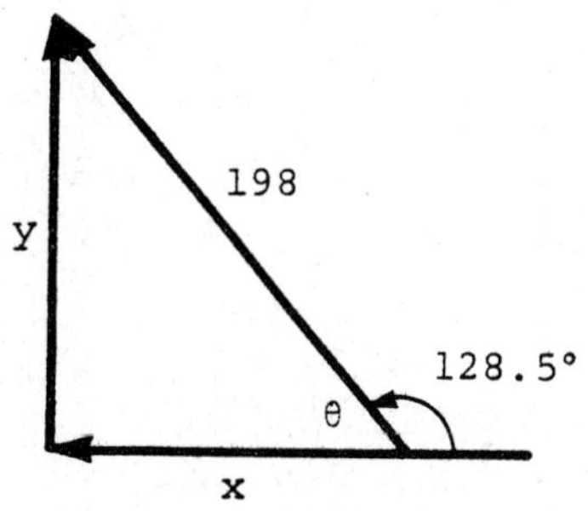

$$\theta = 180° - 128.5° = 51.5°$$

$$\cos 51.5° = \frac{|\mathbf{x}|}{198}$$

$$|\mathbf{x}| = 198 \cos 51.5° = 123$$

$\sin 51.5° = \frac{|\mathbf{y}|}{198}$

$|\mathbf{y}| = 198 \sin 51.5°$

$= 155$

39. Forces of 250 newtons and 450 newtons, forming an angle of 85°

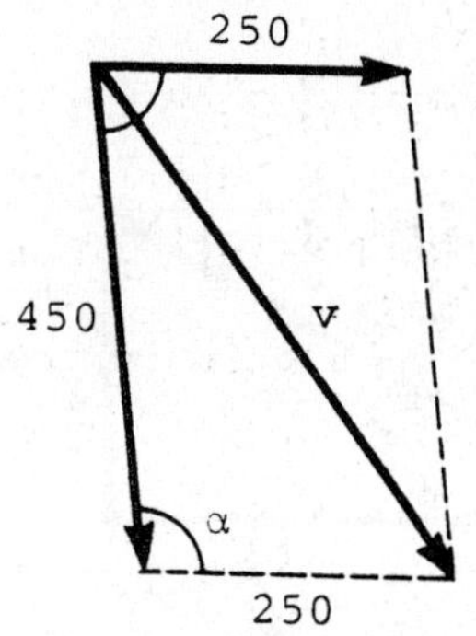

$\alpha = 180° - 85° = 95°$

$|\mathbf{v}|^2 = (250)^2 + (450)^2$

$- 2(250)(450) \cos 95°$

$= 284{,}610$

$|\mathbf{v}| = 530$ newtons

41. Forces of 17.9 newtons and 25.8 newtons, forming an angle of 105° 30′

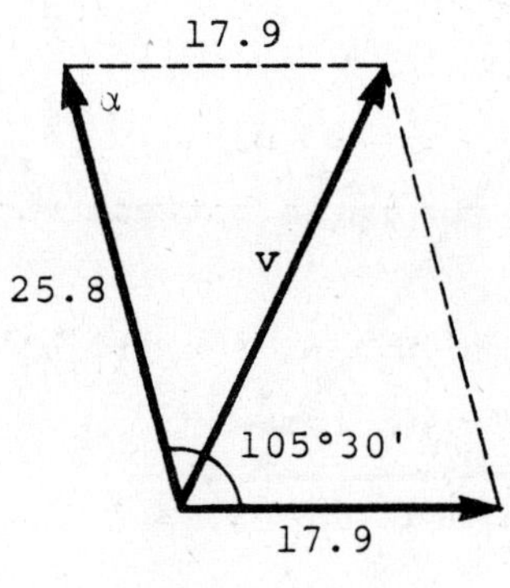

$\alpha = 180° - 105° \ 30'$

$= 74° \ 30'$

$|\mathbf{v}|^2 = (25.8)^2 + (17.9)^2$

$- 2(25.8)(17.9) \cos 74° \ 30'$

$= 739.218$

$|\mathbf{v}| = 27.2$ lb

43. Forces of 116 lb and 139 lb, forming an angle of 140° 50′

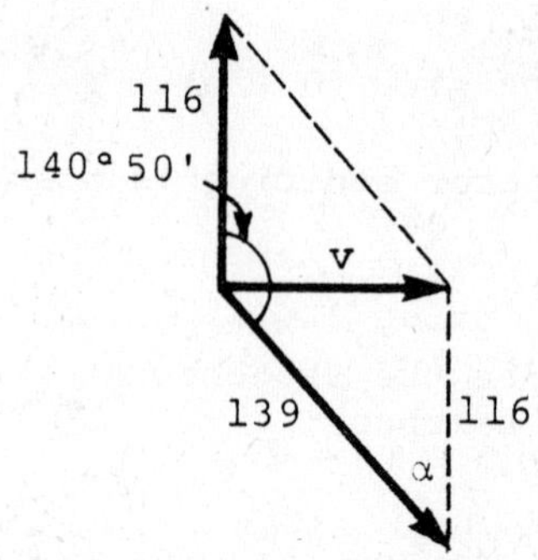

$\alpha = 180° - 140° \ 50' = 39° \ 10'$

$|\mathbf{v}|^2 = (139)^2 + (116)^2$

$- 2(139)(116) \cos 39° \ 10'$

$= 7774.74$

$|\mathbf{v}| = 88.2$ lb

45. Verify $\mathbf{u} + \mathbf{v} = \mathbf{v} + \mathbf{u}$.

$\mathbf{u} + \mathbf{v} = \langle a_1, b_1 \rangle + \langle a_2, b_2 \rangle$

$= \langle a_1 + a_2, b_1 + b_2 \rangle$

$= \langle a_2 + a_1, b_2 + b_1 \rangle$

$= \langle a_2, b_2 \rangle + \langle a_1, b_1 \rangle$

$= \mathbf{v} + \mathbf{u}$

Therefore, $\mathbf{u} + \mathbf{v} = \mathbf{v} + \mathbf{u}$.

47. Verify $-1(\mathbf{u}) = -\mathbf{u}$.

$-1(\mathbf{u}) = -1\langle a_1, b_1 \rangle$

$= \langle -a_1, -b_1 \rangle$

$= -\mathbf{u}$

Therefore, $-1(\mathbf{u}) = -\mathbf{u}$.

49. Verify $\mathbf{u} + 0 = \mathbf{u}$.

$$\begin{aligned}\mathbf{u} + 0 &= \langle a_1, b_1\rangle + \langle 0, 0\rangle \\ &= \langle a_1 + 0, b_1 + 0\rangle \\ &= \langle a_1, b_1\rangle \\ &= \mathbf{u}\end{aligned}$$

Therefore, $\mathbf{u} + 0 = \mathbf{u}$.

51. $\mathbf{u} = 6\mathbf{i} - 2\mathbf{j}; \mathbf{v} = -3\mathbf{i} + 2\mathbf{j}$

$$\begin{aligned}\mathbf{u}\cdot\mathbf{v} &= (6)(-3) + (-2)(2) \\ &= (-18) + (-4) \\ &= -22\end{aligned}$$

53. $\mathbf{u} = \langle -6, 8\rangle$, $\mathbf{v} = \langle 3, -4\rangle$

$\mathbf{u} = -6\mathbf{i} + 8\mathbf{j}$, $\mathbf{v} = 3\mathbf{i} - 4\mathbf{j}$

$$\begin{aligned}\mathbf{u}\cdot\mathbf{v} &= (-6)(3) + 8(-4) \\ &= -18 + (-32) \\ &= -50\end{aligned}$$

55. Verify $\mathbf{u}\cdot\mathbf{v} = |\mathbf{u}||\mathbf{v}| \cos \alpha$.

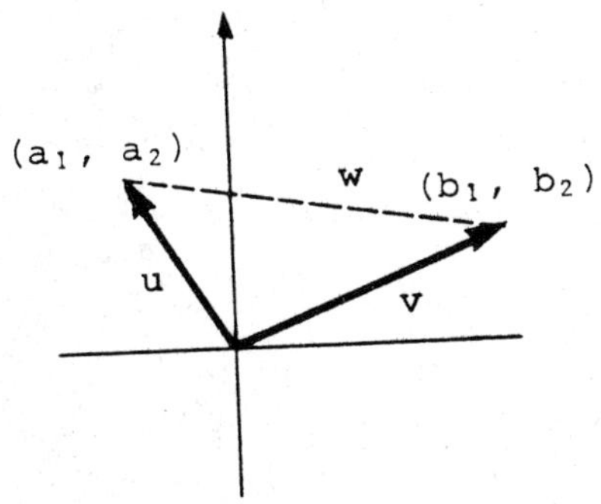

By the law of cosines,

$$|\mathbf{w}|^2 = |\mathbf{u}|^2 + |\mathbf{v}|^2 - 2|\mathbf{u}||\mathbf{v}| \cos \alpha.$$

Let $\mathbf{w} = \langle b_1 - a_1, b_2 - a_2\rangle$.

Therefore,

$$\mathbf{w}\cdot\mathbf{w} = \mathbf{u}\cdot\mathbf{u} + \mathbf{v}\cdot\mathbf{v} - 2|\mathbf{u}||\mathbf{v}| \cos \alpha$$

$$\begin{aligned}(b_1 - a_1)^2 + (b_2 - a_2)^2 &= a_1{}^2 + a_2{}^2 + b_1{}^2 + b_2{}^2 \\ &\quad - 2|\mathbf{u}||\mathbf{v}| \cos \alpha\end{aligned}$$

$$\begin{aligned}b_1{}^2 - 2a_1b_1 + a_1{}^2 + b_2{}^2 - 2b_2a_2 + a_2{}^2 &= a_1{}^2 + a_2{}^2 + b_1{}^2 + b_2{}^2 \\ &\quad - 2|\mathbf{u}||\mathbf{v}| \cos \alpha\end{aligned}$$

$$\begin{aligned}-2a_1b_1 - 2a_2b_2 &= -2|\mathbf{u}||\mathbf{v}| \cos \alpha \\ a_1b_1 + a_2b_2 &= |\mathbf{u}||\mathbf{v}| \cos \alpha \\ \mathbf{u}\cdot\mathbf{v} &= |\mathbf{u}||\mathbf{v}| \cos \alpha.\end{aligned}$$

Therefore, $\mathbf{u}\cdot\mathbf{v} = |\mathbf{u}||\mathbf{v}| \cos \alpha$.

57. $\mathbf{u} = \langle 1, 8\rangle = \mathbf{i} + 8\mathbf{j}$

$\mathbf{v} = \langle 2, -5\rangle = 2\mathbf{i} - 5\mathbf{j}$

$$\begin{aligned}|\mathbf{u}| &= \sqrt{1^2 + 8^2} = \sqrt{1 + 64} = \sqrt{65} \\ |\mathbf{v}| &= \sqrt{2^2 + (-5)^2} = \sqrt{4 + 25} = \sqrt{29} \\ \mathbf{u}\cdot\mathbf{v} &= (1)(2) + 8(-5) \\ &= 2 - 40 = -38 \\ \mathbf{u}\cdot\mathbf{v} &= |\mathbf{u}||\mathbf{v}| \cos \theta \\ -38 &= \sqrt{65}\sqrt{29} \cos \theta \\ \frac{-38}{\sqrt{65}\sqrt{29}} &= \cos \theta \\ 151° &= \theta\end{aligned}$$

59. Verify $\mathbf{u}\cdot\mathbf{v} = \mathbf{v}\cdot\mathbf{u}$.

Let θ = the angle between vectors $\mathbf{u}$ and $\mathbf{v}$.

$$\mathbf{u}\cdot\mathbf{v} = |\mathbf{u}||\mathbf{v}| \cos \theta = |\mathbf{v}||\mathbf{u}| \cos \theta$$

$$\mathbf{v}\cdot\mathbf{u} = |\mathbf{v}||\mathbf{u}| \cos \theta$$

Therefore, $\mathbf{u}\cdot\mathbf{v} = \mathbf{v}\cdot\mathbf{u}$.

61. Verify $\mathbf{u}\cdot(\mathbf{v} + \mathbf{w}) = \mathbf{u}\cdot\mathbf{v} + \mathbf{u}\cdot\mathbf{w}$

Let $\mathbf{u} = a\mathbf{i} + b\mathbf{j}$, $\mathbf{v} = c\mathbf{i} + d\mathbf{j}$, and $\mathbf{w} = e\mathbf{i} + f\mathbf{j}$.

$$\mathbf{u} \cdot (\mathbf{v} + \mathbf{w}) = \mathbf{u} \cdot [(c + e)\mathbf{i} + (d + f)\mathbf{j}]$$
$$= a(c + e) + b(d + f)$$
$$= ac + ae + bd + bf$$
$$\mathbf{u} \cdot \mathbf{v} + \mathbf{u} \cdot \mathbf{w} = (ac + bd) + (ae + bf)$$
$$= ac + ae + bd + bf$$

Therefore, $\mathbf{u} \cdot (\mathbf{v} + \mathbf{w}) = \mathbf{u} \cdot \mathbf{v} + \mathbf{u} \cdot \mathbf{w}$.

63. Given **u** and **v** are not 0, $\mathbf{u} \cdot \mathbf{v} = 0$. Verify that **u** and **v** are perpendicular.

$$\mathbf{u} \cdot \mathbf{v} = |\mathbf{u}||\mathbf{v}| \cos \theta,$$
$$0 = |\mathbf{u}||\mathbf{v}| \cos \theta$$

Since **u** and **v** are not 0,

$$0 = \cos \theta$$
$$90° = \theta.$$

Since θ is the angle between **u** and **v**, and $\theta = 90°$, **u** and **v** are perpendicular.

Section 7.5

1. Let α = the angle between the forces.

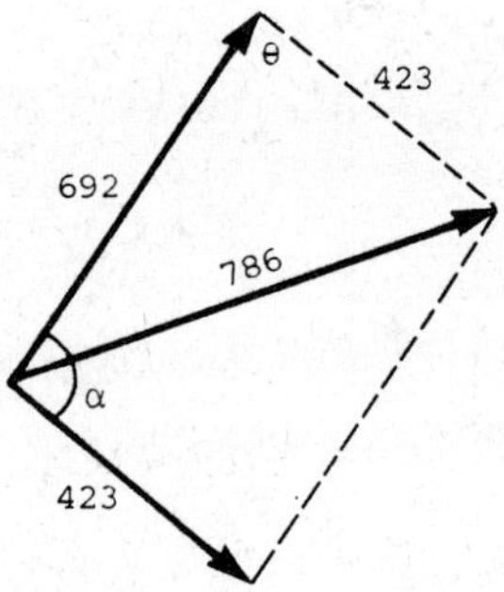

To find α, use the law of cosines to find θ.

$$786^2 = 692^2 + 423^2 - 2(692)(423) \cos \theta$$
$$\cos \theta = \frac{786^2 - 692^2 - 423^2}{-2(692)(423)}$$
$$\theta = 86.1°$$
$$\alpha = 180° - 86.1° = 93.9°$$

3. Let θ = the angle that the hill makes with the horizontal.

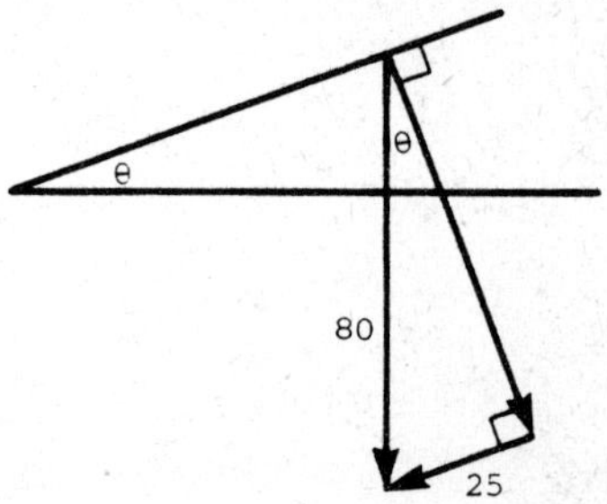

The 80-lb downward force has a 25-lb component parallel to the hill. The two right triangles are similar and have congruent angles.

$$\sin \theta = \frac{25}{80}$$
$$\theta = 18°$$

5. Find the force needed to pull a 60-ton monolith along the causeway.

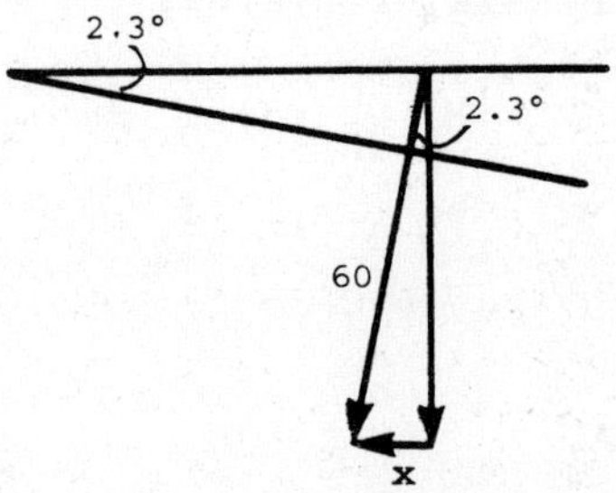

The force needed to pull 60 tons is equal to the magnitude of **x**, the component parallel to the causeway.

$$\sin 2.3° = \frac{|\mathbf{x}|}{60}$$

$$|\mathbf{x}| = 60 \sin 2.3°$$

$$= 2.4$$

The force needed is 2.4 tons.

7. Find the direction and magnitude of the equilibrant.

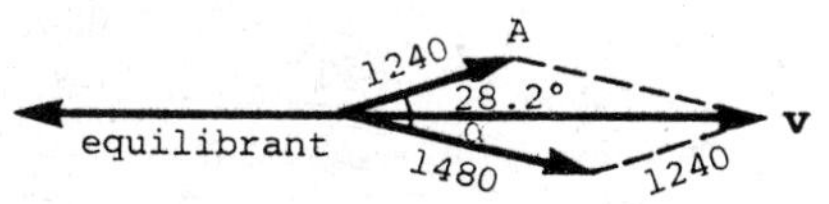

$A = 180° - 28.2° = 151.8°$

Use the law of cosines to find the magnitude of the resultant **v**.

$$|\mathbf{v}|^2 = 1240^2 + 1480^2 - 2(1240)(1480) \cos 151.8°$$

$$|\mathbf{v}| = 2640 \text{ lb}$$

Use the law of sines to find α.

$$\frac{\sin \alpha}{1240} = \frac{\sin 151.8°}{2640}$$

$$\sin \alpha = \frac{1240 \sin 151.8°}{2640}$$

$$\alpha = 12.82°$$

$$\theta = 180° - 12.82° = 167.2°$$

The equilibrant has a magnitude of 2640 lb and makes an angle of 167.2° with the 1480-lb force. The weight is 2640 lb at an angle of 167.2° with the 1480-lb force.

9. Find the weight of the crate and the tension on the horizontal rope.

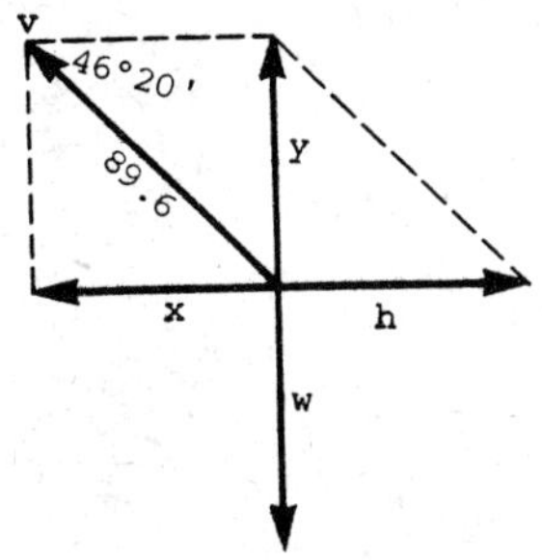

v has horizontal component **x** and vertical component **y**. The resultant **v** + **h** also has vertical component **y**. The resultant balances the weight of the crate, so its vertical component is the equilibrant of the crate's weight.

$$|\mathbf{w}| = |\mathbf{y}| = 89.6 \sin 46° \, 20'$$

$$|\mathbf{w}| = 64.8 \text{ lb}$$

Since the crate is not moving side-to-side, **h**, the horizontal tension on the rope, is the opposite of **x**.

$$|\mathbf{h}| = |\mathbf{x}| = 89.6 \cos 46° \, 20'$$

$$|\mathbf{h}| = 61.9 \text{ lb}$$

The weight of the crate is 64.8 lb; the tension is 61.9 lb.

11. Find the magnitude of the second force and of the resultant. Use the parallelogram rule; **v** is the resultant and **x** is the second force.

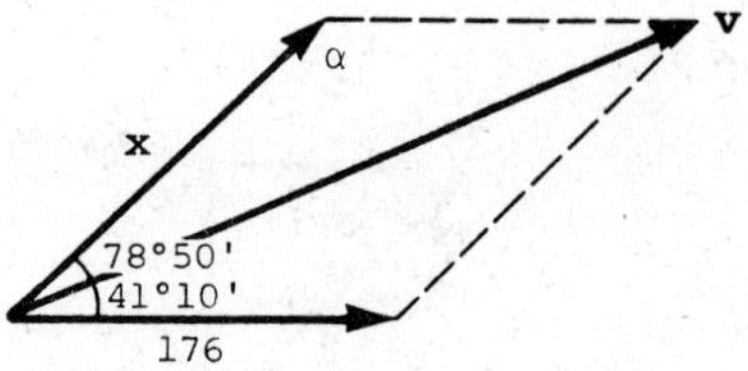

$\alpha = 180° - 78°\ 50'$

The angle between the second force and the resultant is

$\beta = 78°\ 50' - 41°\ 10' = 37°\ 40'$.

Use the law of sines to find $|\mathbf{v}|$.

$$\frac{|\mathbf{v}|}{\sin \alpha} = \frac{176}{\sin \beta}$$

$$|\mathbf{v}| = \frac{176 \sin 101°\ 10'}{\sin 37°\ 40'}$$

$$|\mathbf{v}| = 283 \text{ lb}$$

Use the law of sines to find $|\mathbf{x}|$.

$$\frac{|\mathbf{x}|}{\sin 41°\ 10'} = \frac{176}{\sin 37°\ 40'}$$

$$|\mathbf{x}| = \frac{176 \sin 41°\ 10'}{\sin 37°\ 40'}$$

$$|\mathbf{x}| = 190 \text{ lb}$$

13. Let $\mathbf{v}$ = the groundspeed vector of the plane.

Find the actual bearing of the plane.

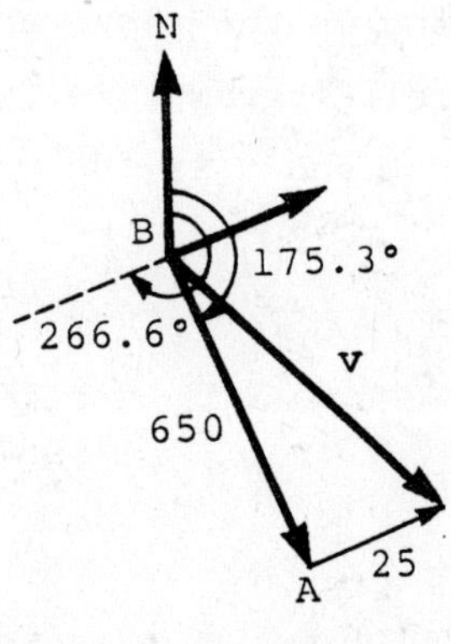

Angle A = 266.6° − 175.3° = 91.3°

Use the law of cosines to find $|\mathbf{v}|$.

$$|\mathbf{v}|^2 = 25^2 + 650^2 - 2(25)(650) \cos 91.3°$$

$$= 423{,}862$$

$$|\mathbf{v}| = 651$$

Use the law of sines to find the drift angle, B, the angle that $\mathbf{v}$ makes with the airspeed vector of the plane.

$$\frac{\sin B}{25} = \frac{\sin A}{651}$$

$$\sin B = .038392573$$

$$B = 2.2°$$

The bearing is 175.3° − 2.2°

= 173.1°

15. Find the distance of the strip from point A.

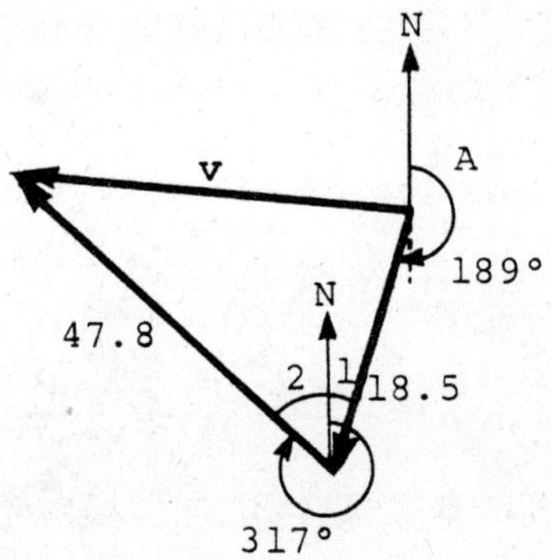

Angle 1 = 189° − 180° = 9°

Angle 2 = 360° − 317° = 43°

Angle 1 + Angle 2 = 9° + 43°

= 52°

Use the law of cosines to find $|\mathbf{v}|$.

$$|\mathbf{v}|^2 = (47.8)^2 + (18.5)^2 - 2(47.8)(18.5) \cos 52°$$

$$|\mathbf{v}|^2 = 1538.2311$$

$$|\mathbf{v}| = 39.2 \text{ km}$$

17. Let $\mathbf{v}$ = the groundspeed vector. Find the bearing and groundspeed of the plane.

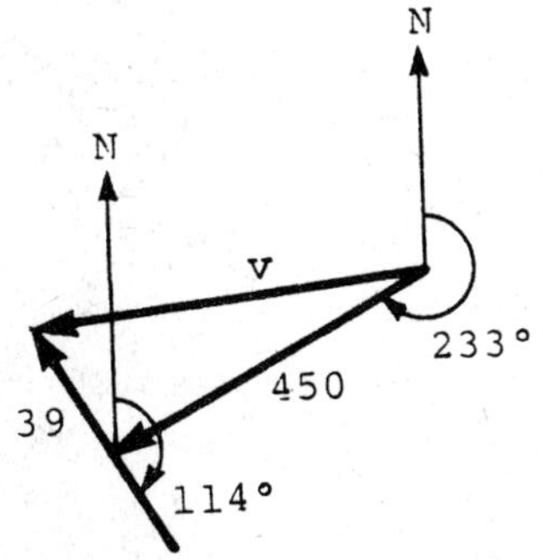

Angle A = 233° − 114° = 119°

Use the law of cosines to find $|\mathbf{v}|$.

$$|\mathbf{v}|^2 = 39^2 + 450^2 - 2(39)(450)\cos 119°$$

$$|\mathbf{v}|^2 = 221{,}038$$

$$|\mathbf{v}| = 470$$

The groundspeed is 470 mph.

Use the law of sines to find angle B, the drift angle.

$$\frac{\sin B}{39} = \frac{\sin 119°}{470}$$

$$\sin B = .0726$$

$$B = 4°$$

The bearing is 4° + 233° = 237°.

19. Let $\mathbf{v}$ = the airspeed vector. The groundspeed must be

$$\frac{400}{2.5} = 160 \text{ mph.}$$

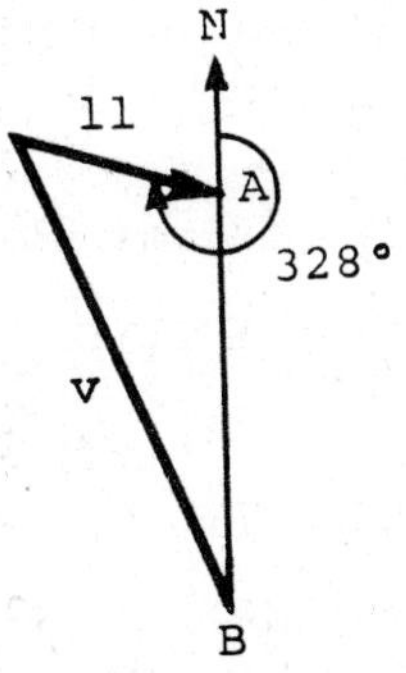

Angle A = 328° − 180° = 148°

Use the law of cosines to find $|\mathbf{v}|$.

$$|\mathbf{v}|^2 = 11^2 + 160^2 - 2(11)(160)\cos 148°$$

$$|\mathbf{v}|^2 = 28{,}706$$

$$|\mathbf{v}| = 170$$

The airspeed must be 170 mph.

Use the law of sines to find B, the drift angle.

$$\frac{\sin B}{11} = \frac{\sin 148°}{170}$$

$$\sin B = \frac{11 \sin 148°}{170}$$

$$B = 2.0°$$

The bearing must be

$$360° - 2.0° = 358°.$$

21. Find the groundspeed and resulting bearing.

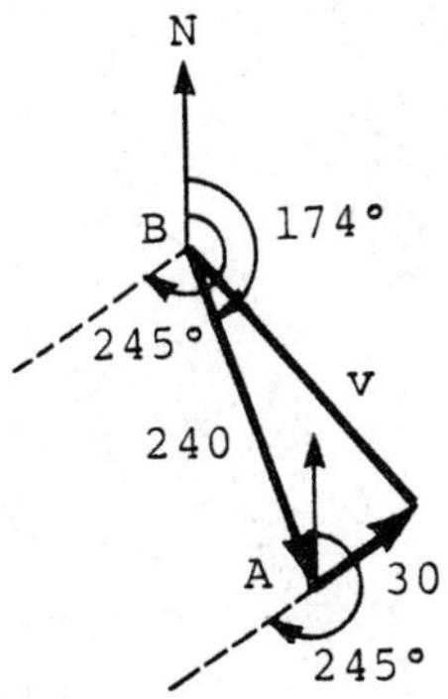

Angle A = 245° − 174° = 71°

Use the law of cosines to find $|\mathbf{v}|$.

$$|\mathbf{v}|^2 = 30^2 + 240^2 - 2(30)(240)\cos 71°$$

$$|\mathbf{v}|^2 \approx 53{,}812$$

$$|\mathbf{v}| \approx 230$$

The groundspeed is 230 km per hr.

Use the law of sines to find angle B.

$$\frac{\sin B}{30} = \frac{\sin 71°}{230}$$

$$\sin B = .1233$$

$$B = 7°$$

The resulting bearing is

$$174° - 7° = 167°.$$

23. At what time did the pilot turn? The plane will fly 2.6 hr before it runs out of fuel. In 2.6 hr, the carrier will travel (2.6)(30) = 83.2 mi and the plane will travel a total of (2.6)(520) = 1352 mi. Suppose it travels x miles on its initial bearing; then it travels 1352 - x miles after having turned.

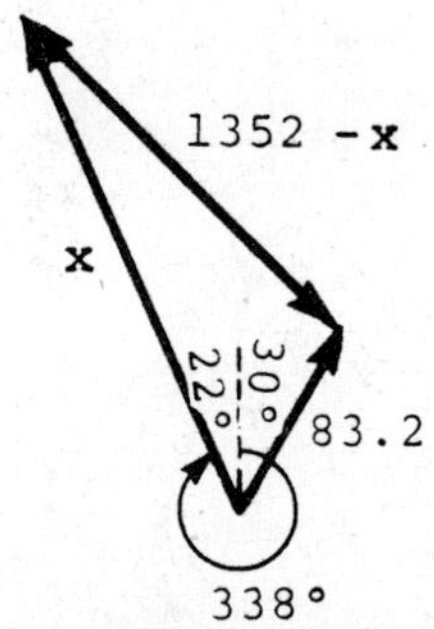

Use the law of cosines to get an equation in x.

$$(1352 - x)^2 = x^2 + (83.2)^2 - 2(x)(83.2) \cos 52°$$

$$1{,}827{,}904 - 2704x + x^2 = x^2 + 6922.24 - 102.45x$$

$$1{,}820{,}982 = 2601.55x$$

$$700 = x$$

To travel 700 miles at 520 mph requires $\frac{700}{520} = 1.35$ hr, or 1 hr and 21 min. The pilot turned at 1 hr 21 min after 2 P.M., or at 3:21 P.M.

Chapter 7 Review Exercises

1. Find b, given C = 74° 10′, c = 96.3 m, B = 39° 30′.

Use the law of sines to find b.

$$\frac{b}{\sin B} = \frac{c}{\sin C}$$

$$b = \frac{c \sin B}{\sin C}$$

$$= \frac{96.3 \sin 39° 30'}{\sin 74° 10'}$$

$$= 63.7 \text{ m}$$

3. Find B, given C = 51° 20′, c = 68.3 m, b = 58.2 m.

Use the law of sines to find B.

$$\frac{\sin B}{b} = \frac{\sin C}{c}$$

$$\sin B = \frac{b \sin C}{c}$$

$$= \frac{58.2 \sin 51° 20'}{68.3}$$

$$\sin B = .66533254$$

$$B = 41° 40'$$

Angle B cannot be obtuse, since $b < c$, $B < C$, and C is acute.

5. Given $B = 39° 50'$, $b = 268$ m, $a = 340$ m; find A.

$a > b > a \sin B$ tells us that two triangles are possible and A can have two value

Use the law of sines to find A.

$$\frac{\sin A}{a} = \frac{\sin B}{b}$$

$$\sin A = \frac{a \sin B}{b}$$

$$= \frac{340 \sin 39° 50'}{268}$$

$$\sin A = .81264638$$

$$A = 54° 20' \text{ or } A = 125° 40'$$

7. If we are given a, A, and C in a triangle ABC, B also is known since $B = 180° - A - C$. We now have the case where we know two angles and the included side. Since we do not have two sides and the angle opposite one of the sides, the possibility of the ambiguous case does not exist.

9. Given $a = 86.14$ in, $b = 253.2$ in, $c = 241.9$ in, find A.

Use the law of cosines to find A.

$$a^2 = b^2 + a^2 - 2bc \cos A$$

$$\cos A = \frac{b^2 + c^2 - a^2}{2bc}$$

$$= \frac{253.2^2 + 241.9^2 - 86.14^2}{2(253.2)(241.9)}$$

$$\cos A = .94046923$$

$$A = 19.87° \quad \text{or} \quad 19° 51'$$

11. Given $A = 51° 20'$, $c = 68.3$ m, $b = 58.2$ m; find a.

Use the law of cosines to find a.

$$a^2 = b^2 + c^2 - 2bc \cos A$$

$$a = \sqrt{(58.2)^2 + (68.3)^2 - 2(58.2)(68.3) \cos 51° 20'}$$

$$a = 55.5 \text{ m}$$

13. Given $A = 46° 10'$, $b = 184$ cm, $c = 192$ cm; find a.

Use the law of cosines to find a.

$$a^2 = b^2 + c^2 - 2bc \cos A$$

$$a = \sqrt{(184)^2 + (192)^2 - 2(184)(192) \cos 46° 10'}$$

$$a = 148 \text{ cm}$$

15. Solve the triangle, given $A = 25° 10'$, $a = 6.92$ yd, $b = 4.82$ yd.

$a > b$ tells us that only one triangle is possible.

Use the law of sines to find b.

$$\frac{\sin B}{b} = \frac{\sin A}{a}$$

$$\sin B = \frac{b \sin A}{a}$$

$$= \frac{4.82 \sin 25° 10'}{6.92}$$

$$B = 17° 10'$$

(Angle B cannot be obtuse, since $b < a$, $B < A$, and A is acute.)

$$C = 180° - A - B$$

$$= 137° 40'$$

Use the law of sines to find c.

$$\frac{c}{\sin C} = \frac{a}{\sin A}$$

$$c = \frac{a \sin C}{\sin A}$$

$$= \frac{6.92 \sin 137° 40'}{\sin 25° 10'}$$

$$c = 11.0 \text{ yd}$$

17. Solve the triangle, given $a = 27.6$ cm, $b = 19.8$ cm, $C = 42° 30'$.

Using the law of cosines, we have

$$c^2 = a^2 + b^2 - 2ab \cos C$$

$$c^2 = 27.6^2 + 19.8^2 - 2(27.6)(19.8) \cos 42° 30'$$

$$c = 18.7 \text{ cm.}$$

Using the law of sines, we have

$$\frac{a}{\sin A} = \frac{c}{\sin C}$$

$$\sin A = \frac{27.6 \sin 42° 30'}{18.7}$$

$$A = 88° 20'.$$

Since the sum of the angles of a triangle is 180°,

$$B = 180° - A - C$$

$$= 180° - 88° 20' - 42° 30'$$

$$B = 49° 10'.$$

19. Given $b = 840.6$ m, $c = 715.9$ m, $A = 149° 18'$, find the area.

Angle A is included between sides b and c.

$$\text{area} = \frac{1}{2} bc \sin A$$

$$= \frac{1}{2}(840.6)(715.9) \sin 149° 18'$$

$$\text{area} = 153{,}600 \text{ m}^2$$

21. Given $a = .913$ km, $b = .816$ km, $c = .582$ km, find the area.

$$s = \frac{1}{2}(a + b + c)$$

$$= \frac{1}{2}(.913 + .816 + .582)$$

$$= 1.1555$$

$$\text{area} = \sqrt{(1.1555)(.2425)(.3395)(.5735)}$$

$$\text{area} = .234 \text{ km}^2$$

23. Substitute $a = 7$, $b = 7\sqrt{3}$, $c = 14$, $A = 30°$, $B = 60°$, and $C = 90°$ into

$$\frac{a + b}{c} = \frac{\cos \frac{1}{2}(A - B)}{\sin \frac{1}{2} C}$$

in order to verify the formula.

$$\frac{a+b}{c} = \frac{7+7\sqrt{3}}{14}$$

$$= \frac{7(1+\sqrt{3})}{14} = \frac{1+\sqrt{3}}{2}$$

$$\frac{\cos\frac{1}{2}(A-B)}{\sin\frac{1}{2}C} = \frac{\cos\frac{1}{2}(30° - 60°)}{\sin\frac{1}{2}(90°)}$$

$$= \frac{\cos\frac{1}{2}(-30°)}{\sin 45°}$$

$$= \frac{\cos 15°}{\sin 45°}$$

$$= \frac{\frac{\sqrt{2}+\sqrt{6}}{4}}{\frac{\sqrt{2}}{2}} = \frac{1+\sqrt{3}}{2}$$

(Use half-angle formulas from Section 5.6 to find cos 15°.)

Each expression is equal to $\frac{1+\sqrt{3}}{2}$.

25. $a = 7,\ b = 7,\ c = 6$

$$s = \frac{1}{2}(7+7+6) = 10$$

$$\text{area} = \sqrt{10(3)(3)(4)}$$
$$= \sqrt{360}$$
$$= 18.97 \text{ m}^2$$

$18.97 \text{ m}^2 \div 7.5 \text{ m}^2/\text{can} \approx 2.5$ cans

27. Let AC = the height of the tree.

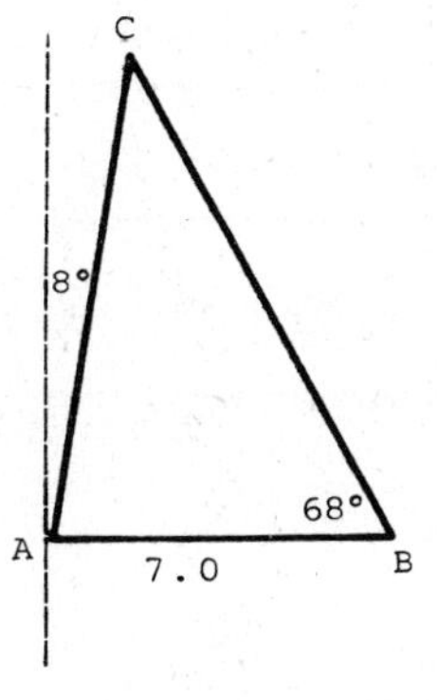

Angle A = 90° − 8.0° = 82°

Angle C = 180° − B − C = 30°

Use the law of sines to find

AC = b.

$$\frac{b}{\sin B} = \frac{c}{\sin C}$$

$$b = \frac{c \sin B}{\sin C}$$

$$= \frac{7.0 \sin 68°}{\sin 30°}$$

$$= 13$$

The tree is 13 m tall.

29. Sketch a triangle showing the situation as follows.

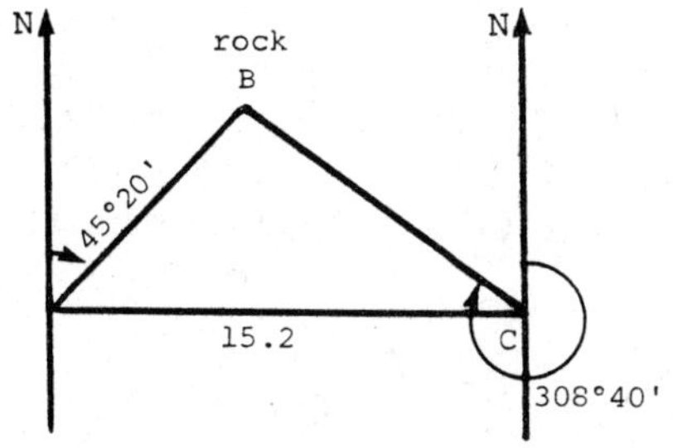

Angle A = 90° − 45° 20′ = 44° 40′

Angle C = 308° 40′ − 270° = 38° 40′

Angle B = 180° − A − C = 96° 40′

Use the law of sines to find BC = a.

$$\frac{a}{\sin A} = \frac{b}{\sin B}$$

$$a = \frac{b \sin A}{\sin B}$$

$$= \frac{15.2 \sin 44° 40'}{\sin 96° 40'}$$

$$a = 10.8$$

The distance between them is 10.8 mi.

31. Let x = the distance between the boats.

In 3 hr, the first boat travels 3(36.2) = 108.6 km and the second travels 3(45.6) = 136.8 km.

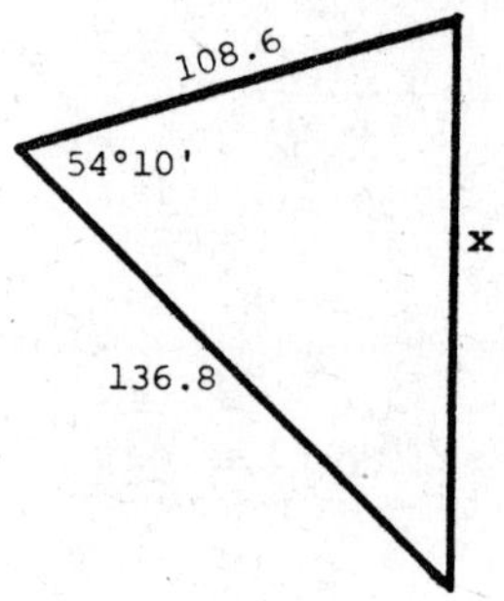

Use the law of cosines to find x.

$$x^2 = (108.6)^2 + (136.8)^2 - 2(108.6)(136.8)\cos 54°\ 10'$$
$$= 13{,}113.359$$
$$x = 114.5 \text{ km}$$

33. Let c = the length of the tunnel.

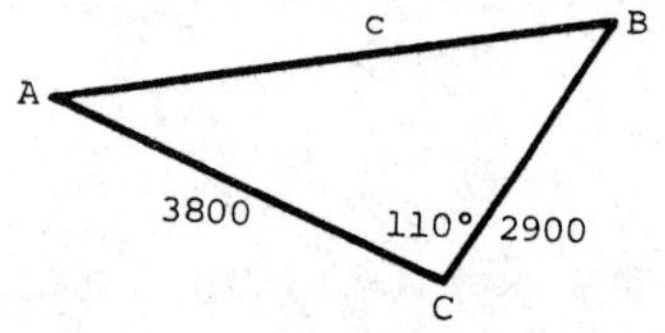

Use the law of cosines to find c.

$$c^2 = 3800^2 + 2900^2 - 2(3800)(2900)\cos 110°$$
$$c^2 = 30{,}388{,}124$$
$$c = 5500$$

The tunnel is 5500 m long.

35. Let x = the distance between the ends of the two equal sides.

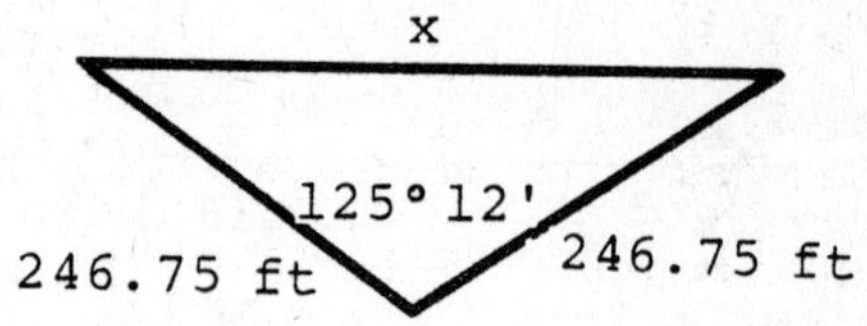

$$x^2 = (246.75)^2 + (246.75)^2 - 2(246.75)(246.75)\cos 125°\ 12'$$
$$x = 438.14$$

The distance between the ends of the two equal sides is 438.14 ft.

37. See the answer sketch in the back of the textbook.

39. See the answer sketch at the back of the textbook.

41. The statement: A diagonal of a parallelogram must bisect the angles of the parallelogram. Draw parallelogram ABCD so that diagonal AC forms θ_1 and θ_2 from angle A and α_1 and α_2 form angle C. Side AB $\neq$ side BC and side AD $\neq$ side CD.

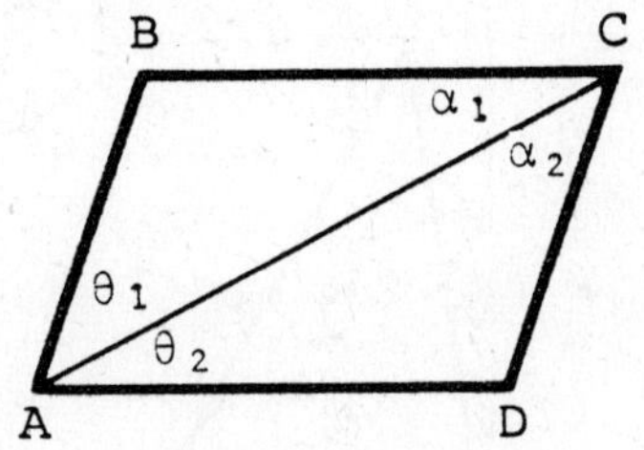

$\theta_1 = \alpha_2$ and $\theta_2 = \alpha_1$ since they are alternate interior angles.

Assume that $\theta_1 = \theta_2$ and $\alpha_1 = \alpha_2$. Then, $\theta_1 = \alpha_1$ and $\theta_2 = \alpha_2$ by substitution triangles ABC and ACD are both isoceles with sides AB = BC and AD = CD. But this statement is a contradiction, so the original statement is false.

43. Find the horizontal and vertical components.

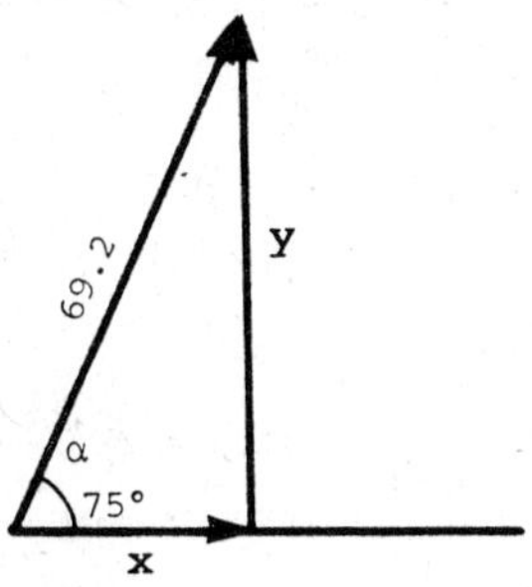

Horizontal:

$$\cos 75° = \frac{|\mathbf{x}|}{69.2}$$

$$|\mathbf{x}| = 69.2 \cos 75° = 17.9$$

Vertical:

$$\sin 75° = \frac{|\mathbf{y}|}{69.2}$$

$$|\mathbf{y}| = 69.2 \sin 75° = 66.8$$

In Exercises 45 and 47 find the magnitude of the resultant force.

45. Forces of 15 lb and 23 lb, forming an angle of 87°

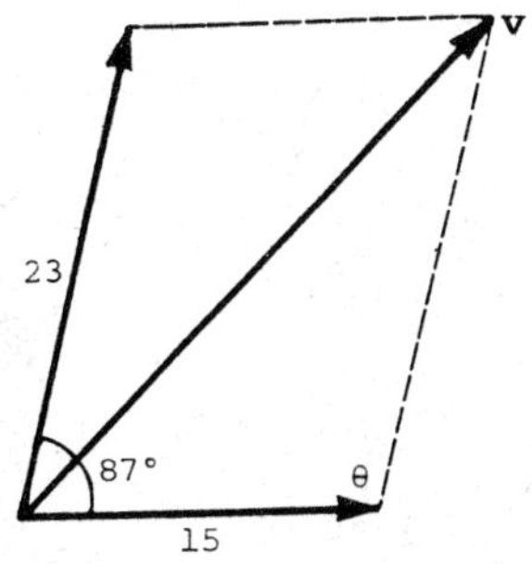

Use the law of cosines to find $|\mathbf{v}|$.

$$|\mathbf{v}|^2 = 15^2 + 23^2 - 2(15)(23) \cos 93°$$

$$|\mathbf{v}|^2 = 790$$

$$|\mathbf{v}| = 28 \text{ lb}$$

47. Forces of 85.2 newtons and 69.4 newtons, forming an angle of 58° 20′

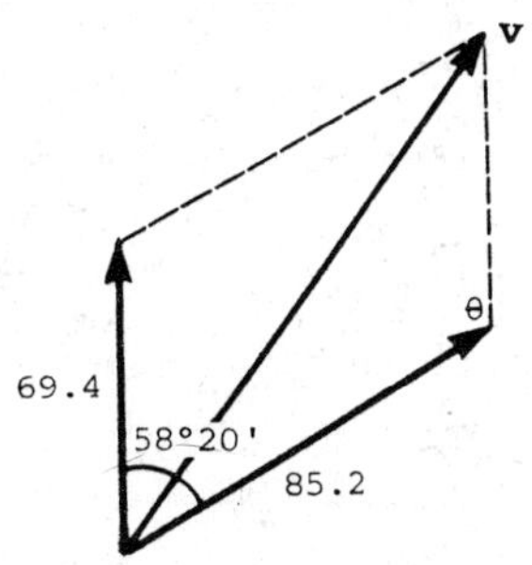

$$\theta = 180° - 58°\,20' = 121°\,40'$$

Use the law of cosines to find $|\mathbf{v}|$.

$$|\mathbf{v}|^2 = (85.2)^2 + (69.4)^2 - 2(85.2)(69.4) \cos 121°\,40'$$

$$|\mathbf{v}|^2 = 18{,}284$$

$$|\mathbf{v}| = 135 \text{ newtons}$$

49. Let $\mathbf{x}$ = the resultant vector.

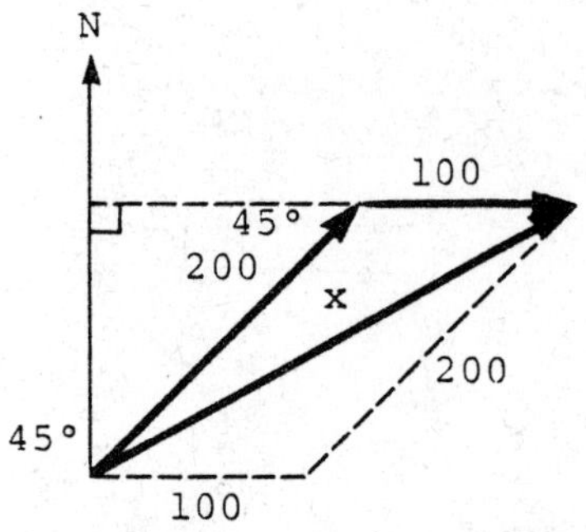

$\alpha = 180° - 45° = 135°$

$|\mathbf{x}|^2 = 200^2 + 100^2 - 2(200)(100)\cos 135°$

$|\mathbf{x}| = 280$

Using the law of cosines again, we get

$$\cos\theta = \frac{200^2 - 100^2 - 280^2}{-2(100)(280)}$$

$$\theta = 30.4°$$

The force is 280 newtons at an angle of 30.4° with the first rope.

51. Let θ = the angle that the hill makes with the horizontal.

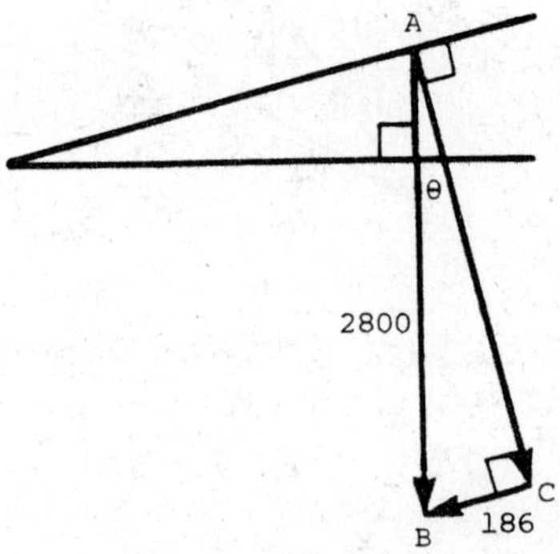

The downward force of 2800 lb has component **AC** perpendicular to the hill and component **CB** parallel to the hill. **AD** represents the force of 186 lb that keeps the car from rolling down the hill. Since vectors **AD** and **BC** are equal, $|\mathbf{BC}| = 186$. Angle B = angle EAB because they are alternate interior angles so the two right triangles are similar. Hence, angle θ = angle BAC.

$$\sin BAC = \frac{186}{2800}$$

$$BAC = 3°\ 50' = \theta$$

The angle that the hill makes with the horizontal is 3° 50′.

53. Let $\mathbf{v}$ = the resultant vector.

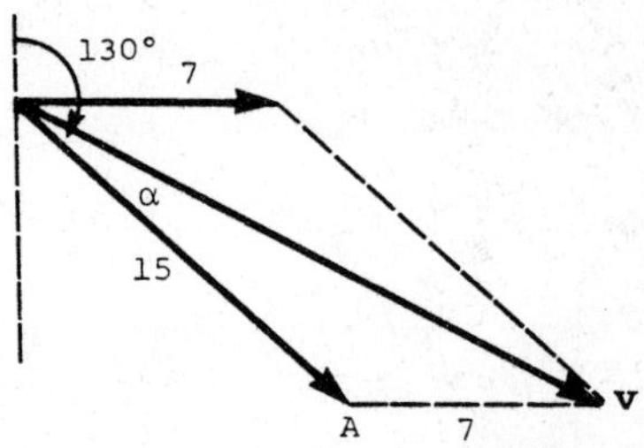

Angle A = 180° − (130° − 90°)
= 140°

Use the law of cosines to find the magnitude of the resultant $\mathbf{v}$.

$|\mathbf{v}|^2 = 15^2 + 7^2 - 2(15)(7)\cos 140°$

$|\mathbf{v}| = 21$

The resulting speed is 21 kph.

Use the law of sines to find α.

$$\frac{\sin \alpha}{7} = \frac{\sin 140°}{21}$$

$$\sin \alpha = \frac{7 \sin 140°}{21}$$

$$\alpha = 12°$$

The bearing is 130° − 12° = 118°.

CHAPTER 3 TEST ANSWERS

1. $\pi/4$ **2.** 2π **3.** $4\pi/3$ **4.** $11\pi/4$ **5.** 270° **6.** −60°

7. 420° **8.** 70° **9.** $\sqrt{3}/2$ **10.** $-\sqrt{2}/2$ **11.** $\sqrt{3}/3$ **12.** −1

13. $\sqrt{3}/3$ **14.** 2 **15.** 3.18 cm **16.** 286 ft^2 **17.** 8900 km

18. .78974917 **19.** .15516683 **20.** −.13744854 **21.** .62632786

22. .38053767 **23.** 1.3074946 **24.** $\pi/6$ cm per min

25. Multiply radians per second by 3600 to get radians per hour.

15.

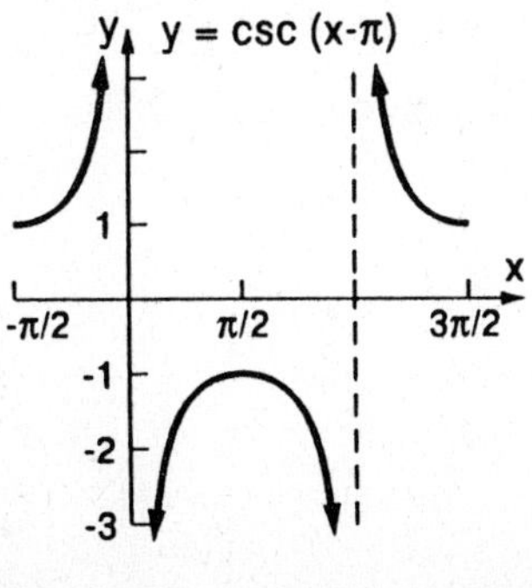

16.

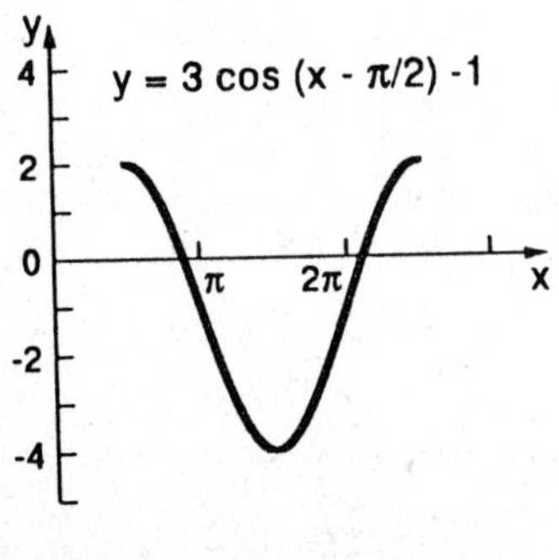

CHAPTER 7 TEST

Find the indicated part of each triangle ABC if the triangle exists.

[7.3] 1. $A = 150.4°$, $b = 31.5$ cm, $c = 22.3$ cm; find a.

[7.1] 2. $A = 21° 20'$, $B = 62° 10'$, $b = 91.2$ m; find a.

[7.2] 3. $A = 35.6°$, $a = 58.2$ m, $b = 104.8$ m; find all possible values of B.

[7.2] 4. $B = 29° 15'$, $b = 16.3$ cm, $c = 20.4$ cm; find all possible values of C.

Find the area of each triangle ABC.

[7.3] 5. $a = 45.5$ yd, $b = 54.2$ yd, $c = 63.7$ yd

[7.1] 6. $B = 50.7°$, $a = 39.2$ cm, $c = 17.3$ cm

[7.2] 7. What happens if you use the law of sines to try to find angle B of triangle ABC if you are given a, b, and angle A, such that $a < b \sin A$?

[7.4] **Sketch the indicated vectors.**

8. **a** − **b**

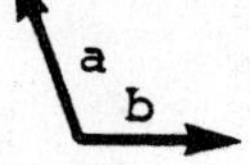

9. 2**m** + **n**

[7.4] **Find the vertical component of each of the following vectors. In each case, α is the angle that the vector makes with the horizontal.**

10. $\alpha = 27°$, magnitude 55

11. $\alpha = 98° 30'$, magnitude 38.6

Solve each of the following.

[7.1] **12.** Two ships on a north–south line are 11 miles apart. The northernmost ship sights a lighthouse at a bearing of 105°. Meanwhile, the other ship sights the lighthouse at a bearing of 45°. How far is the northernmost ship from the lighthouse?

[7.3] **13.** A painter needs to cover a triangular region that is 21 m by 32 m by 15 m. A can of paint covers 50 m^2 of area. How many cans (to the next higher number of cans) will he need?

[7.5] **14.** Find the force required to pull a 65-lb weight up a ramp inclined at 15° to the horizontal.

[7.5] **15.** A force of 32 newtons makes an angle of 55° with a force of 48 newtons. Find the magnitude of the equilibrant.

[7.5] **16.** A plane flying 550 mph on a bearing of 75° encounters a 25-mph wind blowing from due south. Find the reulting bearing of the plane.

CHAPTER 6 TEST ANSWERS

1. $\pi/3$ **2.** $-\pi/2$ **3.** $-\pi/3$ **4.** $\pi/4$ **5.** 4/5 **6.** $\pi/3$

7. $\frac{2 - 5\sqrt{3}}{12}$ **8.** $2\pi/3$; $-\pi/3$ is not in the range of $\text{Cot}^{-1}\ x$.

9. 45°, 225° **10.** 0°, 45°, 225° **11.** 45°, 135°, 225°, 315°

12. 90° **13.** $\pi/3$, $4\pi/3$ **14.** $\pi/6$, $5\pi/6$, $7\pi/6$, $11\pi/6$ **15.** $\pi/2$

16. $\pi/6$, $\pi/2$, $5\pi/6$, $3\pi/2$ **17.** $x = \frac{1}{2}\ \text{Arccos}\ \frac{y}{3}$ **18.** $x = \text{Arctan}\ \frac{3y + 1}{2}$

19. −1 **20.** 4/5

CHAPTER 7 TEST ANSWERS

1. 52.1 cm

2. 37.5 cm

3. The triangle does not exist.

4. 37° 40′ or 142° 20′

5. 1210 yd^2

6. 262 cm^2

7. sin B > 1 results, which means that no such triangle exists.

8.

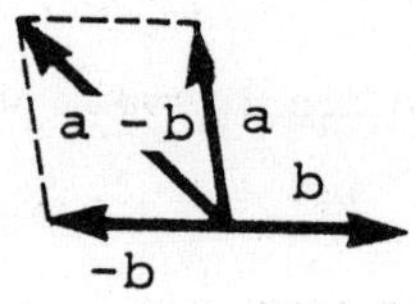

9.

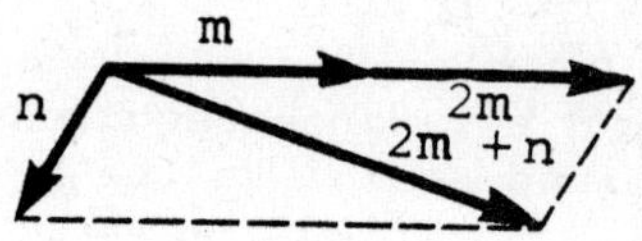

10. 25

11. 38.2

12. 9.0 mi

13. 3 cans

14. 17 lb

15. 71 newtons

16. 73°

CHAPTER 8 COMPLEX NUMBERS AND POLAR EQUATIONS

Section 8.1

1. $\sqrt{-4} = \sqrt{4} \cdot \sqrt{-1} = 2i$

3. $\sqrt{-\frac{25}{9}} = i\sqrt{\frac{25}{9}} = \frac{\sqrt{25}}{\sqrt{9}}i = \frac{5}{3}i$

5. $\sqrt{-150} = i\sqrt{150} = i\sqrt{25 \cdot 6} = i\sqrt{25} \cdot \sqrt{6}$
$= 5i\sqrt{6}$

7. $\sqrt{-80} = i\sqrt{80} = i\sqrt{16 \cdot 5}$
$= 4i\sqrt{5}$

9. $\sqrt{-3} \cdot \sqrt{-3} = i\sqrt{3} \cdot i\sqrt{3}$
$= -3$

11. $\sqrt{-5} \cdot \sqrt{-6} = i\sqrt{5} \cdot i\sqrt{6}$
$= -\sqrt{30}$

13. $\frac{\sqrt{-12}}{\sqrt{-8}} = \frac{i\sqrt{12}}{i\sqrt{8}}$
$= \sqrt{\frac{3}{2}} = \frac{\sqrt{6}}{2}$

15. $\frac{\sqrt{-24}}{\sqrt{72}} = \frac{i\sqrt{24}}{\sqrt{72}}$
$= i\sqrt{\frac{1}{3}} = i\frac{1}{\sqrt{3}} \cdot \frac{\sqrt{3}}{\sqrt{3}}$
$= \frac{\sqrt{3}}{3}i$

17. $x^2 = -16$
$x = \pm\sqrt{-16}$
$x = \pm i\sqrt{16}$
$x = \pm 4i$

The solutions are $4i$ and $-4i$.

19. $z^2 + 12 = 0$
$z^2 = -12$
$z = \pm\sqrt{-12}$
$z = \pm i\sqrt{12}$
$z = \pm i(2\sqrt{3})$
$z = \pm 2i\sqrt{3}$

The solutions are $2i\sqrt{3}$ and $-2i\sqrt{3}$.

21. $3x^2 + 4x + 2 = 0$

Use the quadratic formula with $a = 3$, $b = 4$, and $c = 2$.

$$x = \frac{-4 \pm \sqrt{4^2 - 4 \cdot 3 \cdot 2}}{2 \cdot 3}$$
$$x = \frac{-4 \pm \sqrt{16 - 24}}{6}$$
$$x = \frac{-4 \pm \sqrt{-8}}{6}$$
$$x = \frac{-4 \pm i\sqrt{8}}{6}$$
$$x = \frac{2(-2 \pm i\sqrt{2})}{6}$$
$$x = \frac{-2 \pm i\sqrt{2}}{3}$$

The solutions are $-\frac{2}{3} + \frac{\sqrt{2}}{3}i$ and $-\frac{2}{3} - \frac{\sqrt{2}}{3}i$.

23. $m^2 - 6m + 14 = 0$

Use the quadratic formula with $a = 1$, $b = -6$, and $c = 14$.

$$m = \frac{-(-6) \pm \sqrt{(-6)^2 - 4 \cdot 1 \cdot 14}}{2 \cdot 1}$$
$$m = \frac{6 \pm \sqrt{-20}}{2}$$
$$m = \frac{6 \pm i\sqrt{20}}{2}$$

$m = \frac{2(3 \pm i\sqrt{5})}{2}$

$m = 3 \pm i\sqrt{5}$

The solutions are $3 + i\sqrt{5}$ and $3 - i\sqrt{5}$.

25. $4z^2 = 4z - 7$

$4z^2 - 4z + 7 = 0$

Use the quadratic formula with $a = 4$, $b = -4$, and $c = 7$.

$z = \frac{-(-4) \pm \sqrt{(-4)^2 - 4 \cdot 4 \cdot 7}}{2 \cdot 4}$

$z = \frac{4 \pm \sqrt{-96}}{8}$

$z = \frac{4 \pm i\sqrt{96}}{8}$

$z = \frac{4(1 \pm i\sqrt{6})}{8}$

$z = \frac{1 \pm i\sqrt{6}}{2}$

The solutions are $\frac{1}{2} + \frac{\sqrt{6}}{2}i$ and $\frac{1}{2} - \frac{\sqrt{6}}{2}i$.

27. $m^2 + 1 = -m$

$m^2 + m + 1 = 0$

Use the quadratic formula with $a = 1$, $b = 1$, $c = 1$.

$m = \frac{-1 \pm \sqrt{1^2 - 4 \cdot 1 \cdot 1}}{2 \cdot 1}$

$m = \frac{-1 \pm \sqrt{-3}}{2}$

$m = \frac{-1 \pm i\sqrt{3}}{2}$

The solutions are $-\frac{1}{2} + \frac{\sqrt{3}}{2}i$ and $-\frac{1}{2} - \frac{\sqrt{3}}{2}i$.

29. A quadratic equation of the form $ax^2 + bx + c = 0$ with a, b, and c real numbers and with $a \neq 0$ will have nonreal complex solutions if the discriminant, $b^2 - 4ac$, is negative.

31. If $ax^2 + bx + c = 0$ has all real coefficients and if $4 + \frac{1}{2}i$ is a solution, then $4 - \frac{1}{2}i$ is also a solution, because the solution $4 + \frac{1}{2}i$ is a simplification of the solution $\frac{-b + \sqrt{b^2 - 4ac}}{2a}$, where $-\frac{b}{2a}$ is the real part, or 4, and $\frac{\sqrt{b^2 - 4ac}}{2a}$, or $\frac{1}{2}i$, is the imaginary part. Since $\frac{-b - \sqrt{b^2 - 4ac}}{2a}$ is also a solution, $4 - \frac{1}{2}i$ is also a solution.

33. $(2 - 5i) + (3 + 2i)$

$= (2 + 3) + (2 - 5)i$

$= 5 - 3i$

35. $(-2 + 3i) - (3 + i)$

$= (-2 - 3) + (3 - 1)i$

$= -5 + 2i$

37. $(1 - i) - (5 - 2i)$

$= (1 - 5) + [-1 - (-2)]i$

$= -4 + i$

39. $(2 + i)(3 - 2i)$

$= 6 - 4i + 3i - 2i^2$

$= 6 - i - 2(-1)$

$= 8 - i$

41. $(2 + 4i)(-1 + 3i)$

$= -2 + 6i - 4i + 12i^2$

$= -2 + 2i + 12(-1)$

$= -14 + 2i$

43. $(2 - i)(2 + i) = 4 - 2i + 2i - i^2$

$= 4 - (-1)$

$= 5$ or $5 + 0i$

Note that the product of conjugates is the difference of squares.

$(2 - i)(2 + i) = 2^2 - i^2$

$= 4 - (-1)$

$= 5$

45. $\frac{5}{i} = \frac{5}{0 + i} \cdot \frac{0 - i}{0 - i}$

$= \frac{5}{i} \cdot \frac{-i}{-i}$

$= \frac{-5i}{-i^2}$

$= \frac{-5i}{-(-1)}$

$= \frac{-5i}{1}$

$= -5i$

or $0 - 5i$

47. $\frac{4 - 3i}{4 + 3i} = \frac{4 - 3i}{4 + 3i} \cdot \frac{4 - 3i}{4 - 3i}$

$= \frac{16 - 24i + 9i^2}{16 - 9i^2}$

$= \frac{16 - 24i + 9(-1)}{16 - 9(-1)}$

$= \frac{7 - 24i}{25}$

$= \frac{7}{25} - \frac{24}{25}i$

49. $\frac{4 + i}{6 + 2i} = \frac{4 + i}{6 + 2i} \cdot \frac{6 - 2i}{6 - 2i}$

$= \frac{24 - 2i - 2i^2}{36 - 4i^2}$

$= \frac{24 - 2i - 2(-1)}{36 - 4(-1)}$

$= \frac{26 - 2i}{40}$

$= \frac{13 - i}{20}$

$= \frac{13}{20} - \frac{1}{20}i$

51. $i^{12} = (i^4)^3$

$= 1^3$

$= 1$

53. $i^{18} = i^{16} \cdot i^2$

$= (i^4)^4 \cdot (-1)$

$= 1^4 \cdot (-1)$

$= 1 \cdot (-1)$

$= -1$

55. $i^{-3} = i^{-4} \cdot i^1$

$= (i^4)^{-1} \cdot i$

$= 1^{-1} \cdot i$

$= 1 \cdot i$

$= i$

57. $i^{-10} = i^{-12} \cdot i^2$

$= (i^4)^{-3} \cdot (-1)$

$= 1^{-3} \cdot (-1)$

$= 1 \cdot (-1)$

$= -1$

59. To simplify i^n, divide n by 4 and obtain a quotient q with a remainder r, which must be 0, 1, 2, or 3. Thus,

$$n = 4q + r.$$

Therefore,

$$i^n = i^{4q+r}$$
$$= i^{4q} \cdot i^r.$$

Since $i^{4q} = 1$,

$$i^n = 1 \cdot i^r$$
$$= i^r.$$

Thus, if $r = 0$, $i^n = i^0 = 1$;
if $r = 1$, $i^n = i^1 = i$;
if $r = 2$, $i^n = i^2 = -1$; and
if $r = 3$, $i^n = i^3 = -i$.

61. $2x + yi = 4 - 3i$

$2x = 4$
$x = 2$
$y = -3$

63. $7 - 2yi = 14x - 30i$

$7 = 14x$
$\frac{1}{2} = x$

$-2y = -30$
$y = 15$

65. $x + yi = (2 + 3i)(4 - 2i)$
$= 8 - 4i + 12i - 6i^2$
$= 8 + 8i - 6(-1)$
$= 14 + 8i$

$x = 14$, $y = 8$

67. $I = 8 + 6i$, $Z = 6 + 3i$

$E = IZ$
$= (8 + 6i)(6 + 3i)$
$= 48 + 24i + 36i + 18i^2$
$= 48 + 60i + 18(-1)$
$= 30 + 60i$

69. $I = 7 + 5i$, $E = 28 + 54i$

$E = IZ$

$Z = \frac{E}{I}$

$= \frac{28 + 54i}{7 + 5i}$

$= \frac{28 + 54i}{7 + 5i} \cdot \frac{7 - 5i}{7 - 5i}$,

$= \frac{196 + 378i - 140i - 270i^2}{49 - 25i^2}$

$= \frac{466 + 238i}{74}$

$= \frac{233}{37} + \frac{119}{37}i$

71. $\left(\frac{\sqrt{2}}{2} + \frac{\sqrt{2}}{2}i\right)^2$

$= \left(\frac{\sqrt{2}}{2}\right)^2 + 2\frac{\sqrt{2}}{2} \cdot \frac{\sqrt{2}}{2}i + \left(\frac{\sqrt{2}}{2}i\right)^2$

$= \frac{2}{4} + \frac{4}{4}i + \frac{2}{4}i^2$

$= \frac{1}{2} + i + \frac{1}{2}(-1)$

$= \frac{1}{2} + i - \frac{1}{2}$

Since $\left(\frac{\sqrt{2}}{2} + \frac{\sqrt{2}}{2}i\right)^2 = i$, $\frac{\sqrt{2}}{2} + \frac{\sqrt{2}}{2}i$ is a square root of i.

Section 8.2

For Exercises 1–9, see the answer graphs in the back of your textbook.

11. In order for a complex number $a + bi$ also to be a real number, its imaginary part b must be zero.

13. A complex number of the form a + bi will have its corresponding vector lying on the y-axis provided a = <u>0</u>.

15. $4 - 3i, -1 + 2i$

$(4 - 3i) + (-1 + 2i) = 3 - i$

17. $5 - 6i, -2 + 3i$

$(5 - 6i) + (-2 + 3i) = 3 - 3i$

19. $-3, 3i$

$(-3 + 0i) + (0 + 3i) = -3 + 3i$

21. $2 + 6i, -2i$

$(2 + 6i) + (0 - 2i) = 2 + 4i$

23. $7 + 6i, 3i$

$(7 + 6i) + (0 + 3i) = 7 + 9i$

25. $2(\cos 45° + i \sin 45°)$

$= 2\left(\frac{\sqrt{2}}{2} + \frac{i\sqrt{2}}{2}\right) = \sqrt{2} + i\sqrt{2}$

27. $10(\cos 90° + i \sin 90°)$

$= 10(0 + i)$

$= 0 + 10i = 10i$

29. $4(\cos 240° + i \sin 240°)$

$= 4\left(-\frac{1}{2} - i\frac{\sqrt{3}}{2}\right)$

$= -2 - 2i\sqrt{3}$

31. $(\cos 30° + i \sin 30°)$

$= \frac{\sqrt{3}}{2} + \frac{1}{2}i$

33. $5 \text{ cis } 300°$

$= 5(\cos 300° + i \sin 300°)$

$= 5\left[\frac{1}{2} + i\left(-\frac{\sqrt{3}}{2}\right)\right]$

$= \frac{5}{2} - \frac{5\sqrt{3}}{2}i$

35. $\sqrt{2} \text{ cis } 180°$

$= \sqrt{2}(\cos 180° + i \sin 180°)$

$= \sqrt{2}(-1 + i \cdot 0)$

$= -\sqrt{2} + 0i$

$= -\sqrt{2}$

37. $3 - 3i = x + yi$

Find r and θ.

$x = 3, y = -3$

$r = \sqrt{x^2 + y^2}$

$= \sqrt{(3)^2 + (-3)^2}$

$= \sqrt{18}$

$= 3\sqrt{2}$

$\tan \theta = \frac{y}{x} = \frac{-3}{3} = -1$

$3 - 3i$ is in quadrant IV, so

$\theta = 315°$.

$3 - 3i = 3\sqrt{2}(\cos 315° + i \sin 315°)$

39. $-3 - 3i\sqrt{3}$

$x = -3, y = -3\sqrt{3}$

$r = \sqrt{(-3)^2 + (-3\sqrt{3})^2} = 6$

$\tan \theta = \frac{-3\sqrt{3}}{-3} = \sqrt{3}$

$-3 - 3i\sqrt{3}$ is in quadrant III, so

$\theta = 240°$.

$-3 - 3i\sqrt{3} = 6(\cos 240° + i \sin 240°)$

41. $\sqrt{3} - i$

$x = \sqrt{3}$, $y = -1$

$r = \sqrt{(\sqrt{3})^2 + (-1)^2} = 2$

$\tan \theta = \frac{-1}{\sqrt{3}} = -\frac{\sqrt{3}}{3}$

$\sqrt{3} - i$ is in quadrant IV, so

$\theta = 330°$.

$\sqrt{3} - i = 2(\cos 330° + i \sin 330°)$

43. $-5 - 5i$

$x = -5$, $y = -5$

$r = \sqrt{(-5)^2 + (-5)^2}$
$= \sqrt{50} = 5\sqrt{2}$

$\tan \theta = \frac{-5}{-5} = 1$

$-5 - 5i$ is in quadrant III, so

$\theta = 225°$.

$-5 - 5i = 5\sqrt{2}(\cos 225° + i \sin 225°)$

45. $2 + 2i$

$x = 2$, $y = 2$

$r = \sqrt{(2)^2 + (2)^2} = \sqrt{8} = 2\sqrt{2}$

$\tan \theta = \frac{2}{2} = 1$

$2 + 2i$ is in quadrant I, so

$\theta = 45°$.

$2 + 2i = 2\sqrt{2}(\cos 45° + i \sin 45°)$

47. $-4 = -4 + 0i$

$x = -4$, $y = 0$

$r = \sqrt{(-4)^2 + 0^2} = 4$

The graph of -4 is on the negative x-axis, so $\theta = 180°$.

$-4 = 4(\cos 180° + i \sin 180°)$

49. $-2i = 0 - 2i$

$x = 0$, $y = -2$

$r = \sqrt{0^2 + (-2)^2} = 2$

The graph of $-2i$ is on the negative y-axis, so $\theta = 270°$.

$-2i = 2(\cos 270° + i \sin 270°)$

51. $2 + 3i$

$x = 2$, $y = 3$

$r = \sqrt{(2)^2 + (3)^2} = \sqrt{13}$

$\tan \theta = \frac{3}{2}$

$2 + 3i$ is in quadrant I, so

$\theta = 56.31°$.

$2 + 3i$
$= \sqrt{13}(\cos 56.31° + i \sin 56.31°)$

53. $3(\cos 250° + i \sin 250°)$
$= 3[-.34202014 + (-.93969262)i]$
$= -1.0260604 - 2.8190779i$

55. $12i = 0 + 12i$

$x = 0$, $y = 12$

$r = \sqrt{0^2 + 12^2} = 12$

$12i$ is on the positive y-axis, so $\theta = 90°$.

$12i = 12(\cos 90° + i \sin 90°)$

57. $3 + 5i$

$x = 3$, $y = 5$

$r = \sqrt{3^2 + 5^2} = \sqrt{34}$

$\tan \theta = \frac{5}{3}$

$3 + 5i$ is in quadrant I, so

$\theta = 59.04°$.

$3 + 5i$

$= \sqrt{34}(\cos 59.04° + i \sin 59.04°)$

Section 8.3

1. $[3(\cos 60° + i \sin 60°)]$
$\cdot [2(\cos 90° + i \sin 90°)]$
$= 6(\cos 150° + i \sin 150°)$
$= 6(-\frac{\sqrt{3}}{2} + i\frac{1}{2})$
$= -3\sqrt{3} + 3i$

3. $[2(\cos 45° + i \sin 45°)]$
$\cdot [2(\cos 225° + i \sin 225°)]$
$= 4(\cos 270° + i \sin 270°)$
$= 4(0 - i)$
$= 0 - 4i$ or $-4i$

5. $[4(\cos 60° + i \sin 60°)]$
$\cdot [6(\cos 330° + i \sin 330°)]$
$= 24(\cos 390° + i \sin 390°)$
$= 24(\cos 30° + i \sin 30°)$
$= 24(\frac{\sqrt{3}}{2} + i\frac{1}{2})$
$= 12\sqrt{3} + 12i$

7. $[5 \text{ cis } 90°][3 \text{ cis } 45°]$
$= 15 \text{ cis } 135°$
$= 15(\cos 135° + i \sin 135°)$
$= 15(-\frac{\sqrt{2}}{2} + \frac{\sqrt{2}}{2}i)$
$= -\frac{15\sqrt{2}}{2} + \frac{15\sqrt{2}}{2}i$

9. $[\sqrt{3} \text{ cis } 45°][\sqrt{3} \text{ cis } 225°]$
$= 3 \text{ cis } 270°$
$= 3(\cos 270° + i \sin 270°)$
$= 3(0 - i)$
$= 0 - 3i$ or $-3i$

11. $\frac{4(\cos 120° + i \sin 120°)}{2(\cos 150° + i \sin 150°)}$
$= 2[\cos (-30°) + i \sin (-30°)]$
$= 2(\frac{\sqrt{3}}{2} - \frac{1}{2}i)$
$= \sqrt{3} - i$

13. $\frac{16(\cos 300° + i \sin 300°)}{8(\cos 60° + i \sin 60°)}$
$= 2(\cos 240° + i \sin 240°)$
$= 2(-\frac{1}{2} - \frac{\sqrt{3}}{2}i)$
$= -1 - i\sqrt{3}$

15. $\frac{3 \text{ cis } 305°}{9 \text{ cis } 65°}$
$= \frac{1}{3} \text{ cis } 240°$
$= \frac{1}{3}(\cos 240° + i \sin 240°)$
$= \frac{1}{3}(-\frac{1}{2} - \frac{\sqrt{3}}{2}i)$
$= -\frac{1}{6} - \frac{\sqrt{3}}{6}i$

17. $\frac{8}{\sqrt{3} + i} = \frac{8(\sqrt{3} - i)}{(\sqrt{3} + i)(\sqrt{3} - i)}$
$= \frac{8\sqrt{3} - 8i}{3 + 1} = \frac{8\sqrt{3} - 8i}{4}$
$= 2\sqrt{3} - 2i$

19. $\frac{-i}{1 + i} = \frac{-i(1 - i)}{(1 + i)(1 - i)}$
$= \frac{-i - 1}{1 + 1} = \frac{-1 - i}{2}$
$= -\frac{1}{2} - \frac{1}{2}i$

21. $\dfrac{2\sqrt{6} - 2i\sqrt{2}}{\sqrt{2} - i\sqrt{6}}$

$= \dfrac{(2\sqrt{6} - 2i\sqrt{2})(\sqrt{2} + i\sqrt{6})}{(\sqrt{2} - i\sqrt{6})(\sqrt{2}+ i\sqrt{6})}$

$= \dfrac{4\sqrt{3} + 12i - 4i + 4\sqrt{3}}{2 + 6}$

$= \dfrac{8\sqrt{3} + 8i}{8}$

$= \sqrt{3} + i$

23. $[2.5(\cos 35° + i \sin 35°)]$

$\cdot [3.0(\cos 50° + i \sin 50°)]$

$= 7.5(\cos 85° + i \sin 85°)$

$= .65366807 + 7.4714602i$

25. $(12 \text{ cis } 18.5°)(3 \text{ cis } 12.5°)$

$= 36 \text{ cis } 31°$

$= 36(\cos 31° + i \sin 31°)$

$= 30.858023 + 18.541371i$

27. $\dfrac{45(\cos 127° + i \sin 127°)}{22.5(\cos 43° + i \sin 43°)}$

$= 2(\cos 84° + i \sin 84°)$

$= .20905693 + 1.9890438i$

29. $\left[2 \text{ cis } \dfrac{5\pi}{9}\right]^2$

$= \left[2 \text{ cis } \dfrac{5\pi}{9}\right]\left[2 \text{ cis } \dfrac{5\pi}{9}\right]$

$= 4 \text{ cis } \dfrac{10\pi}{9}$

$= 4\left(\cos \dfrac{10\pi}{9} + i \sin \dfrac{10\pi}{9}\right)$

$= -3.7587705 - 1.3680806i$

31. $(2 \text{ cis } 45°)(5 \text{ cis } 90°)$ and $[2 \text{ cis } (-315°)][5 \text{ cis } (-270°)]$ are the same because 45° and -315° are coterminal angles and 90° and -270° are coterminal angles. Thus, the factors of the two products are the same numbers.

33. $E = 8(\cos 20° + i \sin 20°)$, $R = 6$, $X_L = 3$

$I = \dfrac{E}{Z}$, $Z = R + X_L i$

Write $Z = 6 + 3i$ in trigonometric form

$x = 6$, $y = 3$, so $r = \sqrt{6^2 + 3^2} = \sqrt{45}$.

$\tan \theta = \dfrac{3}{6} = \dfrac{1}{2}$, so $\theta = 26.6°$

$Z = \sqrt{45} \text{ cis } 26.6°$

$I = \dfrac{8 \text{ cis } 20°}{\sqrt{45} \text{ cis } 26.6°}$

$= 1.19 \text{ cis}(-6.6°)$

$= 1.2 - .14i$

Section 8.4

1. $[3(\cos 30° + i \sin 30°)]^3$

$= 3^3[\cos (3)(30°) + i \sin (3)(30°)]$

$= 27(\cos 90° + i \sin 90°)$

$= 0 + 27i$ or $27i$

3. $(\cos 45° + i \sin 45°)$

$= [\cos (45°)(8) + i \sin (8)(45°)]$

$= 1 + 0i$ or 1

5. $[3 \text{ cis } 100°]^3$

$= 3^3 \text{ cis}(3 \cdot 100°)$

$= 27 \text{ cis } 300°$

$= 27(\cos 300° + i \sin 300°)$

$= 27\left(\frac{1}{2} - \frac{\sqrt{3}}{2}i\right)$

$= \frac{27}{2} - \frac{27\sqrt{3}}{2}i$

7. $(\sqrt{3} + i)^5$

$x = \sqrt{3},\ y = 1$

$r = \sqrt{(\sqrt{3})^2 + (1)^2} = 2$

$\tan \theta = \frac{1}{\sqrt{3}}$ so $\theta = 30°$.

$(\sqrt{3} + i)^5$

$= [2(\cos 30° + i \sin 30°)]^5$

$= 2^5(\cos 150° + i \sin 150°)$

$= 32\left(-\frac{\sqrt{3}}{2} + i\frac{1}{2}\right)$

$= -16\sqrt{3} + 16i$

9. $(2 - 2i\sqrt{3})^4$

$r = \sqrt{(2)^2 + (-2\sqrt{3})^2} = 4$

$\tan \theta = \frac{-2\sqrt{3}}{2} = -\sqrt{3}$ so $\theta = 300°$.

$(2 - 2i\sqrt{3})^4$

$= [4(\cos 300° + i \sin 300°)]^4$

$= 4^4(\cos 1200° + i \sin 1200°)$

$= 256\left(-\frac{1}{2} + \frac{i\sqrt{3}}{2}\right)$

$= -128 + 128i\sqrt{3}$

11. $(-2 - 2i)^5$

$r = \sqrt{(-2)^2 + (-2)^2}$

$= \sqrt{8} = 2\sqrt{2}$

$\tan \theta = \frac{-2}{-2} = 1$ so $\theta = 225°$.

$(-2 - 2i)^5$

$= (2\sqrt{2})^5(\cos 225° + i \sin 225°)^5$

$= 32\sqrt{32}(\cos 1125° + i \sin 1125°)$

$= 128\sqrt{2}(\cos 45° + i \sin 45°)$

$= 128\sqrt{2}\left(\frac{\sqrt{2}}{2} + \frac{\sqrt{2}}{2}i\right)$

$= 128 + 128i$

In Exercises 13–29, see the answer graphs in the back of your textbook.

13. $(\cos 0° + i \sin 0°)$

$= 1(\cos 0° + i \sin 0°)$

$r = 1,\ \theta = 0°,\ n = 3$

$r^{1/3} = 1^{1/3} = 1$

$\alpha = \frac{\theta + 360° \cdot k}{n}$

$= \frac{0° + 360° \cdot k}{3}$

If $k = 0$, $\alpha = \frac{0° + 360° \cdot 0}{3} = 0°$.

If $k = 1$, $\alpha = \frac{0° + 360° \cdot 1}{3} = 120°$.

If $k = 2$, $\alpha = \frac{0° + 360° \cdot 2}{3} = 240°$.

So the cube roots are

$(\cos 0° + i \sin 0°)$

$(\cos 120° + i \sin 120°)$

$(\cos 240° + i \sin 240°)$.

15. $8 \text{ cis } 60°$

$r = 8,\ \theta = 60°,\ n = 3$

$r^{1/3} = 8^{1/3} = 2$

$\alpha = \frac{60° + 360 \cdot k}{3}$; $k = 0, 1, 2$

$\alpha = 20°, 140°, 260°$

So the cube roots are

2 cis 20°

2 cis 140°

2 cis 260°.

17. $-8i = 8(\cos 270° + i \sin 270°)$

$r = 8,\ \theta = 270°,\ n = 3$

$r^{1/3} = 8^{1/3} = 2$

$\alpha = \dfrac{270° + 360° \cdot k}{3};\ k = 0, 1, 2$

$\alpha = 90°, 210°, 330°$

So the cube roots are

$2(\cos 90° + i \sin 90°)$

$2(\cos 210° + i \sin 210°)$

$2 \cos 330° + i \sin 330°).$

19. $-64 = 64(\cos 180° + i \sin 180°)$

$r = 64,\ \theta = 180°,\ n = 3$

$r^{1/3} = 64^{1/3} = 4$

$\alpha = \dfrac{180° + 360° \cdot k}{3};\ k = 0, 1, 2$

$\alpha = 60°, 180°, 300°$

So the cube roots are

$4(\cos 60° + i \sin 60°),$

$4(\cos 180° + i \sin 180°),$

$4(\cos 300° + i \sin 300°).$

21. $1 + i\sqrt{3}$

$r = \sqrt{(1)^2 + (\sqrt{3})^2} = 2$

$\tan \theta = \dfrac{\sqrt{3}}{1}$ so $\theta = 60°.$

$1 + i\sqrt{3} = 2(\cos 60° + i \sin 60°)$

$\alpha = \dfrac{60° + 360° \cdot k}{3};\ k = 0, 1, 2$

$\alpha = 20°, 140°, 260°$

The cube roots are

$\sqrt[3]{2}(\cos 20° + i \sin 20°),$

$\sqrt[3]{2}(\cos 140° + i \sin 140°),$

$\sqrt[3]{2}(\cos 260° + i \sin 260°).$

23. $-2\sqrt{3} + 2i$

$r = \sqrt{(-2\sqrt{3})^2 + (2)^2} = 4$

$\tan \theta = \dfrac{2}{-2\sqrt{3}}$ so $\theta = 150°.$

$-2\sqrt{3} + 2i = 4(\cos 150° + i \sin 150°)$

$r^{1/3} = 4^{1/3} = \sqrt[3]{4}$

$\alpha = \dfrac{150° + 360° \cdot k}{3};\ k = 0, 1, 2$

$\alpha = 50°, 170°, 290°$

The cube roots are

$\sqrt[3]{4}(\cos 50° + i \sin 50°),$

$\sqrt[3]{4}(\cos 170° + i \sin 170°),$

$\sqrt[3]{4}(\cos 290° + i \sin 290°).$

25. Find all the second (or square) roots of 1.

$1 = 1(\cos 0° + i \sin 0°)$

$r = 1,\ \theta = 0°,\ n = 2$

$r^{1/2} = 1^{1/2} = 1$

$\alpha = \dfrac{0° + 360° \cdot k}{2};\ k = 0, 1$

$\alpha = 0°, 180°$

The second roots of 1 are

$(\cos 0° + i \sin 0°),$

$(\cos 180° + i \sin 180°).$

27. Find all the sixth roots of 1.

$1 = 1(\cos 0° + i \sin 0°)$

$r = 1,\ \theta = 0°,\ n = 6$

$r^{1/6} = 1^{1/6} = 1$

$\alpha = \frac{0° + 360° \cdot k}{6}$;

$k = 0, 1, 2, 3, 4, 5$

$\alpha = 0°, 60°, 120°, 180°, 240°, 300°$

The sixth roots of 1 are

$(\cos 0° + i \sin 0°)$,

$(\cos 60° + i \sin 60°)$,

$(\cos 120° + i \sin 120°)$,

$(\cos 180° + i \sin 180°)$,

$(\cos 240° + i \sin 240°)$,

$(\cos 300° + i \sin 300°)$.

29. Find all the second (square) roots of i.

$i = \cos 90° + i \sin 90°$

$r = 1, \theta = 90°, n = 2$

$r^{1/2} = 1^{1/2} = 1$

$\alpha = \frac{90° + 360° \cdot k}{2}$; $k = 0, 1$

$\alpha = 45°, 225°$

The second roots of i are

$(\cos 45° + i \sin 45°)$,

$(\cos 225° + i \sin 225°)$.

31. If a is any real number, then a will have one real nth root if n is odd for the following reasons.
If a is positive, $\sqrt[n]{a}$ for n odd will be a positive real number.
If a is zero, $\sqrt[n]{a} = 0$ for n odd.
If a is negative, $\sqrt[n]{a}$ for n odd will be a negative real number.
In any case, a will have a real nth root if n is odd.

33. The statement, "Every real number must have two real square roots," is false. Consider, for example, the real number −4. Its two square roots are 2i and −2i which are not real.

35. $x^3 - 1 = 0$

$x^3 = 1$

$= 1 + 0i$

$= 1(\cos 0° + i \sin 0°)$

The modulus of the solutions is

$1^{1/3} = 1$.

The arguments of the solutions are

$\alpha = \frac{0° + 360° \cdot k}{3}$; $k = 0, 1, 2$.

If $k = 0$, $\frac{0° + 360° \cdot 0}{3} = 0°$.

If $k = 1$, $\frac{0° + 360° \cdot 1}{3} = 120°$.

If $k = 2$, $\frac{0° + 360° \cdot 2}{3} = 240°$.

$x = (\cos 0° + i \sin 0°)$,

$(\cos 120° + i \sin 120°)$,

$(\cos 240° + i \sin 240°)$

37. $x^2 + i = 0$

$x^3 = -i$

$= 0 - i$

$= 1(\cos 270° + i \sin 270°)$

$1^{1/3} = 1$

$\alpha = \frac{270° + 360° \cdot k}{3}$; $k = 0, 1, 2$

$\alpha = 90°, 210°, 330°$

$x = (\cos 90° + i \sin 90°)$,

$(\cos 210° + i \sin 210°)$,

$(\cos 330° + i \sin 330°)$

39. $x^3 - 8 = 0$

$x^3 = 8$

$= 8 + 0i$

$= 8(\cos 0° + i \sin 0°)$

$8^{1/3} = 2$

$\alpha = \dfrac{0° + 360° \cdot k}{3}$; $k = 0, 1, 2$

$\alpha = 0°, 120°, 240°$

$x = 2(\cos 0° + i \sin 0°),$

$2(\cos 120° + i \text{ sub } 120°),$

$2(\cos 240° + i \sin 240°)$

41. $x^4 + 1 = 0$

$x^4 = -1$

$= -1 + 0i$

$= 1(\cos 180° + i \sin 180°)$

$1^{1/4} = 1$

$\alpha = \dfrac{180° + 360° \cdot k}{4}$; $k = 0, 1, 2, 3$

$\alpha = 45°, 135°, 225°, 315°$

$x = (\cos 45° + i \sin 45°),$

$(\cos 135° + i \sin 135°),$

$(\cos 225° + i \sin 225°),$

$(\cos 315° + i \sin 315°)$

43. $x^4 - i = 0$

$x^4 = i$

$= 1 + i$

$= 1(\cos 90° + i \sin 90°)$

$1^{1/4} = 1$

$\alpha = \dfrac{90° + 360° \cdot k}{4}$; $k = 0, 1, 2, 3$

$\alpha = 22.5°, 112.5°, 202.5°, 292.5°$

$x = (\cos 22.5° + i \sin 22.5°),$

$(\cos 112.5° + i \sin 112.5°),$

$(\cos 202.5° + i \sin 202.5°),$

$(\cos 292.5° + i \sin 292.5°)$

45. $x^3 - (4 + 4i\sqrt{3}) = 0$

$x^3 = 4 + 4i\sqrt{3}$

$r = \sqrt{(4)^2 + (4\sqrt{3})^2} = 8$

$\tan \theta = \dfrac{4\sqrt{3}}{4}$; $\theta = 60°$

$x^3 = 8(\cos 60° + i \sin 60°)$

$8^{1/3} = 2$

$\alpha = \dfrac{60° + 360° \cdot k}{3}$; $k = 0, 1, 2$

$\alpha = 20°, 140°, 260°$

$x = 2(\cos 20° + i \sin 20°),$

$2(\cos 140° + i \sin 140°),$

$2(\cos 260° + i \sin 260°)$

47. $x^3 + 4 - 5i = 0$

$x^3 = -4 + 5i$

$r = \sqrt{(-4)^2 + (5)^2} = \sqrt{41}$

$\tan \theta = -\dfrac{5}{4}$, θ is in quadrant II.

$\theta = 128.65981°$

$x^3 = \sqrt{41}(\cos 128.65981°$

$+ i \sin 128.65981°)$

$(\sqrt{41})^{1/3} = 1.8569376$

$\alpha = \dfrac{\theta + 360° \cdot k}{3}$; $k = 0, 1, 2$

If $k = 0$, $\alpha = \dfrac{128.65981° + 0 \cdot 360°}{3}$

$= 42.886603°.$

If $k = 1$, $\alpha = \dfrac{128.65981° + 1 \cdot 360°}{3}$

$= 162.8866°.$

If $k = 2$, $\alpha = \dfrac{128.65981° + 2 \cdot 360°}{3}$

$= 282.8866°.$

$x = 1.8569376$
$\cdot (\cos 42.886603° + i \sin 42.886603°)$,
1.8569376
$\cdot (\cos 162.8866° + i \sin 162.8866°)$,
1.8569376
$\cdot (\cos 282.8866° + i \sin 282.8866°)$

In standard form,

$x = 1.3606 + 1.2637i$,
$-1.7747 + .5464i$,
$.4141 - 1.8102i$.

49.
$$x^3 - 1 = 0$$
$$(x - 1)(x^2 + x + 1) = 0$$
$$x - 1 = 0$$
$$x = 1$$
or
$$x^2 + x + 1 = 0$$

Use the quadratic formula with $a = 1$, $b = 1$, and $c = 1$.

$$x = \frac{-1 \pm \sqrt{1^2 - 4 \cdot 1 \cdot 1}}{2 \cdot 1}$$
$$x = \frac{-1 \pm \sqrt{-3}}{2}$$
$$x = \frac{-1 \pm i\sqrt{3}}{2}$$
$$x = -\frac{1}{2} \pm \frac{\sqrt{3}}{2}i$$
$$x = 1, -\frac{1}{2} + \frac{\sqrt{3}}{2}i, -\frac{1}{2} - \frac{\sqrt{3}}{2}i$$

We see that the solutions are the same as Exercise 35.
For Exercise 35, the solutions are

$(\cos 0° + i \sin 0°)$
$(\cos 120° + i \sin 120°)$
$(\cos 240° + i \sin 240°)$.

In standard form, these are

$$1, -\frac{1}{2} + \frac{\sqrt{3}}{2}i, -\frac{1}{2} - \frac{\sqrt{3}}{2}i.$$

Section 8.5

For Exercises 1–35 see the answer graphs in the back of your textbook (except for Exercises 11 and 25). Answer may vary for Exercises 1-9.

1. Two other pairs of polar coordinates for (1, 45°) are (1, 405°) and (−1, 225°).

3. Two other pairs of polar coordinates for (−2, 135°) are (−2, 495°) and (2, 315°).

5. Two other pairs of polar coordinates for (5, −60°) are (5, 300°) and (−5, 120°).

7. Two other pairs of polar coordinates for (−3, −210°) are (−3, 150°) and (3, −30°).

9. Two other pairs of polar coordinates for (3, 300°) are (3, 660°) and (−3, 120°).

11. If point P lies on an axis in the Cartesian plane, then θ must be a quadrantal angle if (r, θ) represents the point P in polar coordinates.

13. $r = 2 + 2 \cos \theta$

θ	0°	30°	60°	90°	120°	150°
$\cos \theta$	1	.9	.5	0	−.5	−.9
r	4	3.8	3	2	1	.2

θ	180°	210°	240°	270°	300°	330°
$\cos \theta$	−1	−.9	−.5	0	.5	.9
r	0	.3	1	2	3	3.7

15. $r = 3 + \cos \theta$ (limaçon)

θ	0°	30°	60°	90°	120°	150°
r	4	3.9	3.5	3	2.5	2.1

θ	180°	210°	240°	270°	300°	330°
r	2	2.1	2.5	3	3.5	3.9

17. $r = 4 \cos 2\theta$ (four-leaved rose)

θ	0°	30°	45°	60°	90°	120°	135°	150°
r	4	2	0	−2	−4	−2	0	2

θ	180°	210°	225°	240°	270°	300°	315°	330°
r	4	2	0	−2	−4	−2	0	2

19. $r^2 = 4 \cos 2\theta$ (lemniscate)

$r = \pm 2\sqrt{\cos 2\theta}$

Graph only exists for [0°, 45°], [135°, 225°], and [315°, 360°] because $\cos 2\theta$ must be positive.

θ	0°	30°	45°	135°	150°
r	±2	±1.4	0	0	±1.4

θ	180°	210°	225°	315°	330°
r	±2	±1.4	0	0	±1.4

21. $r = 4(1 - \cos \theta)$ (cardioid)

θ	0°	30°	60°	90°	120°	150°
r	0	.5	2	4	6	7.5

θ	180°	210°	240°	270°	300°	330°
r	8	7.5	6	4	2	.5

23. $r = 2 \sin \theta \tan \theta$ (cissoid)

r is undefined at $\theta = 90°$ and $\theta = 270°$.

θ	0°	30°	45°	60°	90°	120°	135°	150°	180°
r	0	.6	1.4	3	–	−3	−1.4	−.6	0

Notice that for [180°, 360°), the graph retraces the path traced for [0°, 180°).

25. To plot a point (r, θ) in polar coordinates with $r < 0$, move $|r|$ units in the direction opposite the ray for θ.

27. $r = 2 \sin \theta$

Multiply both sides by r.

$$r^2 = 2r \sin \theta$$

Since $r^2 = x^2 + y^2$ and $y = r \sin \theta$, $x^2 + y^2 = 2y$.

Complete the square on y.

$$x^2 + y^2 - 2y + 1 = 1$$
$$x + (y - 1)^2 = 1$$

The graph is a circle with center at (0, 1) and radius 1.

29. $r = \dfrac{2}{1 - \cos\theta}$

Multiply both sides by $1 - \cos\theta$.

$r - r\cos\theta = 2$

$\sqrt{x^2 + y^2} - x = 2$

$\sqrt{x^2 + y^2} = 2 + x$

$x^2 + y^2 = (2 + x)^2$

$x^2 + y^2 = 4 + 4x + x^2$

$y^2 = 4(1 - x)$

The graph is a parabola with vertex at (-1, 0) and axis y = 0.

31. $r + 2\cos\theta = -2\sin\theta$

$r^2 = -2r\sin\theta - 2r\cos\theta$

$x^2 + y^2 = -2y - 2x$

$x^2 + y^2 + 2y + 2x = 0$

$(x + 1)^2 + (y + 1)^2 = 2$

The graph is a circle with center (-1, -1) and radius $\sqrt{2}$.

33. $r = 2\sec\theta$

$r = \dfrac{2}{\cos\theta}$

$r\cos\theta = 2$

$x = 2$

The graph is a vertical line.

35. $r(\cos\theta + \sin\theta) = 2$

$r\cos\theta + r\sin\theta = 2$

$x + y = 2$

The graph is the line with intercepts (0, 2) and (2, 0).

37. $x + y = 4$

$r\cos\theta + r\sin\theta = 4$

$r(\cos\theta + \sin\theta) = 4$

39. $x^2 + y^2 = 16$

$r^2 = 16$

$r = 4$

41. $y = 2$

$r\sin\theta = 2$ or

$r = \dfrac{2}{\sin\theta} = 2\csc\theta$

43. Graph $r = \theta$, a spiral of Archimedes.

θ	-360°	-270°	-180°	-90°
θ (radians)	-6.3	-4.7	-3.1	-1.6
r	-6.3	-4.7	-3.1	-1.6

θ	0°	90°	180°	270°	360°
θ (radians)	0	1.6	3.1	4.7	6.3
r	0	1.6	3.1	4.7	6.3

See the answer graph at the back of the textbook.

Chapter 8 Review Exercises

1. $\sqrt{-9} = i\sqrt{9} = 3i$

3. $x^2 = -81$

$x = \pm\sqrt{-81}$

$x = \pm i\sqrt{81}$

$x = \pm i(9)$

$x = \pm 9i$

The solutions are 9i and -9i.

5. $(1 - i) - (3 + 4i) + 2i$

$= (1 - 3) + (-1 - 4 + 2)i$

$= -2 - 3i$

7. $(6 - 5i) + (2 + 7i) - (3 - 2i)$

$= (6 + 2 - 3) + (-5 + 7 + 2)i$

$= 5 + 4i$

9. $(3 + 5i)(8 - i)$

$= 24 - 3i + 40i - 5i^2$

$= 24 + 37i - 5(-1)$

$= 29 + 37i$

11. $(2 + 6i)^2 = 4 + 24i + 36i^2$

$= 4 + 24i + 36(-1)$

$= -32 + 24i$

13. $(1 - i)^3 = 1 - i^3 + 3i^2 - 3i$

$= -2 - 2i$

15. $\frac{6 + 2i}{3 - i} = \frac{6 + 2i}{3 - i} \cdot \frac{3 + i}{3 + i}$

$= \frac{18 + 12i + 2i^2}{9 - i^2}$

$= \frac{18 + 12i + 2(-1)}{9 - (-1)}$

$= \frac{16 + 12i}{10}$

$= \frac{8 + 6i}{5}$

$= \frac{8}{5} + \frac{6}{5}i$

17. $\frac{2 + i}{1 - 5i} = \frac{2 + i}{1 - 5i} \cdot \frac{1 + 5i}{1 + 5i}$

$= \frac{2 + 11i + 5i^2}{1 - 25i^2}$

$= \frac{2 + 11i + 5(-1)}{1 - 25(-1)}$

$= \frac{-3 + 11i}{26}$

$= -\frac{3}{26} + \frac{11}{26}i$

19. $i^{53} = i^{52} \cdot i$

$= (i^4)^{13}i$

$= i^{13}i$

$= 1 \cdot i$

$= i$

or $0 + i$

21. $[5(\cos 90° + i \sin 90°)]$

$\cdot [6(\cos 180° + i \sin 180°)]$

$= 5 \cdot 6[\cos (90° + 180°)$

$+ i \sin (90° + 180°)]$

$= 30(\cos 270° + i \sin 270°)$

$= 30(0 - i)$

$= 0 - 30i$ or $-30i$

23. $\frac{2(\cos 60° + i \sin 60°)}{8(\cos 300° + i \sin 300°)}$

$= \frac{1}{4}[\cos (-240°) + i \sin (-240°)]$

$= -\frac{1}{8} + \frac{\sqrt{3}}{8}i$

25. $(\sqrt{3} + i)^3$

$r = \sqrt{(\sqrt{3})^2 + (1)^2} = 2$

$\tan \theta = \frac{1}{\sqrt{3}} = \frac{\sqrt{3}}{3}, \ \theta = 30°$

$(\sqrt{3} + i)^3 = [2(\cos 30° + i \sin 30°)]^3$

$= 2^3(\cos 90° + i \sin 90°)$

$= 0 + 8i$ or $8i$

27. $(\cos 100° + i \sin 100°)^6$

$= \cos 600° + i \sin 600°$

$= -\frac{1}{2} - \frac{\sqrt{3}}{2}i$

29. The vector representing a real number will lie on the x-axis in the complex plane.

For Exercises 31–35 see the answer graphs in the back of your textbook.

35. $7 + 3i$ and $-2 + i$

$(7 + 3i) + (-2 + i) = 5 + 4i$

37. $-2 + 2i$

$r = \sqrt{(-2)^2 + (2)^2}$
$= \sqrt{8} = 2\sqrt{2}$

$\tan\theta = \frac{2}{-2} = -1$

θ is in quadrant II so $\theta = 135°$.

$-2 + 2i$
$= 2\sqrt{2}(\cos 135° + i \sin 135°)$

39. $2(\cos 225° + i \sin 225°)$

$= 2(-\frac{\sqrt{2}}{2} - \frac{i\sqrt{2}}{2})$
$= -\sqrt{2} - i\sqrt{2}$

41. $1 - i$

$r = \sqrt{(1)^2 + (-1)^2} = \sqrt{2}$

$\tan\theta = \frac{-1}{1} = -1$

θ is in quadrant IV so $\theta = 315°$.

$1 - i = \sqrt{2}(\cos 315° + i \sin 315°)$

43. $-4i$

$r = 4$
$\theta = 270°$
$-4i = 4(\cos 270° + i \sin 270°)$

45. Find the fifth roots of $-2 + 2i$.

$r = \sqrt{(-2)^2 + 2^2} = \sqrt{8}$

$\tan\theta = \frac{2}{-2} = -1$

θ is in quadrant II so $\theta = 135°$.

$-2 + 2i = \sqrt{8}(\cos 135° + i \sin 135°)$

$(\sqrt{8})^{1/5} = (8^{1/2})^{1/5} = 8^{1/10} = \sqrt[10]{8}$

$\alpha = \frac{135° + 360 \cdot k}{5}$; $k = 0, 1, 2, 3, 4$

$\alpha = 27°, 99°, 171°, 243°, 315°$

The fifth roots are

$\sqrt[10]{8}(\cos 27° + i \sin 27°)$,
$\sqrt[10]{8}(\cos 99° + i \sin 99°)$,
$\sqrt[10]{8}(\cos 171° + i \sin 171°)$,
$\sqrt[10]{8}(\cos 243° + i \sin 243°)$,
$\sqrt[10]{8}(\cos 315° + i \sin 315°)$.

47. The real number -32 has one real fifth root. The one real fifth root is -2, and all other fifth roots are not real.

49. $x^3 + 125 = 0$

$x^3 = -125$
$-125 = -125 + 0i$
$-125 = 125(\cos 180° + i \sin 180°)$
$125^{1/3} = 5$

$\alpha = \frac{180° + 360° \cdot k}{3}$; $k = 0, 1, 2$

$\alpha = 60°, 180°, 300°$

$x = 5(\cos 60° + i \sin 60°)$,
$5(\cos 180° + i \sin 180°)$,
$5(\cos 300° + i \sin 300°)$

51. $x^2 + i = 0$

$x^2 = -i$
$= 0 - i$
$= 1(\cos 270° + i \sin 270°)$
$\sqrt[2]{1} = 1$

$\alpha = \frac{270° + 360° \cdot k}{2}$; $k = 0, 1$

$\alpha = 135°, 315°$

$x = (\cos 135° + i \sin 135°),$

$(\cos 315° + i \sin 315°)$

For Exercises 53–55 see the answer graphs in the back of your textbook.

53. $r = -1 + \cos \theta$

θ	0°	30°	45°	60°	90°
r	0	-.7	-.3	-.5	0

θ	120°	135°	150°	180°	270°	315°
r	-1.5	-1.7	-1.9	-2	-1	-.3

55. $r = 2 \sin 4\theta$

θ	0°	7.5°	15°	22.5°	0°	37.5°	45°
r	0	.1	$\sqrt{3}$	2	$\sqrt{3}$	1	0

θ	52.5°	60°	67.5°	75°	82.5°	90°
r	-1	$-\sqrt{3}$	-2	$-\sqrt{3}$	-1	0

The graph continues to form eight petals for the interval [0°, 360°).

57.

$$r = \frac{3}{1 + \cos \theta}$$

$$r(1 + \cos \theta) = 3$$

$$r + r \cos \theta = 3$$

$$\sqrt{x^2 + y^2} + x = 3$$

$$x^2 + y^2 = (3 - x)^2$$

$$x^2 + y^2 = 9 - 6x + x^2$$

$$y^2 = 9 - 6x$$

$$y^2 + 6x - 9 = 0$$

59.

$$r = \sin \theta + \cos \theta$$

$$r^2 = r \sin \theta + r \cos \theta$$

$$x^2 + y^2 = x + y$$

61.

$$x = -3$$

$$r \cos \theta = -3$$

63.

$$y = x^2$$

$$r \sin \theta = (r \cos \theta)^2$$

$$r = \frac{\sin \theta}{\cos^2 \theta} = \tan \theta \sec \theta$$

or $r = \frac{\tan \theta}{\cos \theta}$

CHAPTER 8 TEST

[8.1] **Write as a multiple of i.**

1. $\sqrt{-256}$

2. $\sqrt{-\frac{100}{27}}$

[8.1] 3. Simplify i^{47}.

[8.1] 4. Solve $4x^2 + 3x + 1 = 0$.

[8.1] 5. Find the resultant of $3 - 3i$ and $-6 + 2i$.

[8.2] **Write in standard form.**

6. $4(\cos 120° + i \sin 120°)$

7. $6 \text{ cis } (-30°)$

[8.2] **Write in trigonometric form.**

8. $-\frac{3\sqrt{2}}{2} + \frac{3\sqrt{2}}{2}i$

9. $-1 - i\sqrt{3}$

Perform the indicated operations. Write answers in standard form.

[8.1] 10. $(9 - 2i) + i$

[8.1] 11. $(3 - 2i)(7 + 4i)$

[8.1] 12. $\frac{4 - 3i}{2 + i}$

[8.1] 13. $\frac{4 - 5i}{-i}$

[8.3] 14. $(5 \text{ cis } 15°)(8 \text{ cis } 45°)$

[8.3] 15. $\frac{\cos 240° + i \sin 240°}{\cos 60° + i \sin 60°}$

[8.3] 16. $(3 - 3i\sqrt{3})^3$

[8.4] 17. $[7(\cos 30° + i \sin 30°)]^2$

[8.4] **Find all solutions of each equation. Write answer in trigonometric form.**

18. $x^4 - 1 = 0$

19. $x^2 + 9i = 0$

20. $x^3 = 6\sqrt{2} - 6i\sqrt{2}$

[8.4] 21. Find all third roots of $-8i$. Write answers in standard form.

[8.4] 22. How may real fifth roots does −31 have?

[8.5] 23. Write an equivalent equation in polar coordinates for $x^2 + y^2 = 25$.

[8.5] 24. Write an equivalent equation in rectangular coordinates for $r = \frac{4}{1 - \sin \theta}$, and describe its graph.

[8.5] 25. Write an equivalent equation in rectangular coordinates for $r = -7 \sec \theta$, and describe its graph.

CHAPTER 8 TEST ANSWERS

1. $16i$ **2.** $\frac{10\sqrt{3}}{9}i$ **3.** $-i$ **4.** $-\frac{3}{8} + \frac{\sqrt{7}}{8}i, -\frac{3}{8} - \frac{\sqrt{7}}{8}i$

5. $-3 - i$ **6.** $-2 + 2i\sqrt{3}$ **7.** $\sqrt{3} - 3i$ **8.** $3(\cos 135° + i \sin 135°)$

9. $2(\cos 240° + i \sin 240°)$ **10.** $9 - i$ **11.** $29 - 2i$ **12.** $1 - 2i$

13. $5 + 4i$ **14.** $20 + 20i\sqrt{3}$ **15.** -1 or $-1 + 0i$ **16.** -216 or $-216 + 0i$

17. $\frac{49}{2} + \frac{49\sqrt{3}}{2}i$ **18.** $(\cos 0° + i \sin 0°)$, $(\cos 90° + i \sin 90°)$,

$(\cos 180° + i \sin 180°)$, $(\cos 270° + i \sin 270°)$ **19.** $3(\cos 135° + i \sin 135°)$,

$3(\cos 315° + i \sin 315°)$ **20.** $\sqrt[3]{12}(\cos 105° + i \sin 105°)$, $\sqrt[3]{12}(\cos 225° + i \sin 225°)$,

$\sqrt[3]{12}(\cos 345° + i \sin 345°)$ **21.** $2i$ (or $0 + 2i$), $-\sqrt{3} - i$, $\sqrt{3} - i$ **22.** 1

23. $r = 5$ **24.** $x^2 = 8(2 + y)$, a parabola with vertex $(0, -2)$ and axis $x = 0$

25. $x = -7$; a vertical line with intercept $(-7, 0)$

CHAPTER 9 LOGARITHMS

Section 9.1

1. $3^4 \cdot 3^7 = 3^{4+7}$

$= 3^{11}$

3. $\frac{7^4}{7^3} = 7^{4-3}$

$= 7^1$

$= 7$

5. $\frac{6^4 \cdot 6^5}{6^6} = \frac{6^{4+5}}{6^6}$

$= 6^{9-6}$

$= 6^3$

7. $(8^9)^2 = 8^{9 \cdot 2} = 8^{18}$

9. $\frac{(2^5)^2 \cdot 2^4}{2^6} = \frac{2^{(5 \cdot 2)+4}}{2^6}$

$= 2^{14-6}$

$= 2^8$

11. $2^{-3} = \frac{1}{2^3}$

13. $3^{-7} \cdot 3^8 = 3^{-7+8}$

$= 3^1$

$= 3$

15. $\frac{9^{-8} \cdot 9^2}{(9^2)^{-2}} = \frac{9^{-8+2}}{9^{(2)(-2)}}$

$= \frac{9^{-6}}{9^{-4}}$

$= 9^{-6-(-4)}$

$= 9^{-2}$

$= \frac{1}{9^2}$

17. $25^{1/2} = \sqrt{25}$

$= 5$

19. $16^{3/4}$

$16 = 2^4$, so

$16^{3/4} = (\sqrt[4]{16})^3$

$= (\sqrt[4]{2^4})^3$

$= 2^3$

$= 8.$

For Exercises 21–27 see the answer graphs at the back of the textbook.

21. $y = \left(\frac{1}{3}\right)^x$

Make a table of values.

x	−2	−1	0	1	2
$y = \left(\frac{1}{3}\right)^x$	9	3	1	$\frac{1}{3}$	$\frac{1}{9}$

23. $y = 2^{x+1}$

Make a table of values.

x	−2	−1	0	1	2
$y = 2^{x+1}$	$\frac{1}{2}$	1	2	4	8

25. $y = 2^{-x}$

Make a table of values.

x	−2	−1	0	1	2
$y = 2^{-x}$	4	2	1	$\frac{1}{2}$	$\frac{1}{4}$

27. $y = 3^x - 3$

Make a table of values.

x	-2	-1	0	1	2
$y = 3^x - 3$	$-\frac{26}{9}$	$-\frac{8}{3}$	-2	0	6

29. $3^5 = 243$

$a^y = x$ means

$\log_a x = y$, so

$3^5 = 243$ means

$\log_3 243 = 5$.

31. $10^4 = 10{,}000$ means

$\log_{10} 10{,}000 = 4$.

33. $6^{-2} = \frac{1}{36}$ means

$\log_6 \frac{1}{36} = -2$.

35. $\log_4 16 = 2$ means

$4^2 = 16$.

37. $\log_{10} 1000 = 3$ means

$10^3 = 1000$.

39. $\log_{3/4} \frac{9}{16} = 2$ means

$\left(\frac{3}{4}\right)^2 = \frac{9}{16}$.

41. Why are logarithms of negative numbers not defined?

Remember that the base of a logarithm is a positive number. When we say $\log_a x = y$, that means that $a^y = x$. Since a is a positive number, a raised to any number (positive or negative or zero) will result in a positive number. Therefore, x is always posi- tive. Hence, logarithms of negative numbers are not defined.

43. The reason why $\log_a 1 = 0$ is that any positive number, which the base a must be, raised to the power 0 is equal to 1.

For Exercises 45–51, see the answer graphs at the back of the textbook.

45. $y = \log_4 x$

Use the fact that $y = \log_4 x$ means $4^y = x$. Make a table of values.

$x = 4^y$	$\frac{1}{4}$	1	4	16
y	-1	0	1	2

47. $y = \log_{1/3} x$ means

$\left(\frac{1}{3}\right)^y = x$.

$x = \left(\frac{1}{3}\right)^y$	9	3	1	$\frac{1}{3}$
y	-2	-1	0	1

49. $y = \log_2 (x - 1)$ means

$2^y = x - 1$

$2^y + 1 = x$.

$x = 2^y + 1$	$\frac{17}{16}$	$\frac{5}{4}$	$\frac{3}{2}$	2	3	5
y	-4	-2	-1	0	1	2

51. $y = 1 + \log_3 x$

$y - 1 = \log_3 x$ means

$3^{y-1} = x.$

$x = 3^{y-1}$	$\frac{1}{9}$	$\frac{1}{3}$	1	3	9
y	-1	0	1	2	3

53. $\log_6 9 + \log_6 5 = \log_6 9 \cdot 5$

$= \log_6 45$

55. $\log_5 12 - \log_5 7 = \log_5 \frac{12}{7}$

57. $\log_2 7 + \log_2 4 - \log_2 5$

$= \log_2 7 \cdot 4 - \log_2 5$

$= \log_2 \frac{28}{5}$

59. $4 \cdot \log_3 2 = \log_3 2^4$

$= \log_3 16$

61. $\frac{1}{2} \cdot \log_8 7 = \log_8 7^{1/2} = \log_8 \sqrt{7}$

63. $4 \cdot \log_2 3 - 2 \cdot \log_2 5$

$= \log_2 3^4 - \log_2 5^2$

$= \log_2 \frac{3^4}{5^2}$

$= \log_2 \frac{81}{25}$

65. $\log_{10} 14 = \log_{10} 2 \cdot 7$

$= \log_{10} 2 + \log_{10} 7$

$= .3010 + .8451$

$= 1.1461$

67. $\log_{10} 28 = \log_{10} 2^2 \cdot 7$

$= 2 \log_{10} 2 + \log_{10} 7$

$= 2(.3010) + .8451$

$= 1.4471$

69. $\log_{10} 2^8 = 8 \log_{10} 2$

$= 8(.3010)$

$= 2.4080$

71. $\log_{10} 11^3 = 3 \log_{10} 11$

$= 3(1.0414)$

$= 3.1242$

73. $\log_{10} \sqrt[3]{11} = \frac{1}{3} \log_{10} 11$

$= \frac{1}{3}(1.0414)$

$= .3471$

75. Prove $\log_a \frac{x}{y} = \log_a x - \log_a y$.

Let $\log_a x = M$ and $\log_a y = N$. Then

$a^M = x$ and $a^N = y$.

$$\frac{a^M}{a^N} = \frac{x}{y}$$

$$a^{M-N} = \frac{x}{y}$$

$$\log_a \frac{x}{y} = M - N$$

Substituting for M and N gives the result

$$\log_a \frac{x}{y} = \log_a x - \log_a y.$$

Section 9.2

1. log 278

Enter: 278 [log]

Display: 2.4440

3. log 3.28
Enter: 3.28 [log]
Display: .5159

5. log 9.83
Enter: 9.83 [log]
Display: .9926

7. log .327
Enter: .327 [log]
Display: -.4855

9. log .000672
Enter: .000672 [log]
Display: -3.1726

11. log 675,000
Enter: 675,000 [log]
Display: 5.8293

13. ln 5
Enter: 5 [ln]
Display: 1.6094

15. ln 697
Enter: 697 [ln]
Display: 6.5468

17. ln 1.72
Enter: 1.72 [ln]
Display: .5423

19. ln 97.6
Enter: 97.6 [ln]
Display: 4.5809

21. $\ln e^{3.1} = (3.1) \ln e$
$= (3.1)(1)$
$= 3.1$

23. $\ln 4e^2 = \ln 4 + \ln e^2$
$= \ln 4 + 2 \ln e$
$= \ln 4 + 2(1)$
$= \ln 4 + 2$
$= 1.3863 + 2$
$= 3.3863$

25. When -1 is entered into a scientific calculator and the log key is pressed, an error message is displayed. This result has two explanations: (a) The domain of $y = \log x$ is $(0, \infty)$ and -1 is not in this interval and (b) there is no solution to the equation $10^x = -1$.

27. (a) $\log 1 = 0 \neq 1$ since $10^0 = 1$.

(b) $\log 10 = 1$ since $10^1 = 10$.

(c) $\ln e = 1$ since $e^1 = e$.

(d) $(\ln 10)^0 = 1$ since ln 10 is a real number and any real number raised to the 0 power is 1.

Answer: (a)

29. To find the common antilogarithm of .5340 with a calculator with a 10^x key,

Enter: .5340 [10^x]

Display: 3.4198.

Or, use the INV and log keys.

Enter: .5340 [INV] [log]

Display: 3.4198

If necessary, consult the instructions for your calculator.

$$.5340 = \log N$$
$$N = 3.4198$$

31. Find N such that log N = 2.6571.

Enter; 2.6571 [INV] [log]

Display: 454.0462

N = 454.0462

33. Find N such that log N = -1.118.

Enter: 1.118 [+/-] [INV] [log]

Display: .0762

N = .0762

35. To find the natural antilogarithm of 3.8494, use a calculator to compute $e^{3.8494} = 46.9649$, as follows.

Enter: 3.8494 [INV] [ln]

Display: 46.9649

$$3.8494 = \ln N$$
$$N = 46.9649$$

37. $1.3962 = \ln N$

$$N = 4.0398$$

39. $-.4168 = \ln N$

$$N = .6592$$

41. $\log_5 11 = \dfrac{\ln 11}{\ln 5}$

$$= 1.4899$$

43. $\log_3 1.89 = \dfrac{\ln 1.89}{\ln 3}$

$$= .5794$$

45. $\log_8 9.63 = \dfrac{\ln 9.63}{\ln 8}$

$$= 1.0892$$

47. $\log_{11} 47.3 = \dfrac{\ln 47.3}{\ln 11}$

$$= 1.6083$$

49. $\log_{50} 31.3 = \dfrac{\ln 31.3}{\ln 50}$

$$= .8803$$

51. Find the pH of milk, 4×10^{-7}.

$$\begin{aligned} pH &= -\log [H_3O^+] \\ pH &= -\log (4 \times 10^{-7}) \\ &= -(\log 4 + \log 10^{-7}) \\ &= -[\log 4 + (-7)] \\ &= -(-6.4) \\ &= 6.4 \end{aligned}$$

53. Find the pH of limes, 1.6×10^{-2}.

$$\begin{aligned} pH &= -\log [H_3O^+] \\ pH &= -\log (1.6 \times 10^{-2}) \\ &= -(\log 1.6 + \log 10^{-2}) \\ &= -[\log 1.6 + (-2)] \\ &= -(-1.8) \\ &= 1.8 \end{aligned}$$

55. Find the hydronium ion concentration of shampoo with a pH of 5.5.

$$\begin{aligned} pH &= -\log [H_3O^+] \\ 5.5 &= -\log [H_3O^+] \\ -5.5 &= \log [H_3O^+] \\ .0000032 &= H_3O^+ \\ 3.2 \times 10^{-6} &= H_3O^+ \end{aligned}$$

57. Find the hydronium ion concentration of soda pop with a pH of 2.7.

$$\begin{aligned} pH &= -\log [H_3O^+] \\ 2.7 &= -\log [H_3O^+] \\ -2.7 &= \log [H_3O^+] \\ .0020 &= H_3O^+ \\ 2.0 \times 10^{-3} &= H_3O^+ \end{aligned}$$

59. Find N(r) for r = 90% = .9.

$$\begin{aligned} N(r) &= -5000 \ln r \\ N(.9) &= -5000 \ln (.9) \\ &= -5000(-.1054) \\ &= 527 \text{ years.} \end{aligned}$$

61. Find r if N(r) = 1000 years.

$$\begin{aligned} N(r) &= -5000 \ln r \\ 1000 &= -5000 \ln r \\ -.2 &= \ln r \\ .82 &= r \end{aligned}$$

r is 82%.

63. Since a sample of a small community shows two species with 50 individuals each, the proportions, P_1 and P_2, of both species are .5.

$$H = -[P_1 \log_2 P_1 + P_2 \log_2 P_2]$$

$$\begin{aligned} H &= -[.5 \log_2 .5 + .5 \log_2 .5] \\ &= -[.5(-1) + .5(-1)] \\ &= -[-.5 + (-.5)] \\ &= -(-1) \\ &= 1 \end{aligned}$$

65. $S(n) = .36 \ln \left(1 + \frac{n}{.36}\right)$

(a) n = 100

$$S(100) = .36 \ln \left(1 + \frac{100}{.36}\right) \approx 2$$

(b) n = 500

$$S(500) = .36 \ln \left(1 + \frac{500}{.36}\right) \approx 3$$

(c) n = 2000

$$S(2000) = .36 \ln \left(1 + \frac{2000}{.36}\right) \approx 3$$

(d) n = 10

$$S(10) = .36 \ln \left(1 + \frac{10}{.36}\right) \approx 1$$

67. $p = 86.3 \ln h - 680$

(a) h = 3000 ft

$$\begin{aligned} p &= 86.3 \ln 3000 - 680 \\ &= 11 \end{aligned}$$

The percent of moisture that falls as snow is 11%.

(b) $h = 4000$ ft

$$p = 86.3 \ln 4000 - 680$$
$$= 36$$

The percent of moisture that falls as snow is 36%.

(c) $h = 7000$ ft

$$p = 86.3 \ln 7000 - 680$$
$$= 84$$

The percent of moisture that falls as snow is 84%.

Section 9.3

1. $5^x = 125$

Write the right side as a power of 5.

$$5^x = 5^3$$

Use property 1 to set exponents equal.

$$x = 3$$

3.
$$32^k = 2$$
$$(2^5)^k = 2$$
$$2^{5k} = 2^1$$

Use property 1.

$$5k = 1$$
$$k = \frac{1}{5}$$

5. $100^r = 1000$

Write both sides as a power of 10.

$$(10^2)^r = 10^3$$
$$10^{2r} = 10^3$$

Use property 1.

$$2r = 3$$
$$r = \frac{3}{2}$$

7.
$$27^z = 81$$
$$(3^3)^z = 3^4$$
$$3^{3z} = 3^4$$
$$3z = 4$$
$$z = \frac{4}{3}$$

9.
$$2^{5x} = 16^{x+2}$$
$$2^{5x} = (2^4)^{x+2}$$
$$2^{5x} = 2^{4x+8}$$
$$5x = 4x + 8$$
$$x = 8$$

11.
$$25^{2k} = 125^{k+1}$$
$$(5^2)^{2k} = (5^3)^{k+1}$$
$$5^{4k} = 5^{3k+3}$$
$$4k = 3k + 3$$
$$k = 3$$

13.
$$10^p = 3$$

Use property 2 to take logarithms of both sides.

$$\log 10^p = \log 3$$
$$p \log 10 = \log 3$$
$$p = \frac{\log 3}{\log 10}$$
$$p = \log 3$$
$$p = .477$$

15.
$$10^{r+2} = 15$$
$$(r + 2) \log 10 = \log 15$$
$$r + 2 = \log 15$$
$$r = \log 15 - 2$$
$$r = -.824$$

17. $6k = 10$

$k \log 6 = \log 10$

$k = \dfrac{1}{\log 6}$

$k = 1.285$

19. $14^z = 28$

$z \log 14 = \log 28$

$z = \dfrac{\log 28}{\log 14}$

$z = 1.263$

21. $7^a = 32$

$a \log 7 = \log 32$

$a = \dfrac{\log 32}{\log 7}$

$a \approx 1.781$

23. $15^x = 8^{2x+1}$

$x \log 15 = (2x + 1) \log 8$

$x \log 15 - 2x \log 8 = \log 8$

$x(\log 15 - 2 \log 8) = \log 8$

$x = \dfrac{\log 8}{\log 15 - 2 \log 8}$

$x = -1.433$

25. $16^{z+3} = 28^{2z-1}$

$(z + 3) \log 16$

$= (2z - 1) \log 28$

$3 \log 16 + \log 28$

$= 2z \log 28 - z \log 16$

$z = \dfrac{3 \log 16 + \log 28}{2 \log 28 - \log 16}$

$z = 2.993$

27. $\log x - \log (x - 14) = \log 8$

$\log \left(\dfrac{x}{x - 14}\right) = \log 8$

$\dfrac{x}{x - 14} = 8$

$x = 8x - 112$

$7x = 112$

$x = 16$

29. $\log z = 1 - \log (3z - 13)$

$\log z + \log (3z - 13) = 1$

$\log [z(3z - 13)] = \log 10$

$z(3z - 13) = 10$

$3z^2 - 13z - 10 = 0$

$(z - 5)(3z + 2) = 0$

$z = 5$ or $z = -\dfrac{2}{3}$

Logarithms of negative numbers are not defined, so the only solution of the equation is 5.

31. $\log_x 10 = 3$

$x^3 = 10$

$x = \sqrt[3]{10}$

$x = 2.154$

33. $2 + \log x = 0$

$\log x = -2$

$x = 10^{-2}$

$x = \dfrac{1}{100}$

$x = .01$

35. If $a^x = a^y$, then $x = y$. What if a could be 1?

Since $1^4 = 1$ and $1^5 = 1$, then

$$1^4 = 1^5.$$

But $4 \neq 5$, which is why the base a cannot equal one.

37. It would be an error to reject any negative answer in solving a logarithmic equation.
As an example, the equation $\log(x + 2) = 0$ has a negative solution.

$$\log(x + 2) = 0$$
$$10^0 = x + 2$$
$$1 = x + 2$$
$$-1 = x$$

39. For the statement

$$\log x^2 = 2 \log x$$

to be true, x must be positive.

41. (a) $P = P_0 e^{-.04t}$, $P_0 = 1{,}000{,}000$, $t = 1$

$$P = 1{,}000{,}000\, e^{-.04}$$
$$\ln P = \ln 1{,}000{,}000 + (-.04)$$
$$P \approx 961{,}000$$

(b) $750{,}000 = 1{,}000{,}000\, e^{-.04t}$

$$.75 = e^{-.04t}$$
$$\ln .75 = -.04t$$
$$t = \frac{-\ln .75}{.04}$$
$$\approx 7 \text{ yr}$$

(c) $\frac{P_0}{2} = P_0\, e^{-.04t}$

$$\frac{1}{2} = e^{-.04t}$$
$$\ln \frac{1}{2} = -.04t$$
$$t = \frac{\ln \frac{1}{2}}{-.04}$$
$$t \approx 17.3 \text{ yr}$$

43. Use $A = P\left(1 + \frac{r}{n}\right)^{nt}$ for (a)-(d) to find the amount of money in an account at the end of 12 years if \$5000 is deposited at 7% annual interest compounded the following ways.

(a) annually, let n = 1

$$A = 5000\left(1 + \frac{.07}{1}\right)^{1 \cdot 12}$$
$$= \$11{,}260.96$$

(b) semiannually, let n = 2

$$A = 5000\left(1 + \frac{.07}{2}\right)^{2 \cdot 12}$$
$$= \$11{,}416.64$$

(c) quarterly, let n = 4

$$A = 5000\left(1 + \frac{.07}{4}\right)^{4 \cdot 12}$$
$$= \$11{,}497.99$$

(d) daily, let n = 365

$$A = 5000\left(1 + \frac{.07}{365}\right)^{365 \cdot 12}$$
$$= \$11{,}580.90$$

(e) Use the formula $A = Pe^{rt}$ to find the amount if \$5000 is compounded continuously.

$$A = 5000e^{(.07)(12)}$$
$$= \$11{,}581.83$$

45. How much money must be deposited today to become \$1000 in 10 years at 5% compounded continuously?

Use the formula $A = Pe^{rt}$ and solve for P.

$$1000 = Pe^{(.05)(10)}$$
$$1000 = Pe^{.5}$$
$$P = \frac{1000}{e^{.5}}$$
$$P = 606.53$$

One must invest \$606.53.

Chapter 9 Review Exercises

1. $\frac{3^4 \cdot 3^5}{3^6} = \frac{3^{4+5}}{3^6}$
$= 3^{9-6}$
$= 3^3 = 27$

3. $(2^3)^2 \cdot 2^5 = 2^{3 \cdot 2+5}$
$= 2^{11}$

5. $16^{3/4} = (\sqrt[4]{2^4})^3$
$= 2^3$
$= 8$

For Exercises 7–13, see the answer graphs at the back of the textbook.

7. $y = 5^x$

x	-1	0	1
$y = 5^x$	$\frac{1}{5}$	1	5

9. $y = 2^{-x}$

x	-2	-1	0	1
$y = 2^{-x}$	4	2	1	$\frac{1}{2}$

11. $y = \log_3 x$ means
$3^y = x$.

$x = 3^y$	$\frac{1}{3}$	1	3
y	-1	0	1

13. $y = \log_2 (2 - x)$ means
$2^y = 2 - x$
$x = 2 - 2^y$.

$x = 2 - 2^y$	$\frac{7}{4}$	1	-2	-6
y	-2	0	2	3

15. $2^9 = 512$ means
$\log_2 512 = 9$.

17. $\left(\frac{2}{3}\right)^{-1} = \frac{3}{2}$ means
$\log_{2/3} \frac{3}{2} = -1$.

19. $\log_3 81 = 4$ means
$3^4 = 81$.

21. $\log_{1/3} 3 = -1$ means
$\left(\frac{1}{3}\right)^{-1} = 3$.

23. $\log_5 4 + \log_5 3 = \log_5 4 \cdot 3$
$= \log_5 12$

25. $5 \log_2 x = \log_2 x^5$

27. $2 \log_3 x + 4 \log_3 y$
$= \log_3 x^2 + \log_3 y^4$
$= \log_3 x^2y^4$

29. $\log 2.48 = .3945$

31. $\log 1240 = 3.0934$

33. $\log 5{,}270{,}000 = 6.7218$

35. Find N such that $\log N = 3.4314$.

Enter: 3.4314 [INV] [log]

Display: 2700.2253

$N = 2700.2253$

37. Find N such that $\log N = -1.7721$.

Enter: 1.7721 [+/-] [INV] [log]

Display: .0169

$N = .0169$

39. $\ln 15 = 2.7081$

41. $\ln .12 = -2.1203$

43. $3^x = 27$

Write the right side as a power of 3.

$3^x = 3^3$

Use property 1 to set the exponents equal.

$x = 3$

45.

$$2^{m-4} = 8^{2m}$$
$$2^{m-4} = (2^3)^{2m}$$
$$2^{m-4} = 2^{6m}$$

Use property 1.

$$m - 4 = 6m$$
$$5m = -4$$
$$m = -\frac{4}{5}$$

47. $5^n = 33$

Use property 2 to take logarithms of both sides.

$$\log 5^n = \log 33$$
$$n \log 5 = \log 33$$
$$n = \frac{\log 33}{\log 5}$$
$$= 2.173$$

49.

$$\log x = 2 - \log (x - 3)$$
$$\log x + \log (x - 3) = 2$$
$$\log x(x - 3) = 2$$
$$x^2 - 3x = 10^2$$
$$= 100$$
$$x^2 - 3x - 100 = 0$$

$$x = \frac{3 \pm \sqrt{9 + 400}}{2}$$
$$x = \frac{3 \pm \sqrt{409}}{2}$$

Since $\frac{3 - \sqrt{409}}{2}$ is negative and the domain of the common logarithm function does not include negative numbers, the only solution is

$$x = \frac{3 + \sqrt{409}}{2}$$
$$= 11.612.$$

51. $\log_t 16 = 2$

$$t^2 = 16$$
$$t = \pm 4$$

The base cannot be negative, so

$$t = 4.$$

53. $A(t) = 5000(3)^{-.02t}$

$A(0) = 5000$ is the initial amount.

So we find t when $A(t) = 2500$.

$$2500 = 5000(3)^{-.02t}$$

$$3^{-.02t} = \frac{1}{2}$$

$$\log 3^{-.02t} = \log \frac{1}{2}$$

$$\log 3^{-.02t} = -\log 2$$

$$-.02t \log 3 = -\log 2$$

$$.02t = \frac{\log 2}{\log 3}$$

$$t = \left(\frac{1}{.02}\right)\left(\frac{\log 2}{\log 3}\right)$$

$$t \approx 31.5$$

The half-life is about 31.5 days.

55. If \$20,000 is deposited at 7% annual interest compounded quarterly, how much will be in the account after 5 years?

Use $A = P\left(1 + \frac{r}{n}\right)^{nt}$ with $t = 4$.

$$A = P\left(1 + \frac{r}{n}\right)^{nt}$$

$$= 20{,}000\left(1 + \frac{.07}{4}\right)^{4 \cdot 5}$$

$$= \$28{,}295.56$$

CHAPTER 9 TEST

[9.1] **Use the properties of exponents to simplify each of the following. Leave answers in exponential form, with positive exponents only.**

1. $4^9 \cdot 4^4$

2. $\frac{(3^4)^2 \cdot 3^3}{3^{15}}$

[9.1] **Write a table of values of x and y that could be used to graph each of the following.**

3. $y = 2 + 3^x$

4. $y = -1 + \log_3 x$

[9.1] **Write each of the following in logarithmic form.**

5. $4^{-1} = \frac{1}{4}$

6. $25^{1/2} = 5$

[9.1] **Write in exponential form.**

7. $\log_3 243 = 5$

8. $\log_{16} 2 = .25$

[9.1] **Use the properties of logarithms to simplify each expression.**

9. $\log_4 10 + \log_4 3$

10. $\frac{1}{4} \log_6 41$

11. $5 \log_a 2 - \log_a 8$

12. $\log_3 x - 4 \log_3 y$

[9.2] **Use a calculator to find each of the following logarithms. Give answers to four decimal places.**

13. log 13.7

14. log 56,700

15. ln 4.35

16. ln .007

[9.2] **Use a calculator to find the common antilogarithm of each number. Give answers to four decimal places.**

17. 3.9841 **18.** −.2116

[9.3] **Solve each equation. Use logarithms as necessary. Give answers to the nearest thousandth.**

19. $3^{2x-1} = 42$ **20.** $\log_x 175 = 5$

[9.3] **Use common logarithms or natural logarithms and a calculator to find each of the following to four decimal places.**

21. $\log_5 36$ **22.** $\log_{.2} 1.7$

[9.3] **Solve each equation without using a calculator.**

23. $729^x = 27$ **24.** $5^{3x+1} = 625^x$

[9.3] **25.** What values of x could not possibly be solutions of the following equation? (Do not solve.)

$$\log (2x - 3) + \log x^2 = 0$$

CHAPTER 9 TEST ANSWERS

1. 4^{13}

2. $\frac{1}{3^4}$

3.

x	-2	-1	0	1	2
$y = 2 + 3^x$	$\frac{19}{9}$	$\frac{7}{3}$	3	5	11

4.

$x = 3^{y+1}$	$\frac{1}{9}$	$\frac{1}{3}$	1	3	9
y	-3	-2	-1	0	1

5. $\log_4 \frac{1}{4} = -1$

6. $\log_{25} 5 = \frac{1}{2}$

7. $3^5 = 243$

8. $16^{.25} = 2$

9. $\log_4 30$

10. $\log_6 41^{1/4}$

11. $\log_a 4$

12. $\log_3 \frac{x}{y^4}$

13. 1.1367

14. 4.7536

15. 1.4702

16. -4.9618

17. 9640.5098

18. .6143

19. 2.2011

20. 2.8094

21. 2.2266

22. -.3297

23. 1/2

24. 1

25. Any x less than or equal to 3/2

NOTES

NOTES

NOTES

NOTES

NOTES

NOTES

NOTES

NOTES

NOTES

NOTES